To Richard.

THE COMPLETE METALSMITH

AN ILLUSTRATED HANDBOOK

Hope it will give you years of enjoyment and insight into the wonderful world of jewellery making.

Your friend.

Wolfgang

THE SMITH ALSO SITTETH BY THE ANVIL,

AND FIGHTETH WITH THE HEAT OF
THE FURNACE,

AND NOISE OF THE HAMMER AND THE
ANVIL IS EVER IN HIS EARS,

AND HIS EYES LOOK STILL UPON THE
PATTERN OF THE THING THAT
HE MAKETH.

HE SETTETH HIS MIND TO FINISH HIS WORK,

AND WAITETH TO POLISH IT PERFECTLY.

-ECCLESIASTICUS

THE COMPLETE METALSMITH

AN ILLUSTRATED HANDBOOK

Tim McCreight

DAVIS PUBLICATIONS, INC.
Worcester, Massachusetts

ACKNOWLEDGMENTS

So many people contributed to this book that it is impossible to mention them all. My students, who have sampled much of the manuscript, made suggestions that clarified many descriptions. Chuck Evans, Gary Griffin, and Bob Ebendorf reviewed the material meticulously, contributing greatly to the technical accuracy of the information. For help on specific topics I'd like to thank Peter Handler, John Pirtle, Paula Dinneen, Will Earley, and John Cogswell. Wyatt Wade of Davis Publications who supervised the project has been a constant source of support and sound judgment. And for putting up with the late nights and lost weekends I especially thank my family, Jay, Jobie, and Jeff.

Printed in the United States of America
Library of Congress Catalog Number: 81-66573
ISBN: 0-87192-135-9

10 9

CONTENTS

INTRODUCTION

This book represents three years of intensive research and experimentation. Information from hundreds of sources has been collected, distilled, and illustrated. It is intended to be both a text and tool, a blend of instruction and reference. Like any tool, its value increases as you bring to it your own perceptions and skills. It is designed to make the information easily accessible, and built to stand up to years of bench-side use.

Like any craft, metalsmithing involves some chemicals and procedures that are potentially dangerous. Great care has been taken to omit hazards where possible and to give clear warnings wherever they apply. These will be only as effective as you make them. Before using this book, please take the time to read the health and safety suggestions on page 132 which explain the system used throughout the book as a means of providing for your well-being.

ABOUT THIS BOOK

This is an example of a new direction in crafts publishing. THE COMPLETE METALSMITH has been built as a tool with several features that should be explained. Each page has a title panel at the side and top, making it easier to locate specific information. This may be augmented by adding your own index tabs. On the following page, you'll find tabs that can be glued onto the chapter heading pages. To make your book into the Deluxe Model, insert these printed tabs into adhesive plastic tab sleeves which can be purchased at any office supply store.

To meet differing tastes this book is available in both spiral and glued binding. To make the book even more practical for some readers, the pages can be made to fit into a looseleaf binder. To do this, drill ¼ inch holes where indicated on the back cover. Spiral wire may be removed after snipping off the last half inch. Glued bindings may be cut off (after drilling) with handsaw or jeweler's saw.

These do-it-yourself features are intended to make the book practical without making it expensive. You will find THE COMPLETE METALSMITH to be a valuable addition to your workshop.

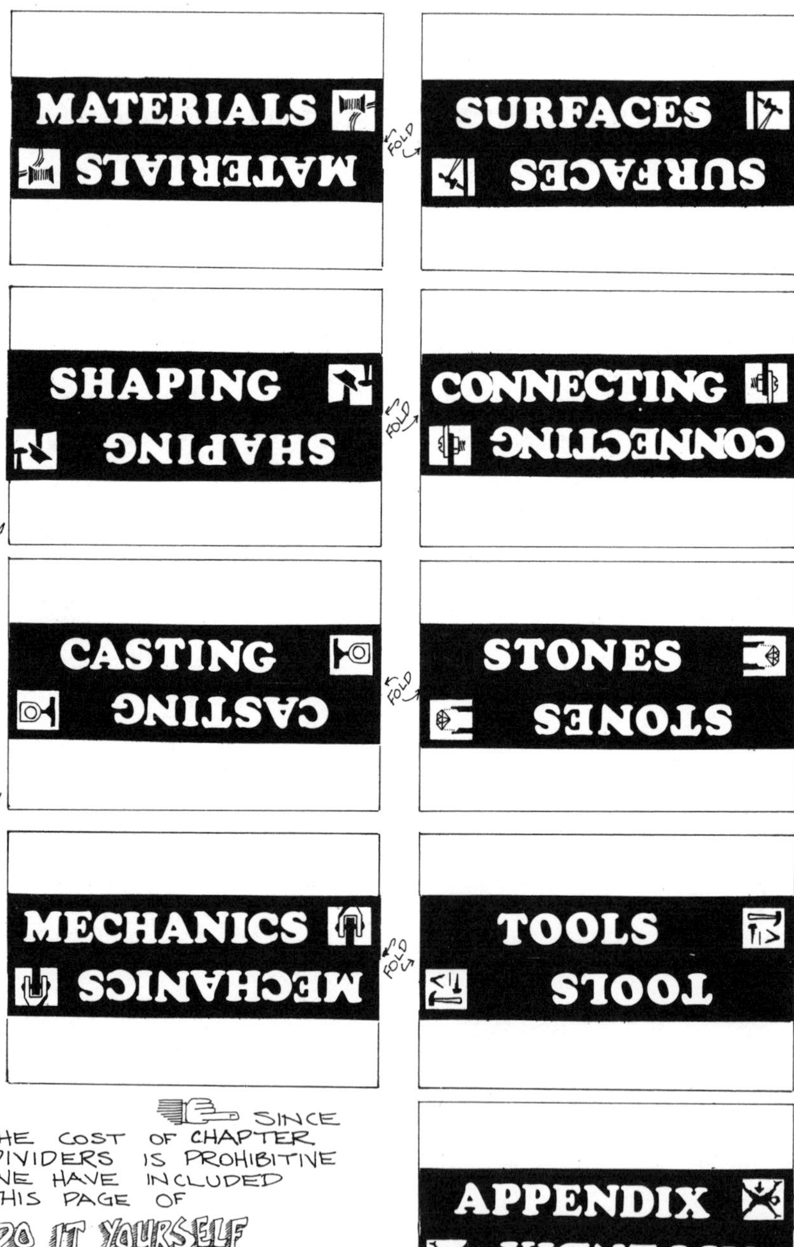

SINCE THE COST OF CHAPTER DIVIDERS IS PROHIBITIVE WE HAVE INCLUDED THIS PAGE OF

DO IT YOURSELF

TABS! CUT AND PASTE AS SHOWN, COVERING UP THE CHAPTER LABEL ON THE FIRST PAGE OF EACH CHAPTER.

MATERIALS

Metallurgy MATERIALS

METALLURGY IS A COMPLEX HIGHLY TECHNICAL FIELD: IT CANNOT BE ACCURATELY SIMPLIFIED. IT IS HELPFUL THOUGH FOR A METALSMITH TO UNDERSTAND AT LEAST IN A GENERAL WAY THE STRUCTURE AND BEHAVIOR OF METAL. THIS UNDERSTANDING CAN HELP TO EXPLAIN PROBLEMS AND RESULTS THAT COME UP IN THE STUDIO.

Crystals

AT ROOM TEMPERATURE METALS EXIST AS CRYSTALS, REGULARLY SHAPED UNITS ARRANGED IN AN ORDERED RECURRING PATTERN CALLED A SPACE LATTICE. THERE ARE 7 CRYSTAL SYSTEMS AND 14 LATTICE CONFIGURATIONS. HERE ARE THOSE ASSOCIATED WITH FAMILIAR METALS.

FACE CENTERED CUBIC

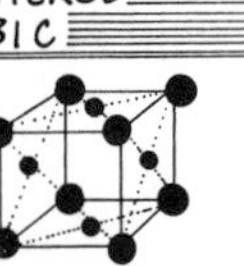

LEAD
COPPER
ALUMINUM
CALCIUM
GOLD
SILVER
NICKEL
IRON (AT HIGH TEMPERATURES)

BODY CENTERED CUBIC

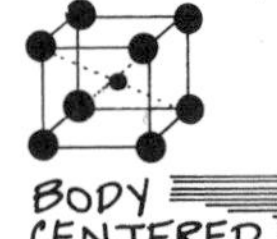

CHROMIUM
LITHIUM
MOLYBDENUM
POTASSIUM
SODIUM
VANADIUM
IRON (AT ROOM TEMPERATURE)

HEXAGONAL CLOSE-PACKED

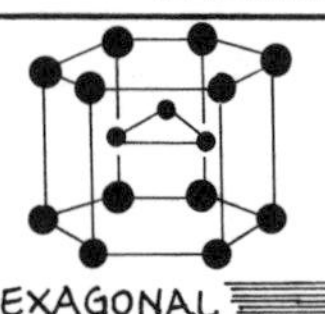

BERYLLIUM
CADMIUM
COBALT
MAGNESIUM
TITANIUM
ZINC

IT IS NOT A COINCIDENCE THAT EASILY WORKED METALS HAVE THE SAME CRYSTAL STRUCTURE: CRYSTAL SHAPE IS ONE FACTOR DETERMINING MALLEABILITY.

Annealing

THIS CAN BE DONE WITH A KILN OR TORCH. COLORS ARE BEST SEEN IN A DIMLY-LIT AREA. HANDY FLUX MAY BE PAINTED ONTO METAL TO SERVE AS A TEMPERATURE INDICATOR: IT IS CLEAR AT 1100° F. IF QUENCHING IN PICKLE, BEWARE OF DANGEROUS FUMES AND SPLASHING OF ACID.

STERLING, 14K GOLD, 10K GOLD } HEAT TO DULL RED; QUENCH AS SOON AS REDNESS DISAPPEARS.

RED GOLDS, COPPER, BRONZE } HEAT TO MEDIUM RED; QUENCH AS SOON AS REDNESS DISAPPEARS.

WHITE GOLD - HEAT TO BRIGHT RED; QUENCH WHEN RED FADES.

BRASS - HEAT TO BRIGHT RED - AIR COOL.

Recrystalization

WHEN HEATED TO ITS MELTING POINT A METAL LOSES ITS CRYSTALINE ORGANIZATION AND BECOMES FLUID. WHEN THE HEAT SOURCE IS REMOVED AND THE METAL COOLS IT REFORMS ITS CRYSTAL PATTERN, STARTING WITH THE FIRST AREAS TO COOL. MANY CLUSTERS OF CRYSTALS START TO FORM SIMULTANEOUSLY, ALL HAVING THE SAME ORDER BUT NOT NECESSARILY THE SAME ORIENTATION.

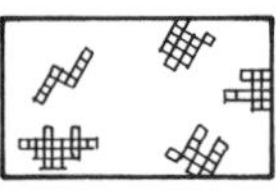

CRYSTALS START TO FORM AS METAL COOLS.

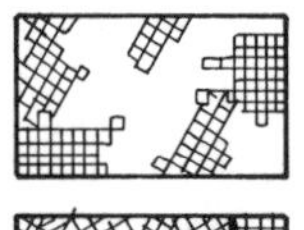

AS THEY GROW, CRYSTALS "BUMP INTO" ONE ANOTHER, FORMING IRREGULAR GRAINS.

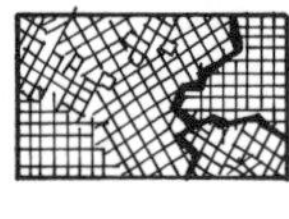

SOLID METAL. THE DARK LINE TRACES A GRAIN BOUNDARY.

CRYSTALS MOVE MOST EASILY WITHIN A SEMI-ORDERED STRUCTURE. CRYSTALS AT A GRAIN BOUNDARY ARE CAUGHT IN A "LOG JAM" WITH THE RESULT THAT THE METAL IS TOUGH, HARD TO WORK.

WHEN WORKED, LARGE CRYSTALS ARE BROKEN INTO SMALLER ONES – CREATING MORE GRAIN BOUNDARIES. THE METAL IS WORK-HARDENED. A SIMILAR CONDITION IS CREATED WHEN METAL IS RAPIDLY COOLED: CRYSTALS DO NOT HAVE TIME TO GROW FROM A FEW CLUSTERS IN AN ORGANIZED PATTERN, BUT INSTEAD FORM MANY SMALL UNORIENTED GRAINS.

IN TIME EVEN AT ROOM TEMPERATURE CRYSTALS WILL REALIGN THEMSELVES INTO AN ORGANIZED LATTICE. BY HEATING METAL WE ACCELERATE THE MOVEMENT OF ATOMS AND THE SUBSEQUENT RECRYSTALLIZATION. THIS PROCESS IS CALLED ANNEALING.

IN ITS ANNEALED STATE THE CRYSTAL ARRANGEMENT CONTAINS IRREGULARITIES CALLED VACANCIES AND DISLOCATIONS. THESE FACILITATE CRYSTAL MOVEMENT AND SO CONTRIBUTE TO MALLEABILITY. IF NONFERROUS METALS ARE HELD AT ANNEALING TEMPERATURES FOR LONG PERIODS THE RESULT IS A MORE PERFECT ARRANGEMENT AND A LESS MALLEABLE MATERIAL. THIS IS CALLED HEAT HARDENING.

General:

- GOLD WAS PROBABLY THE SECOND METAL TO BE WORKED BY EARLY MAN, BEING DISCOVERED AFTER COPPER. QUALITY GOLD WORK CAN BE FOUND FROM AS EARLY AS 3000 BC.
- IF ALL THE GOLD EVER FOUND (ABOUT 120,000 TONS) WERE CAST INTO A SINGLE INGOT IT WOULD MAKE ONLY A 20 YARD CUBE.
- ONE OUNCE OF GOLD CAN BE FLATTENED TO A SHEET THAT WILL COVER 100 SQUARE FEET OR DRAWN TO A WIRE ALMOST A MILE LONG.
- GOLD CAN BE MADE INTO A FOIL THAT IS LESS THAN FIVE MILLIONTHS OF AN INCH THICK — .0000033" OR .00025 MM. AT THIS POINT IT IS SEMI-TRANSPARENT.
- GOLD HAS A FACE-CENTERED CUBIC CRYSTAL.
- PICKLES FOR GOLD WOULD INCLUDE SPAREX #2 OR A MIXTURE OF 1 PART NITRIC ACID (70%) WITH 8 PARTS WATER.
- GOLD DISSOLVES IN AQUA REGIA AND SOLUTIONS OF CHLORINE AND POTASSIUM CYANIDE OR SODIUM CYANIDE.

DATA: Au

MELTING POINT:
1063 °C
1945 °F

HARDNESS: 2-2.5

SPECIFIC GRAVITY:
CAST 19.23
WORKED 19.29-19.34

ATOMIC WEIGHT:
197.2

gold-filled

REFERS TO A MATERIAL ON WHICH A LAYER OF GOLD HAS BEEN BONDED (FUSED). THE RESULTING INGOT IS ROLLED OR DRAWN TO MAKE SHEET AND WIRE. A STANDARD PRACTICE IS TO CLAD THE BASE WITH 10% (WEIGHT) 12K GOLD. SINCE 12K IS HALF PURE IT MEANS THE FINAL RESULT, IF IT WERE MELTED DOWN AND ASSAYED WOULD EQUAL 1/20 OR 5% PURE GOLD. THIS IS MARKED AS: 1/20 G.F. THE ADVANTAGES OF THIS OVER PLATING ARE THAT A THICKER GOLD LAYER MAY BE ACHIEVED, AND THAT BECAUSE OF WORKING THE GOLD IS DENSER.

THE TERM ROLLED GOLD REFERS TO A SIMILAR MATERIAL THAT HAS ONLY HALF AS THICK A GOLD LAYER: 1/40 TH.

ONE NAME FOR THIS CLADDING OF SILVER OVER COPPER IS SHEFFIELD PLATE, FROM AN INDUSTRY CENTERED IN SHEFFIELD, ENGLAND IN THE 1700S.

Fineness of Gold Karats

SINCE FINE (PURE) GOLD IS TOO SOFT FOR MOST USES IT IS ALLOYED WITH OTHER METALS TO ACHIEVE A DESIRED HARDNESS, COLOR, AND MELTING POINT. SILVER AND COPPER ARE THE TWO MOST COMMON ADDITIVES BUT MANY OTHER METALS CAN BE USED.

THE RELATIVE AMOUNT OF GOLD IN AN ALLOY IS CALLED THE KARAT. THIS SIGNIFIES PROPORTION AND SHOULD NOT BE CONFUSED WITH CARAT.

Karat	Fineness		Karat	Fineness	
1K	.0417		13 K	.5417	
2	.0833		14	.5833	
3	.1250		15	.6250	
4	.1667		16	.6667	2/3
5	.2083		17	.7083	
6	.2500	1/4	18	.7500	3/4
7	.2917		19	.7917	
8	.3333	1/3	20	.8333	
9	.3750		21	.8750	
10	.4167		22	.9167	
11	.4583		23	.9583	
12	.5000	1/2	24	1.000	

VOLUNTARY PRODUCT STANDARD 70-76

HAS SET LEGAL TOLERANCES SINCE 1976. IT ALLOWS VARIATION OF 3 PARTS PER 1000 (.072K) ON UNSOLDERED GOODS AND .007 (.168K) ON SOLDERED OBJECTS. THIS IS CALLED PLUMB (I.E. ACCURATE) GOLD. MANUFACTURERS WERE GIVEN UNTIL 1981 TO DISPOSE OF THEIR OLD MERCHANDISE MADE AT LOWER STANDARDS.

COMMON ALLOYS: SEE PAGE 141 FOR OTHERS.

NAME	% Au	% Ag	% Cu	% Ni	% Zn	MELTING POINT °C	MELTING POINT °F	SPEC. GRAV.
22K COINAGE	90	10				940	1724	17.2
18 K YELLOW	75	12½	12½			904	1660	15.5
18K WHITE	75		5	15	5	904	1660	15.7
14K YELLOW	58	25	17			802	1476	13.4
14K WHITE	58		20	14½	7½	927	1700	13.7
10K YELLOW	42	12	41			786	1447	11.6
10K WHITE	42		26	20	11	927	1760	11.8

MATERIALS

Testing

TO BE SCIENTIFICALLY ACCURATE A SAMPLE MUST BE ASSAYED IN A TESTING LABORATORY. THESE TWO TESTS HAVE BEEN USED FOR MANY YEARS AND WITH A LITTLE EXPERIENCE USUALLY WILL BE ACCURATE ENOUGH FOR THE CRAFTSPERSON.

Is it gold?

WITH A SMALL FILE, MAKE A SCRATCH IN AN INCONSPICUOUS SPOT. WITH A WOODEN, GLASS, OR PLASTIC STICK APPLY A DROP OF 70% NITRIC ACID TO THIS SPOT AND OBSERVE THE REACTION. WHEN DONE, RINSE WELL IN RUNNING WATER.

- NO REACTION → IT'S GOLD
- BRIGHT GREEN BUBBLING ALL OVER → IT'S BRASS
- GREEN ONLY IN SCRATCH → GOLD LAYER OVER BRASS
- MILKY IN SCRATCH → GOLD OVER SILVER

EVEN GOLD MAY BE BOUGHT TOO DEAR.

—GERMAN PROVERB

What karat is it?

SAFETY ALERT

DETERMINING KARAT REQUIRES A TESTING KIT:

- NITRIC ACID
- AQUA REGIA
- SAMPLES OF KNOWN KARAT
- TOUCHSTONE (SLATE OR CERAMIC)

THE OBJECT TO BE TESTED IS RUBBED ON THE STONE (I.E. "TOUCHED") TO LEAVE A STREAK. A PARALLEL LINE IS MADE WITH ONE OF THE TEST NEEDLES. ACID IS THEN WASHED ACROSS BOTH MARKS AND THE REACTIONS ARE OBSERVED. WHEN THE SAMPLE COLORS AT THE SAME RATE AS THE TEST STREAK, A MATCH HAS BEEN MADE. NITRIC ACID IS USED FOR LOW KARAT GOLDS; AQUA REGIA IS NEEDED FOR HIGHER KARATS.

alloys

Formulas:

TO RAISE KARAT:

A. AMOUNT TO BE RAISED X ITS KARAT
B. SAME AMOUNT X DESIRED KARAT
C. DIFFERENCE BETWEEN A & B
D. DIFFERENCE BETWEEN PURE (24) & GOAL
E. DIVIDE C BY D

EXAMPLE: IF YOU HAVE 6 DWTS. OF 10K AND WANT 14K

A. 6 × 10 = 60
B. 6 × 14 = 84
C. 84 − 60 = 24
D. 24 − 14 = 10
E. 24 ÷ 10 = 2.4

ADDING 2.4 DWTS OF FINE GOLD WILL RAISE THE INITIAL 10K SAMPLE TO 14K.

TO LOWER KARAT:

A. AMOUNT TO BE LOWERED X ITS KARAT
B. SAME AMOUNT X DESIRED KARAT
C. DIFFERENCE BETWEEN A AND B
D. DIVIDE DIFFERENCE (C) BY THE QUALITY BEING MADE

EXAMPLE: YOU HAVE 9 DWTS OF 18K AND WANT 12K

A. 9 × 18 = 162
B. 9 × 12 = 108
C. 162 − 108 = 54
D. 54 ÷ 12 = 4.5

ADD 4.5 DWTS. OF ALLOY TO CHANGE THE 18 SAMPLE TO 12K.

- A MIXTURE OF ROUGHLY EQUAL PARTS OF GOLD AND SILVER IS CALLED ELECTRUM. MAXIMUM HARDNESS OF THIS ALLOY IS AT 50:50.
- THE HARDEST ALLOY OF GOLD, SILVER, AND COPPER IS REACHED AT 50:25:25. THIS WILL BE 12 KARAT YELLOW.
- ANY INCREASE OF THE COPPER CONTENT IN A GOLD ALLOY WILL LOWER ITS MELTING POINT UP TO 18% COPPER (MELTS: 880°C 1642°F). TO CONTINUE LOWERING, AS WHEN MAKING SOLDER, ADD SILVER.
- MANY KINDS AND COLORS OF GOLD SOLDER ARE COMMERCIALLY AVAILABLE BUT IN A PINCH A GOLD OF A LOWER KARAT MAY BE USED.
- WHITE GOLD USUALLY HAS 10-20% NICKEL AND CAN CONTAIN ZINC, COPPER, TIN, AND MANGANESE. IT HAS NO SILVER.

FLUXES: WHEN POURING GOLD INGOTS:

SAFETY ALERT

AN EVEN MIXTURE OF POWDERED CHARCOAL AND AMMONIUM CHLORIDE (SAL AMMONIAC) IS SPRINKLED ON THE METAL DURING MELTING. IT WILL YIELD A BRIGHT TOUGH INGOT THAT WILL WITHSTAND ROLLING.

IF IRON OR STEEL ARE PRESENT USE A FLUX OF 1 PART POTASSIUM NITRATE (SALTPETRE) AND 2 PARTS POTASSIUM CARBONATE. MELT WITH THIS TO PURIFY; COOL, REMELT WITH THE SAL AMMONIAC FLUX AND POUR INTO WARM (NOT HOT) MOLD.

PLATINUM IS A DENSE WHITE METAL THAT HAS A HIGH RESISTANCE TO CORROSION. IT WAS DISCOVERED BY SPANIARDS IN COLOMBIA, SOUTH AMERICA IN 1538. THEY CALLED IT "PLATINA" BECAUSE OF ITS SIMILARITY TO SILVER ("PLATA"). TODAY WE SPEAK OF THESE SIX METALS COLLECTIVELY AS THE PLATINUM GROUP.

METAL	RELATIVE OCCURRENCE	MELTING POINT		HARDNESS	SPECIFIC GRAVITY
PLATINUM	60 %	3224° F	1773° C	4–4.5	21.45
PALLADIUM	30	2829	1555	4–4.5	12
RHODIUM	4	3571	1966	5	12.4
RUTHENIUM	3	4530	2500	6.5	12.2
IRIDIUM	2	4449	2355	6.5	22.4
OSMIUM	1	5550	3066	7	22.5

BECAUSE OF THEIR HIGH MELTING POINTS THESE METALS REQUIRE AN OXYGEN-AND-FUEL STYLE TORCH FOR SOLDERING OR CASTING.

BECAUSE OF THEIR TOUGHNESS AND GREAT RESISTANCE TO TARNISH THESE METALS ARE WELL-SUITED TO SETTINGS FOR PRECIOUS STONES.

TO TEST, HEAT A SAMPLE TO BRIGHT RED AND AIR COOL. METALS OF THE PLATINUM GROUP WILL REMAIN BRIGHT AND SHINY. BECAUSE OF THIS RESISTANCE NO FLUX IS NEEDED WHEN SOLDERING.

MORE THAN HALF OF ALL PLATINUM METALS MINED ARE USED BY THE JEWELRY INDUSTRY. OTHER USES INCLUDE:

- MEDICAL IMPLANTS.
- ARCHITECTURAL DECORATION, AS LEAF.
- PLATING ON THE TIPS OF FOUNTAIN PENS FOR DURABILITY (ESP. OSMIUM).

RHODIUM IS OFTEN PLATED OVER STERLING ARTICLES TO PROVIDE A BRIGHT TARNISH-RESISTANT OUTER LAYER. ITS REFLECTIVITY INDEX (85 % OF THE VISIBLE SPECTRUM) IS SLIGHTLY LOWER THAN STERLING BUT THIS LOSS OF SHINE IS GENERALLY IMPERCEPTIBLE. WORKED RHODIUM HAS A VICKERS HARDNESS OF 100 BUT ELECTROPLATED RHODIUM HAS A VICKERS OF 775–820, SHOWING THAT IT IS VERY HARD AND DURABLE.

PLATINUM GROUP METALS CAN BE CAST BUT BECAUSE OF THEIR HIGH MELTING POINTS A SPECIAL INVESTMENT MUST BE USED. WHEN BUYING THIS, REQUEST A DATA SHEET AND FOLLOW MIXING DIRECTIONS CAREFULLY. NO FLUX IS NEEDED WHEN MELTING.

CLEANLINESS IS VERY IMPORTANT WHEN HEATING METALS OF THE PLATINUM GROUP. AN OXIDIZING FLAME IS RECOMMENDED. CONTAMINATION BY SILVER, ALUMINUM, IRON, LEAD, ETC. WILL CAUSE INTERCRYSTALINE CRACKING AT THE GRAIN BOUNDARIES. IF CONTAMINATION OCCURS THERE IS NO WAY TO CORRECT THE PROBLEM METALLURGICALLY; THE DAMAGED AREA MUST BE CUT OUT AND REPLACED WITH A PATCH.

SAFETY ALERT

PLATINUM GROUP METALS DISSOLVE SLOWLY IN AQUA REGIA.

RHODIUM WAS SEPARATED FROM PLATINUM IN 1803 AND TAKES ITS NAME FROM THE GREEK WORD "RODON" (ROSE) BECAUSE OF THE COLORS OF THE METALLIC SALTS.

MATERIALS

SILVER, KNOWN IN THE ANCIENT WORLD AS ARGENTUM WAS AT ONE TIME THOUGHT TO BE MORE PRECIOUS THAN GOLD SINCE IT APPEARED IN NATURE LESS COMMONLY.

PURE SILVER, LIKE PURE GOLD, IS TOO SOFT FOR MOST USES AND SO IS ALLOYED. THOUGH MANY METALS MAY BE USED, COPPER IS PREFERED SINCE IT GREATLY TOUGHENS THE ALLOY WITHOUT DETRACTING FROM THE BRIGHT SHINE CHARACTERISTIC OF SILVER.

DATA: Ag

MELTING POINT:
960.5° C
1761° F

HARDNESS: 2-2½

SPECIFIC GRAVITY: 10.5

ATOMIC WEIGHT: 107.88

Sterling IS THE ALLOY MOST COMMONLY USED IN JEWELRY MAKING AND SILVERSMITHING. IT WAS ADOPTED AS A STANDARD ALLOY IN ENGLAND IN THE TWELTH CENTURY WHEN KING HENRY II IMPORTED REFINERS FROM AN AREA OF GERMANY KNOWN AS THE EASTERLING. THE PRODUCT THEY MADE WAS OF A CONSISTENT QUALITY AND CAME INTO USAGE AS CURRENCY BY 1300 WHEN IT WAS KNOWN AS EASTERLING SILVER.

sterling:

MELTING POINT:
893° C
1640° F

HARDNESS: 2½-3

SPECIFIC GRAVITY:
10.41

ANOTHER COMMON ALLOY CONTAINS SLIGHTLY MORE COPPER — 10 TO 20% — AND IS CALLED COIN SILVER. IT MELTS AT A TEMPERATURE SLIGHTLY LOWER THAN STERLING AND IS MORE LIKELY TO TARNISH. A 90% ALLOY WAS USED IN U.S. COINS UNTIL 1966 BUT NOW NO SILVER IS USED. THIS TREND AWAY FROM SILVER COINS HAS BEEN INTERNATIONAL.

Heat Hardening

IN CONVENTIONAL WORK-HARDENING, METAL IS MADE RIGID BY UPSETTING THE ORDERLY ARRANGEMENT OF GRAINS. A SIMILAR RIGIDNESS CAN BE ACHIEVED BY REDUCING THE NUMBER OF DISLOCATIONS AND VACANCIES; I.E. BY CREATING EXTREME REGULARITY. THIS IS ACHIEVED BY WARMING THE METAL SUFFICIENTLY TO BEGIN RECRYSTALIZATION AND HOLDING AT THIS TEMPERATURE LONG ENOUGH TO ALLOW GRADUAL ORDERED CRYSTAL GROWTH.

• FOR STERLING: AFTER ALL SOLDERING IS DONE, HEAT TO 280°C (536°F) AND HOLD FOR 2½ HOURS. QUENCH IN PICKLE OR WATER AND FINISH AS USUAL.

ELECTROLYTIC CLEANING OF STERLING:

THIS PROCEDURE LENDS ITSELF TO THE REMOVAL OF TARNISH FROM FLATWARE OR HOLLOWWARE.

IN AN ALUMINUM POT (OR A POT LINED WITH ALUMINUM FOIL) MIX A DILUTE SOLUTION OF EQUAL PARTS OF BAKING SODA, SALT, AND LIQUID SOAP. A QUARTER CUP OF EACH TO A GALLON OF WATER WOULD BE A TYPICAL MIXTURE. SET THE STERLING INTO THE POT, BRING MIX TO A BOIL, AND ALLOW TO STAND FOR A FEW MINUTES. OXIDES ARE TRANSFERRED TO THE ALUMINUM.

RINSE IN WATER AND, IF A UTENSILE, WASH BEFORE USING.

ANNEALED FINE SILVER HAS A HARDNESS OF VICKERS 26 (TENSILE STRENGTH OF 9 TONS PER SQ. INCH). COLD WORKING INCREASES THE HARDNESS TO VICKERS 95-100, (TENSILE STRENGTH OF 20-22 TONS PER SQUARE INCH).

BRITTANIA SILVER: 958.3 PARTS PER 1000 WAS THE LEGAL ALLOY IN ENGLAND FROM 1697 TO 1719, CONTRIVED TO DISCOURAGE THE MELTING OF COINS. IT IS STILL A LEGAL ALLOY.

SILVER RESISTS AQUA REGIA SINCE HYDROCHLORIC ACID FORMS A DENSE CHLORIDE FILM THAT RESISTS CORROSION.

Fineness— IS TESTED WITH SCHWERTER'S SOLUTION:

1 TSP POTASSIUM DICHROMATE
¾ FL. OZ. 70% NITRIC ACID
½ FL. OZ. DISTILLED WATER

THIS SOLUTION IS APPLIED TO THE METAL AT ROOM TEMPERATURE AND THE COLOR IS OBSERVED

PURE SILVER → BRIGHT BLOOD RED
.925 → DARK RED
.900 → DARK RED
.800 → BROWN
.500 → GREEN
GERMAN (NICKEL) SILVER → DARK BLUE

MATERIALS *Copper & Brass*

COPPER IS AVAILABLE IN MORE THAN 100 ALLOYS. COMPREHENSIVE DATA IS AVAILABLE (FREE) FROM:

COPPER DEVELOPMENT ASSOC. INC.
405 LEXINGTON AVENUE
NEW YORK, N.Y. 10017
ASK FOR THE STANDARDS HANDBOOK

COPPER WAS PROBABLY THE FIRST METAL TO BE PUT TO USE BY OUR ANCESTORS AND REMAINS IMPORTANT TO US TODAY. IT CONDUCTS HEAT AND ELECTRICITY VERY WELL, CAN BE FORMED AND JOINED, AND COMBINES WITH MANY ELEMENTS TO FORM A BROAD RANGE OF ALLOYS.

8000 BC	COPPER DISCOVERED
6000 BC	EGYPTIANS USED COPPER WEAPONS
5000 BC	BEGINNING OF BRONZE AGE
3800 BC	EVIDENCE OF CONTROLLED BRONZE ALLOYING
2750 BC	EGYPTIANS MADE COPPER PIPES

COPPER IS SOLD IN STANDARD SHEETS 36" x 96" (3' x 8') AND IN COILS TWELVE AND EIGHTEEN INCHES WIDE. WHEN ORDERING SPECIFY HARD, HALF-HARD, OR ANNEALED.

DATA: Cu

MELTING POINT:
1083° C
1981° F

HARDNESS: 3

SPECIFIC GRAVITY:
8.96

ATOMIC WEIGHT:
63.54

COPPER MAY BE SOLD BY B&S GAUGE, THOUSANDTHS OF AN INCH, OR OZ. PER SQ. FOOT.

B&S	OZ/SQ.FT.
10	72
12	64
14	48
16	36
18	30
20	24
22	18

MATERIALS

WHEN COPPER IS HOT-ROLLED IT DEVELOPS A SLIGHTLY ROUGH SURFACE. FOR THIS REASON MOST CRAFTSPEOPLE PREFER COLD-ROLLED MATERIAL. COPPER ALLOY # 110 IS A COMMON CHOICE.

WHEN EXPOSED TO MOIST AIR COPPER FORMS POISONOUS ACETATES, SULPHATES, AND CHLORIDES KNOWN COLLECTIVELY AS VERDIGRIS. THE NAME COMES FROM "VERT-DE-GRICE" OLD FRENCH FOR "GREEN OF GREECE," A REFERENCE TO METAL SCULPTURES OF ANTIQUITY. BECAUSE OF THESE COMPOUNDS, WORKERS SHOULD ALWAYS WASH THEIR HANDS AFTER HANDLING COPPER FOR A LONG TIME. COPPER COOKWARE AND SERVING PIECES SHOULD EITHER BE PLATED WITH A NON-CORROSIVE METAL LIKE TIN OR WASHED BEFORE EACH USE.

MOST COPPER IS ELECTROLYTICALLY REFINED; I.E. ELECTRICALLY DEPOSITED ON AN ANODE. THIS PRODUCT IS PURE BUT CONTAINS OXYGEN ATOMS SCATTERED THROUGHOUT THE METAL. WHEN HEATED THIS FORMS CuO_2 THAT BREAKS DOWN THE BOND BETWEEN CRYSTALS AND CAN WEAKEN THE METAL AS MUCH AS 60%. TO ALLEVIATE THIS PROBLEM MOST COPPER IS ALLOYED WITH A DEOXIDIZER LIKE PHOSPHOROUS.

Japanese Alloys

SHAKU-DO → .5 TO 4 % GOLD, BALANCE COPPER
M.P. 1968–1980° F
THIS IS KNOWN FOR THE DEEP PURPLE COLOR ACHIEVED THROUGH OXIDATION.

SHIBU-ICHI → 75 % COPPER, 25 % SILVER
M.P. 1775°F 968°C
THIS IS A SILVERY PINK ALLOY THAT DARKENS AND RETICULATES EASILY.

BRASS & BRONZE

BRASS IS AN ALLOY OF COPPER AND ZINC THAT CAN ACHIEVE A WIDE RANGE OF PROPERTIES AND COLORS. THE BRONZE OF ANTIQUITY WAS A MIX OF 10-20% TIN WITH THE BALANCE BEING COPPER. TODAY THE TERM BRONZE REFERS TO ANY TIN-BEARING BRASS OR GOLDEN COLORED BRASS.

ALPHA BRASSES HAVE LESS THAN 36% Zn
(GOOD FOR COLD WORKING)
BETA BRASSES HAVE MORE THAN 36% Zn
(GOOD FOR HOT WORKING)

BRONZE IS SOMETIMES CALLED BELL METAL OR GUN METAL.

SEE PAGE 141 FOR TECHNICAL DATA ON COMMON BRASSES.

TO DISTINGUISH BRASS FROM BRONZE:

DISSOLVE A SMALL SAMPLE IN 50:50 SOLUTION OF NITRIC ACID AND WATER. TIN WILL BE INDICATED BY A WHITE PRECIPITATE; METASTANNIC ACID.

Properties:

ALUMINUM IS THE MOST ABUNDANT METALLIC ELEMENT ON EARTH. BECAUSE OF ITS LIGHT WEIGHT, RESISTANCE TO CORROSION AND ABILITY TO ALLOY WELL IT IS USED STRUCTURALLY (BUILDINGS, AIRCRAFT, CARS), AS ARCHITECTURAL TRIM (SIDING), AND IN FUNCTIONAL OBJECTS LIKE COOKWARE.

IT IS THE SECOND MOST MALLEABLE AND SIXTH MOST DUCTILE METAL. IT IS USUALLY FOUND IN BAUXITE AS AN OXIDE CALLED ALUMINA: Al_2O_3

DATA: Al

MELTING POINT:
660 °C
1220 °F

HARDNESS: 3

SPECIFIC GRAVITY: 2.7

ATOMIC WEIGHT: 26.97

History

THOUGH THE EXISTENCE OF ALUMINUM WAS THEORIZED IN THE 1700S IT WAS NOT ISOLATED UNTIL 1825. WHEN THE WASHINGTON MONUMENT WAS COMPLETED IN 1884 A 100 OZ. PYRAMID OF ALUMINUM WAS MADE TO CROWN IT. AT THE TIME THIS WAS THE LARGEST MASS OF ALUMINUM EVER MADE; BEFORE PLACEMENT IT WAS DISPLAYED IN TIFFANY'S WINDOW IN NEW YORK.

COMMERCIAL PRODUCTION WAS DEVISED IN 1886 AND MANY ALLOYS HAVE BEEN DEVELOPED SINCE THEN.

PURE ALUMINUM IS CALLED	2S
Al + 1.25% MANGANESE =	3S
Al + 1.2% Mn AND 1% Mg =	4S
Al + 4% Cu, .5Mn, .5 Mg =	17S
Al + 4.5 Cu, .5 Mn, 1.5 Mg =	24S

JOINING-

ALUMINUM CAN BE SOLDERED AND JOINED ONLY WITH SPECIAL SOLDERS, MANY OF WHICH ARE SOLD WITH THEIR OWN FLUX. WELDING CAN BE DONE WITH 43S OR #717 WIRE USED WITH #33 FLUX. MORE DETAILED SPECIFICATIONS ARE AVAILABLE FROM A SUPPLIER. WELDING IS MADE EASIER WITH A TIG (TUNGSTEN INERT GAS) WELDER.

Anodizing

IS A PROCESS OF ELECTRICALLY CAUSING THE FORMATION OF A RESISTANT OXIDE FILM ON THE SURFACE OF ALUMINUM. THE FILM MAY BE COLORED WITH DYES WHICH CAN GIVE FINISHED ALUMINUM PRODUCTS A WIDE RANGE OF COLOR POSSIBILITIES.

Nickel & its Alloys~

THE WORD NICKEL MEANS "DECEIVER" IN GERMAN AND WAS GIVEN TO THE ORE (NICCOLITE) BECAUSE IT WAS EASILY MISTAKEN FOR COPPER ORE.

NICKEL IS A HARD WHITE METAL THAT IS USED PRIMARILY AS AN ALLOYING INGREDIENT: IT INCREASES HARDNESS AND RESISTANCE TO CORROSION WITHOUT IMPAIRING DUCTILITY.

DATA: Ni

MELTING POINT:
1445 °C
2651 °F

SPECIFIC GRAVITY: 8.9

ATOMIC WEIGHT: 58.69

SOME OF THE COMMON ALLOYS ARE:

- NICKEL SILVER (GERMAN SILVER, WHITE BRASS)
  Cu 60%
  Ni 20
  Zn 20
  THIS METAL IS USED IN JEWELRY BECAUSE OF ITS LOW COST AND GENERALLY FAVORABLE WORKING PROPERTIES. IT CAN BE FORGED, STAMPED, SOLDERED, AND POLISHED. THOUGH IT CAN BE CAST THIS IS DIFFICULT BECAUSE OF ITS HIGH MELTING POINT AND TENDENCY TO OXIDIZE.

- MONEL METAL
  Ni 67%
  Cu 30%
  BAL. Fe, Mn, C, Si, S
  THIS TOUGH, OXIDE-RESISTANT METAL HAS MANY USES IN INDUSTRY BUT IS RARELY USED IN THE CRAFTS. IT MELTS AROUND 1300 °C, 2370 °F.

- NICHROME
  Ni 80%
  Cr 20
  BECAUSE OF ITS ABILITY TO REDUCE OXIDATION AND ITS HIGH MELTING POINT (1400°C, 2550°F) THIS WIRE IS USED AS A HEATING ELEMENT IN ELECTRIC KILNS.

- NICKEL ALLOY #752
  Cu 65%
  Ni 18
  Zn 17
  THIS ALLOY WILL "SWELL" WHEN HEATED ABOVE 980 °C (1800°F). WHEN ITS RETICULATED OXIDE SKIN IS REMOVED IN A NITRIC ACID PICKLE THE METAL WILL BE FOUND TO BE DRAMATICALLY PERFORATED. IT MAY BE SOLDERED AND POLISHED.

IRON IS THE WORLD'S MOST WIDELY USED METAL. IT CAN BE ALLOYED WITH A WIDE RANGE OF ELEMENTS TO PRODUCE MANY DIVERSE PROPERTIES.

IRON ORE USUALLY CONTAINS SULPHUR, PHOSPHOROUS, SILICON, AND CARBON. WHEN ALL BUT 3-4 % CARBON HAVE BEEN SMELTED OUT THE RESULTING METAL IS POURED INTO INGOTS AND CALLED CAST IRON OR PIG IRON. FURTHER REFINING IS NECESSARY TO MAKE A STEEL OF GOOD WORKING QUALITIES:

DATA: Fe

MELTING POINT: 1539 °C / 2802 °F

SPECIFIC GRAVITY: 7.87

ATOMIC WEIGHT: 55.85

.15 — .3 % CARBON — MILD (LOW-CARBON) STEEL : CANNOT BE HARDENED
.3 — .5 " — MEDIUM CARBON STEEL } PLAIN CARBON STEEL; USED FOR TOOLS
.5 — 1.6 " — HIGH CARBON STEEL }
2.5 " — MALLEABLE IRON: FOR CAST & MACHINED PARTS

code system: SET BY THE SOCIETY OF AUTOMOTIVE ENGINEERS (SAE) AND AMERICAN IRON AND STEEL INDUSTRY (AISI).

- LETTER INDICATES TYPE OF FURNACE USED IN SMELTING.
- SECOND TWO DIGITS INDICATE MAJOR ALLOYING MATERIAL, IN CODE.
- LAST TWO DIGITS INDICATE THE PER CENT OF THIS ALLOY.

EXAMPLE: **B1065** (CODE FOR CARBON)

THIS IS A PLAIN CARBON STEEL MADE IN AN ACID BESSEMER FURNACE THAT CONTAINS .65 % CARBON. IT WOULD BE USED FOR SPRINGS, TOOLS, AND BLADES.

OTHER METALS USED FOR STEEL ALLOYS ARE:
CHROMIUM: FOR CORROSION RESISTANCE; 10-20% USED IN STAINLESS.
MANGANESE: INCREASES HARDENABILITY AND TENSILE STRENGTH.
MOLYBDENUM: INCREASES CORROSION RESISTANCE; HIGH TEMP. STRENGTH.
TUNGSTEN: FORMS HARD ABRASION-RESISTANT PARTICLES CALLED TUNGSTEN CARBIDE. USED FOR CUTTING EDGES.

Hardening Steel:

(SEE ALSO PAGE 122)

STEEL METALLURGY IS A COMPLEX FIELD AND DESERVES MORE SPACE THAN CAN BE GIVEN HERE. IN A SIMPLIFIED WAY, THOUGH, THIS IS HOW TOOL STEELS MAKE THEIR MAGIC.

ANNEALED CARBON STEEL CONTAINS FERRITE, WHICH IS MALLEABLE, AND HARD PARTICLES OF CARBIDE CALLED CEMENTITE.

WHEN HEATED TO A GLOWING RED THE CARBIDES DISSOLVE INTO THE IRON; THE RESULT IS CALLED AUSTENITE. THE TEMPERATURE AT WHICH THIS OCCURS IS CALLED THE CRITICAL RANGE.

PEARLITE

IF THE STEEL IS COOLED QUICKLY THE RESULT IS A HARD NEEDLE-LIKE STRUCTURE CALLED MARTENSITE. THIS IS WHAT GIVES CARBON STEEL ITS TOUGHNESS. UNFORTUNATELY IT ALSO MAKES IT BRITTLE.

BY HEATING THIS TO A PRESCRIBED TEMPERATURE AND COOLING IT AT A CERTAIN RATE THE STRESS MAY BE RELIEVED WITHOUT REMOVING ALL THE HARDNESS. THE RESULT CONTAINS HARD CEMENTITE PARTICLES HELD IN A TOUGH MATRIX OF MARTENSITE. THIS PROCESS IS CALLED TEMPERING OR DRAWING THE TEMPER, AND USUALLY TAKES PLACE BETWEEN 200-350 °C (400-600 °F).

AUSTENITE

MARTENSITE

IT IS IMPORTANT TO DISTINGUISH BETWEEN WEAR RESISTANCE AND HARDNESS. THE FIRST DEPENDS ON THE NUMBER AND HARDNESS OF THE PARTICLES; THE LATTER ON THE STRENGTH OF THE MATRIX. FOR INSTANCE, GRAVEL IN MUD WILL NOT MAKE A DURABLE MATERIAL EVEN THOUGH THE GRAVEL (PARTICLES) ARE HARD. IN STEEL THIS PROPERTY IS MOSTLY CONTROLLED BY THE ALLOY AND NOT BY HEAT TREATMENT. INCREASED CARBON - UP TO 1.6 % - MEANS MORE PARTICLES BUT LESS MATRIX OR INCREASED WEAR RESISTANCE BUT DECREASED HARDNESS.

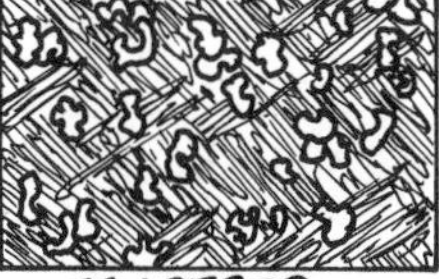
TEMPERED MARTENSITE

White Metal MATERIALS

THE TERM "WHITE METALS" REFERS TO THE SEVERAL MALLEABLE, GRAY-COLORED METALS WITH LOW MELTING POINTS SHOWN HERE, AND TO THE ALLOYS HAVING THESE METALS AS PRIMARY INGREDIENTS. THESE ARE ALSO CALLED EASILY FUSIBLE ALLOYS, POT METAL, AND TYPE METAL, THE LATTER NAME COMING FROM THE USE OF THESE ALLOYS IN MAKING PRINTERS' TYPE.

METAL	MELTING POINT		S.G.
LEAD	621 °F	327 °C	11.3
TIN	450	233	7.3
BISMUTH	520	271	9.8
ANTIMONY	1168	631	6.6
CADMIUM	610	321	8.7

TYPICAL ALLOYS

MELTING RANGE		LEAD	TIN	BISMUTH	CADMIUM
60-68 °C	140-154 °F	26.7 %	13.3 %	50 %	10 %
72	132	27.8	12.4	50.5	9.3
70-80	128-176	34.5	9.3	50	6.2
70-84	1 -153	30.9	14.9	50.8	3.4
94-104	171-189	22	22	56	
94-143	171-259	33.4	33.3	33.3	
94-149	171-270	16	17	67	
143-163	259-295	43	43	14	

BECAUSE OF THEIR LOW MELTING POINTS WHITE METALS CAN BE MELTED WITH ALMOST ANY TORCH. TO HELP REDUCE OXIDATION MELTING IS BEST DONE IN A SMALL-NECKED CRUCIBLE OR LADLE.

DURING MELTING THE METAL MAY BE PROTECTED FROM OXYGEN BY A COATING OF OLIVE OIL, LINSEED OIL, OR LARD. THESE FLOAT ON THE SURFACE OF THE MELT; IN POURING THE METAL WILL SLIDE OUT FROM UNDERNEATH.

CRITICISM COMES EASIER THAN CRAFTSMANSHIP.

ZEUXIUS 400 BC

CAUTION:

SAFETY ALERT

THE FUMES PRODUCED BY THESE METALS ARE DANGEROUS. HEAT UNDER A VENTILATING HOOD OR ARRANGE A FAN OVER YOUR SHOULDER.

LEAD CAN BE ABSORBED THROUGH THE SKIN: WASH WELL AFTER HANDLING ANY LEAD-BEARING ALLOY. IT IS ESPECIALLY UNWISE TO EAT, DRINK, OR SMOKE IN AN AREA WHERE WHITE METAL IS BEING WORKED.

% TIN	% LEAD	MELTING POINT	
100	0	450 °F	233 °C
0	100	621	327
83.4	16.6	381	194
80	20	372	189
75	25	367	186
66.6	33.4	385	196
50	50	466	241
25	75	552	289

Pewter & Brittania

PEWTER AS USED IN ANTIQUITY AND ASSOCIATED WITH COLONIAL AMERICA WAS AN ALLOY OF LEAD AND TIN. IN THE LATE 1700'S A SUBSTITUTE ALLOY WAS DEVELOPED IN ENGLAND AND NAMED BRITTANIA METAL. TODAY THE WORDS PEWTER AND BRITTANIA ARE USED INTERCHANGEABLY AND REFER TO THIS:

91 % TIN
7 % ANTIMONY
2 % COPPER

IT CAN BE SAWN, SOLDERED, FORMED, AND CAST. FINISHING CAN BE DONE WITH FINE STEEL WOOL AND A MIX OF LAMPBLACK (SOOT) AND KEROSENE BLENDED TO A PASTE.

contamination

WHEN HEATED ABOVE THEIR MELTING POINT WHITE METALS WILL BURN PITS INTO GOLD, PLATINUM, SILVER, COPPER, BRASS, ETC. USE SEPARATE FILES AND SOLDERING TOOLS TO KEEP THESE METALS AWAY FROM EACH OTHER.

TO REMOVE WHITE METAL, FILE, SCRAPE, SAND OR USE EITHER OF THESE SOLUTIONS:

3 oz. GLACIAL ACETIC ACID
1 oz. HYDROGEN PEROXIDE

8 oz. FLUOBORIC ACID
1.6 oz 30% HYDROGEN PEROXIDE
22 oz. WATER

ALLOW WORK TO SOAK FOR SEVERAL HOURS.

THERMOSETTING PLASTICS ARE GENERALLY AVAILABLE AS LIQUIDS THAT REACT WITH A CATALYST OR HARDENER TO CROSS-LINK LARGE MOLECULES (POLYMERS) WITH SMALL ONES (MONOMERS) IN A PROCESS CALLED POLYMERIZATION. AFTER CURING THE RESULTANT MATERIAL CANNOT BE RETURNED TO ITS ORIGINAL STATE. THERMOSETTING PLASTICS ARE USUALLY EPOXIES OR POLYESTERS.

	EPOXIES	POLYESTERS
ADHESIVE STRENGTH	VERY GOOD	FAIR
COLOR STABILITY IN SUNLIGHT	FAIR	GOOD
CONTROL OVER CURING RATE	GOOD	FAIR
COST	TWICE POLYESTERS	HALF EPOXIES
SHRINKAGE	½ %	7%

SAFETY ALERT

SAFETY

THERMOSETTING PLASTICS PRODUCE FUMES THAT CAN CAUSE SEVERE DAMAGE, EVEN IN SMALL DOSES. SERIOUS VENTILATION FACILITIES ARE A MUST. SKIN IRRITATION IS ALSO LIKELY TO RESULT FROM CONTACT, SO GLOVES SHOULD BE WORN. SPECIFIC HEALTH HAZARDS ARE GIVEN IN DETAIL IN THE TWO BOOKS LISTED BELOW. ANYONE INTENDING TO WORK WITH THESE MATERIALS SHOULD DO SOME CAREFUL READING BEFORE GETTING STARTED.

Casting:

BECAUSE THEY ARE LIQUID, THERMOSETTING PLASTICS ARE COMMONLY USED TO FILL A MOLD OR ENCASE AN OBJECT. THE SEQUENCE GIVEN HERE PROVIDES A GENERAL INTRODUCTION TO THE PROCESS.

1. CAREFUL MEASUREMENT IS IMPORTANT SO A SENSITIVE SCALE IS NEEDED. WAX-COATED PAPER CUPS MAKE HANDY CONTAINERS FOR MEASURING, MIXING, AND AS MOLDS FOR SMALL SLABS. POUR OUT THE DESIRED AMOUNT OF RESIN; THINNER MAY BE ADDED TO FACILITATE THE REMOVAL OF BUBBLES.
2. WEIGH AND GRADUALLY STIR IN ADDITIVES. ADD PIGMENTS TO ACHIEVE THE DESIRED HUE—USUALLY A LITTLE GOES A LONG WAY.
3. WEIGH AND ADD CATALYST. MIX THOROUGHLY (SEVERAL MINUTES) BUT AVOID WHIPPING UP BUBBLES. SEE CHART BELOW.
4. POUR INTO MOLD. A RELEASE AGENT LIKE POLYVINYL ALCOHOL ON THE MOLD WILL MAKE REMOVAL EASIER. THE MOLD CAN BE MADE OF PLASTIC, RUBBER, WAX, PLASTICENE, OR SEALED PLASTER.
5. CURING WILL USUALLY TAKE ABOUT 24 HOURS, LESS FOR CASTINGS UNDER ½" THICK. EVEN WHEN CURED THE PLASTIC WILL HAVE A GUMMY LAYER ON TOP. TEST CURING BY POKING THROUGH THIS WITH A PIN. WHEN THE PLASTIC IS SOLID THE GUMMY LAYER IS SCRAPED OFF AND THE MATERIAL CAN BE SAWN, FILED, SANDED, AND BUFFED.

Layering:

BECAUSE OF THE SLOW RATE OF CURE IT IS POSSIBLE TO BUILD UP LAYERS OF PLASTIC WITH EXCELLENT BONDING: IN FACT LAYERING IS RECOMMENDED FOR CASTINGS OVER 2" THICK. FOLLOW THE INSTRUCTIONS ABOVE THROUGH #4. AT THIS POINT THE RESIN/CATALYST MIX CAN BE DIVIDED UP AND GIVEN VARIOUS COLORS OR PROPERTIES. EACH LAYER MAY BE ADDED AS SOON AS THE PRECEDING ONE HAS BEGUN TO GEL.

Embedding:

ANY WATER-FREE OBJECT CAN BE EMBEDDED IN PLASTIC. THE PROCESS IS AS ABOVE, WITH THE OBJECT SET INTO PLACE MIDWAY INTO THE POURING. AS LONG AS THE FIRST LAYER IS GOOEY WHEN THE SECOND IS POURED, THERE WILL BE NO DIVISION LINE.

THICKER CASTINGS REQUIRE A SMALLER PROPORTION OF CATALYST:

CASTING THICKNESS	% CATALYST
¼"	2
½"	1
¾"	½
1"	¼

books

PLASTICS FOR JEWELRY - HARRY HOLLANDER WATSON-GUPTILL NYC: 1974

HEALTH HAZARDS MANUAL FOR ARTISTS - M. MCCANN FOUNDATION FOR COMMUNITY OF ARTISTS: NYC

THERMOPLASTICS ARE SOLID AT ROOM TEMPERATURE. THEY ARE COMMONLY AVAILABLE AS SHEETS, RODS, TUBES, AND BLOCKS. THESE SUBDIVISIONS AND BRAND NAMES ARE THERMOPLASTICS:

ACRYLIC – PLEXIGLAS, LUCITE, PERSPEX, ACRYLOID
POLYCARBONATE – LEXAN
POLYSTYRENE – STYROFOAM®

THERMOPLASTICS ARE LONG CHAIN-LIKE MOLECULES (POLYMERS) THAT LIE SIDE BY SIDE. WHEN HEATED THEY CAN BE BENT AND SO ALLOW FOR FORMING. WHEN REHEATED THE POLYMERS WILL RETURN TO THEIR ORIGINAL POSITION. BECAUSE OF THIS WE SAY THAT THERMOPLASTICS HAVE A "MEMORY."

heat forming acrylic

THERMOPLASTICS MAY BE FORMED AT TEMPERATURES AROUND 200°–350° F (100–180° C). SPECIFIC TEMPERATURES WILL DEPEND ON THE MATERIAL, THE DEGREE OF DEFORMATION, AND THE THICKNESS OF THE SECTION. FORMING MAY BE DONE BY HAND, IN FORMS PRESSED TOGETHER, OR WITH VACUUM PRESSURE.

THE FOLLOWING SEQUENCE IS GIVEN TO PROVIDE A GENERAL INTRODUCTION TO THE POSSIBILITIES OF THIS TECHNIQUE. BEFORE TRYING THIS, READ FURTHER AND LOOK FOR ADVICE FROM SOMEONE FAMILIAR WITH PLASTICS. A LOCAL SUPPLIER WILL HAVE MANUFACTURERS' DATA SHEETS AND CAN OFTEN HELP WITH SPECIFIC PROJECTS.

1. AFTER REMOVING THE PROTECTIVE PAPER, SET THE SHEET OR ROD INTO A KITCHEN OVEN AND HEAT TO THE POINT WHERE THE PLASTIC WILL BEND WHEN PUSHED WITH A BLUNT TOOL, (ABOUT 300° F).

2. WEARING CLEAN COTTON GLOVES, PULL THE PLASTIC OUT AND BEND IT OR PUSH IT OVER A RIGID FORM. HOLD IT IN POSITION UNTIL IT COOLS – USUALLY JUST A MINUTE OR TWO. IF THE PLASTIC COOLS BEFORE FORMING IS COMPLETED, RETURN THE PIECE TO THE OVEN AND REWARM IT.

3. A STRIP HEATER IS USED TO ACHIEVE STRAIGHT BENDS. THESE MAY BE BOUGHT AT A HOBBY SHOP OR PLASTICS SUPPLY COMPANY.

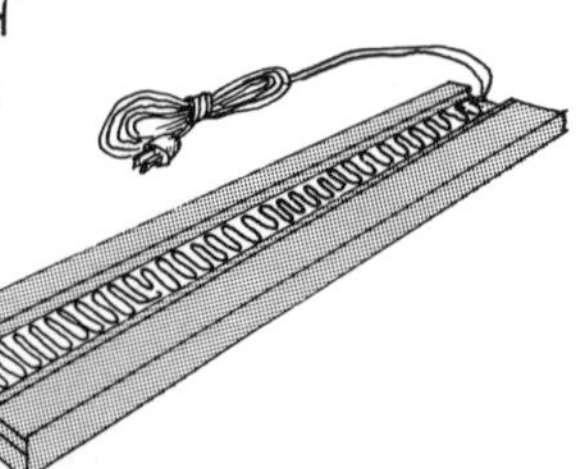

Safety:

SAFETY ALERT

THE BIGGEST PROBLEMS HERE COME FROM:

1. DUST CREATED BY CUTTING & SANDING.
2. TOXIC FUMES RELEASED BY HEAT CREATED BY MACHINING.
3. TOXIC VAPORS GIVEN OFF BY SOLVENTS (GLUES).

WHEN CUTTING THERMOPLASTICS ON A POWER MACHINE, VENTILATE AND WEAR GOGGLES AND A RESPIRATOR. THESE SAME PRECAUTIONS ARE NEEDED WHEN CEMENTING, PROPORTIONATE TO THE SCALE OF THE WORK BEING DONE.

cutting

THERMOPLASTICS CAN BE CUT, DRILLED, AND TURNED LIKE WOOD. WHEN POSSIBLE THE PAPER COATING SHOULD BE LEFT ON FOR THESE OPERATIONS.

SHEET OF 1/4" OR THINNER CAN ALSO BE BROKEN ALONG A STRAIGHT EDGE. MAKE A DEEP GOUGE USING A SCRIBE AND STRAIGHT EDGE. BREAK OVER A TABLE EDGE OR DOWEL; USE PLIERS FOR SMALL PIECES.

joining

THERMOPLASTICS CAN BE HELD TOGETHER WITH EPOXY CEMENT OR CYANO-ACRYLATES (E.G. SUPER GLUE) BUT A STRONGER AND NEATER JOINT IS MADE WITH A GLUE DEVISED JUST FOR THIS PURPOSE. IT IS A SOLVENT THAT PENETRATES A SEAM BY CAPILLARY ACTION AND CHEMICALLY WELDS THE JOINT. THE AREA SHOULD BE SCRAPED AND FILED BUT NOT POLISHED. THE PROTECTIVE PAPER IS REMOVED; THE PIECES CAN BE HELD IN PLACE WITH MASKING TAPE. THE SOLVENT IS APPLIED WITH A BRUSH OR SYRINGE AND WILL DRY QUICKLY.

finishing

EDGES ARE SMOOTHED WITH A FILE THEN SCRAPED WITH A FLAT PIECE OF STEEL LIKE THE BACK OF A HACK SAW BLADE. FINE ABRASIVE PAPERS MAY BE USED. A MUSLIN BUFF WITH WHITE DIAMOND OR A PLASTIC COMPOUND WILL REMOVE SCRATCHES. AVOID BUILDING UP HEAT.

SURFACES

Polishing SURFACES

A POLISHED APPEARANCE IS THE RESULT OF A PERFECTLY FLAT SURFACE. UNDER MAGNIFICATION THE CROSS-SECTION OF SCRATCHES LOOKS LIKE THIS. LIGHT IS REFLECTED AROUND THE SCRATCHES LIKE SOUND BEING ECHOED IN A MOUNTAIN VALLEY.

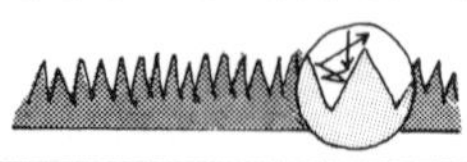

1. GOOD FINISHING BEGINS WHEN YOU FIRST BEGIN TO HANDLE THE METAL. AVOID MAKING UNNECESSARY SCRATCHES BY STORING THE STOCK CAREFULLY. DON'T SCRIBE A LINE UNTIL YOU ARE SURE OF YOUR PLANS.

2. GENERALLY A FILE IS USED TO SMOOTH EDGES OR DEFINE SHAPES. THE NEXT STEP IS USUALLY TO ABRASIVE PAPERS. SILICON CARBIDE (WET/DRY PAPER) IS A POPULAR CHOICE. THIS CAN BE BOUGHT AT A HARDWARE STORE OR JEWELRY SUPPLY COMPANY. OTHER CHOICES WOULD INCLUDE EMERY, GARNET, AND CROCUS (ROUGE) PAPERS. WHATEVER YOUR PREFERENCE, THE PAPER SHOULD BE WRAPPED AROUND A BOARD OR DOWEL TO INCREASE YOUR LEVERAGE. THE CUTTING POWER OF THE PAPER DEPENDS ON THE FORCE BEHIND IT.

CREASE FOR SHARP EDGES.

MASKING TAPE

3. GRITS:

100'S	VERY COARSE
200'S	COARSE
300'S	MEDIUM
400'S	FINE
500'S	VERY FINE

ADVANCE FROM COARSE TO FINE PAPERS, TAKING CARE NOT TO SKIP OR ABBREVIATE ANY STEP. AS YOU SWITCH GRITS CHANGE THE DIRECTION OF YOUR STROKE. THIS WILL MAKE IT EASIER TO TELL WHEN THE MARKS OF THE PREVIOUS ABRASIVE HAVE BEEN WORKED OUT.

IF YOU ARE AFTER A MIRROR FINISH YOU WILL PROBABLY PROGRESS TO A 500 PAPER. KEEP IN MIND THAT THERE IS NO UNIVERSAL 'RIGHT' FINISH. YOU CAN STOP AT ANY POINT THAT COMPLIMENTS THE PIECE.

4. POLISHING STICKS CAN BE MADE BY GLUING LEATHER OR FELT ONTO WOOD. COMPOUND IS THEN RUBBED INTO THIS.

ANOTHER EFFECTIVE METHOD OF HAND POLISHING IS THE USE OF STRINGS OR THONGS, CALLED TRUMMING.

Compounds:

A TYPICAL FINISHING SEQUENCE MOVES FROM FILES TO PAPERS TO THE USE OF TOUGH POWDERS CALLED OXIDES. THESE MAY BE NATURAL SANDS OR MAN-MADE PARTICLES. THEY ARE USUALLY BOUGHT ALREADY MIXED WITH A VEHICLE OR BONDING AGENT, OFTEN GREASE OR TALLOW.

A WIDE RANGE OF COMPOUNDS ARE AVAILABLE: CONSULT A SUPPLY COMPANY CATALOGUE FOR A DETAILED LIST. MANY OF THESE HAVE BEEN CONCOCTED FROM MIXTURES OF SEVERAL OXIDES FOR SPECIFIC APPLICATIONS. TRY A COUPLE AND CHOOSE THOSE THAT MEET YOUR NEEDS.

REMEMBER TO FOLLOW THE MANUFACTURER'S ADVICE, AND TO KEEP A SEPARATE WHEEL OR STICK FOR EACH COMPOUND.

scratchbrushes

THESE BRASS-BRISTLED BRUSHES WILL GIVE A DELICATE SHINE TO GOLD OR STERLING WORK THAT HAS BEEN FINISHED TO A UNIFORM MATTE WITH PAPER OR PUMICE. LUBRICATE THE SCRUBBING WITH SOAP AND WORK IN ALL DIRECTIONS.

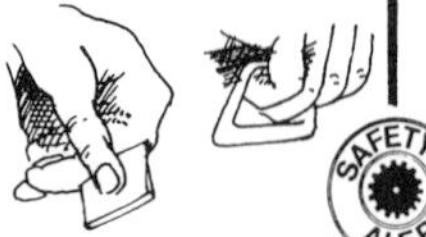

rules for the BUFFING MACHINE

SAFETY ALERT

- PAY ATTENTION! IF YOUR MIND WANDERS, TURN OFF THE MACHINE AND TAKE A BREAK.
- USE A PINCH (BREAKAWAY) GRIP. DON'T ENTWINE YOUR FINGERS INTO THE WORK.
- WEAR GOGGLES. KEEP LONG HAIR AND LOOSE CLOTHING TIED BACK.
- WORK ONLY ON THE LOWER QUARTER OF THE WHEEL.

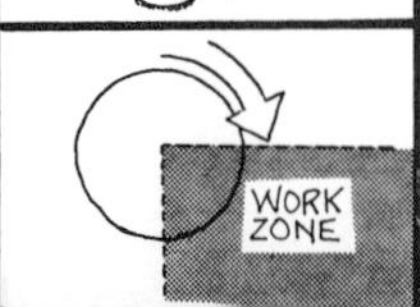

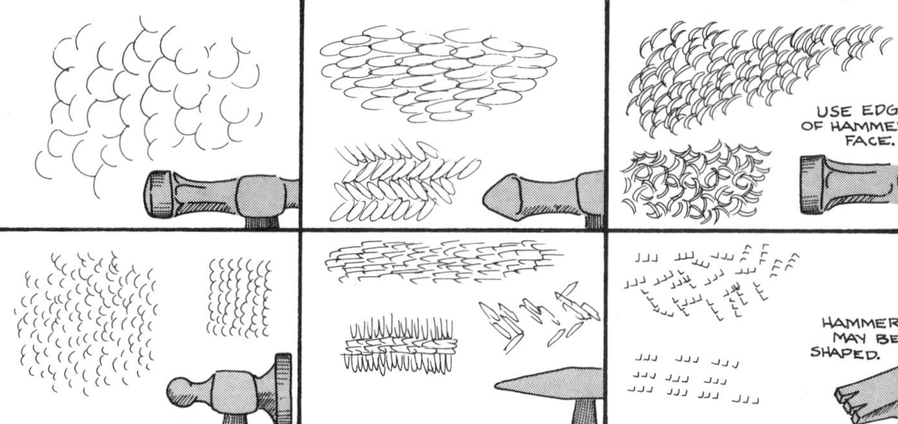

- WORK ON AN ANVIL, PREFERABLY POLISHED.
- ANNEAL METAL BEFORE STARTING.
- USE STOCK THICK ENOUGH TO ABSORB BLOW.
- HOLD HANDLE WHERE IT IS COMFORTABLE. A LOWER GRIP INCREASES POWER.
- YOU MAY ALSO CREATE A RICH SURFACE BY HAMMERING THE METAL ONTO A TEXTURE, LIKE RUSTED STEEL OR CONCRETE.

STAMPING

NOTES

- WORK ON A HARD SURFACE, IF POSSIBLE, POLISHED STEEL.
- METAL SHOULD BE ANNEALED AND THICK ENOUGH TO TOLERATE THE EFFECT. REMEMBER THAT STAMPED AREAS ARE THINNER THAN STARTING SHEET.

DEFINITION: STAMPING IS A NON-CONTINUOUS SERIES OF INDENTATIONS MADE BY A TOOL, USUALLY DRIVEN BY A HAMMER. IT IS LIKE LEATHER TOOLING.

USES:

- AS DECORATION. IT IS VERSATILE, DIRECT, AND PERMANENT.
- TO PROVIDE RECESS FOR
  - SOLDER INLAY
  - NIELLO
  - ENAMELS
  - AMALGAM
  - RESINS
- TO GIVE ILLUSION OF DEPTH.
- TO HALLMARK.

tool design

STAMPS SHOULD HAVE A BEVEL TO GIVE THE DISPLACED METAL SOMEWHERE TO GO.

PUNCHES WITH FACES LARGER THAN 4 MM (3/16") TEND TO TILT AND REQUIRE A VERY HEAVY BLOW. THEY ARE NOT RECOMMENDED FOR HAND CRAFTING.

TO MAKE A CLEAR, STRAIGHT-WALLED IMPRINT IT IS CRITICAL THAT THE TOOL FACE BE FLAT. CHECK WITH MAGNFICATION.

LIKE THIS | NOT THIS

Letters & Numbers

COMMERCIALLY MADE LETTER AND NUMBER STAMPS CAN BE USED FOR SURFACE ENRICHMENT.

FOUND TOOLS –

CARPENTERS' NAIL SETS OR PHILIPS HEAD SCREWDRIVERS WITH THE POINT GROUND OFF MAKE GOOD PUNCHES.

ALSO SEE: CHASING, TOOL-MAKING, HOLDING DEVICES.

Notes:

- THE TOOL IS USUALLY DRAWN TOWARDS THE WORKER, HELD AT SUCH AN ANGLE THAT IT PROPELS ITSELF ALONG.
- THE TOOL MAY BE DIPPED IN OIL TO LUBRICATE ITS TRAVEL.
- USE A LIGHTWEIGHT HAMMER AND SIT COMFORTABLY. THE PROCESS SHOULD BE DELICATE AND CONTROLLED.
- FOR SMALL RADIUS CURVES, TILT TOOL MORE OR SWITCH TO A SMALLER TOOL. SINCE A SHARPER ANGLE MAY CAUSE TOOL TO SLIP A NEW TOOL IS THE BETTER SOLUTION.
- IT IS IMPORTANT THAT THE WORKPIECE BE SECURELY HELD.

DEFINITION & USES

CHASING IS AN ANCIENT AND OFTEN MISUNDERSTOOD TECHNIQUE USED TO INCISE LINES INTO METAL.

THE RESULT LOOKS LIKE ENGRAVING AND THE PROCESS RESEMBLES STAMPING BUT CHASING IS A TECHNIQUE BY ITSELF.

UNLIKE ENGRAVING, NO METAL IS REMOVED. UNLIKE STAMPING, THE TOOL MOVES IN A STEADY, UNBROKEN MOTION.

CHASING CAN BE USED TO CREATE LINEAR PATTERNS ON FLAT OR SHAPED SHEET METAL, AND IS USED TO SHARPEN DETAILS ON CASTINGS.

Holding Devices

TO KEEP CLAMPS AT A DISTANCE, USE A STRIP OF STEEL OR A THIN PIECE OF WOOD. PROTECT AGAINST SCRATCHES WITH A RUBBER BAND.

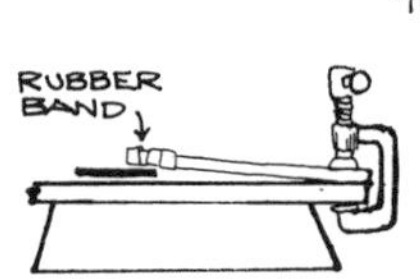

WORKPIECE MAY BE NAILED TO A BLOCK OF WOOD.

INSULATED STAPLES ARE ALREADY CUSHIONED WITH CARDBOARD.

TONGUE ON WOOD BLOCK IS NOT ESSENTIAL, BUT ALLOWS WOOD TO BE CLAMPED MORE SECURELY IN A VISE.

WHEN CLAMPING DIRECTLY ONTO THE BENCH, USE WOOD, LEATHER, OR CARDBOARD PAD TO PREVENT SCRATCHES.

FOR LIGHT WORK, A LARGE PAPER CLAMP IS HANDY. THESE CAN BE BOUGHT AT AN OFFICE SUPPLY STORE.

tools

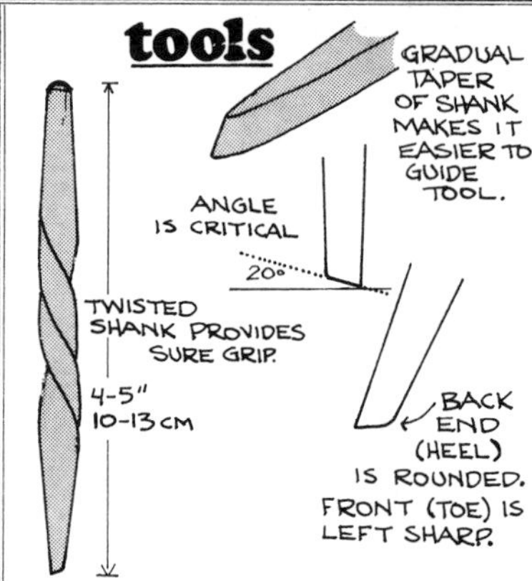

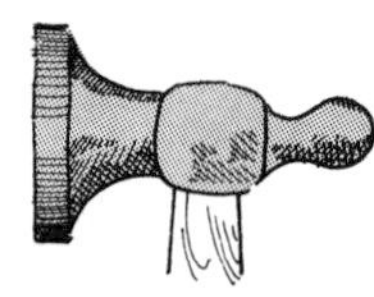

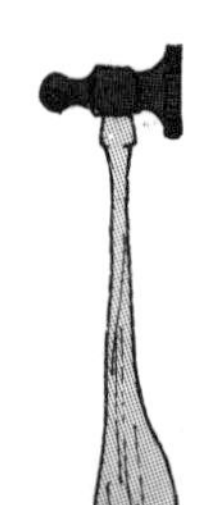

THOUGH ANY LIGHT HAMMER CAN BE USED, THIS ONE HAS EVOLVED OVER THE YEARS JUST FOR THIS TECHNIQUE. IT IS LIGHT ENOUGH TO BE USED FOR HOURS, HAS A LARGE FACE TO "FIND" THE TOOL, AND FITS ON A COMFORTABLE PISTOL GRIP HANDLE. WOOD IS SPRINGY SO HAMMER 'SPANKS' TOOL.

ROLL PRINTING (texture transfer)

DESCRIPTION: A SANDWICH MADE OF SHEET METAL AND A TEXTURING MATERIAL IS PASSED THROUGH THE ROLLING MILL UNDER GREAT PRESSURE. THE METAL IS EMBOSSED, RECEIVING THE REVERSE IMAGE OF THE MATERIAL.

PROCEDURE:

1. WORKPIECE IS ANNEALED AND THOROUGHLY DRIED.
2. WHEN APPROPRIATE, THE TEXTURE MATERIAL IS ANNEALED.*
3. ROLLERS ARE SET BY EYE AND SANDWICH IS TRIED. ADJUST ROLLERS SO PRESSURE IS CORRECT: HANDLE SHOULD BE HARD TO MOVE, BUT SHOULD NOT REQUIRE TWO PEOPLE.
4. PASS SANDWICH BETWEEN ROLLS, TRYING TO KEEP A CONTINUOUS MOVEMENT.

variation:

TO CREATE A RAISED PATTERN, PREPARE A SHEET OF METAL BY MAKING INDENTATIONS. THIS COULD BE DONE BY
- STAMPING
- ENGRAVING
- ETCHING
- ROLL PRINTING
- ETC.

TEMPLATE

A HARD METAL LIKE BRASS IS **PREFERRED.**

THE PRINT IS THEN MADE ON ANNEALED METAL AS ABOVE.

SUGGESTED MATERIALS

BURLAP	SCREEN
SANDPAPER	STRING
LACE	BINDING WIRE
NETTING	STRAIGHT PINS *
DRIED LEAVES	COARSE PAPER
TREE BARK	PAPER CLIPS *
PREPARED TEMPLATES	STAPLES *
	SAWBLADES *

* ANNEAL BY HEATING TO BRIGHT RED AND ALLOWING TO COOL SLOWLY.

tools

EYE PROTECTION REQUIRED!

HOLD HANDPIECE LIKE A PEN AND USE A STROKING OR SCRIBBLING MOTION.

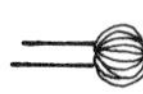
EASIEST TO CONTROL

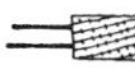
DEEPER CUT

REACHES INTO TIGHT SPOTS

CUTS SHARPEST LINE

TO LUBRICATE AND COOL:

DIP TOOL IN OIL OF WINTERGREEN

FLEX SHAFT CARVING

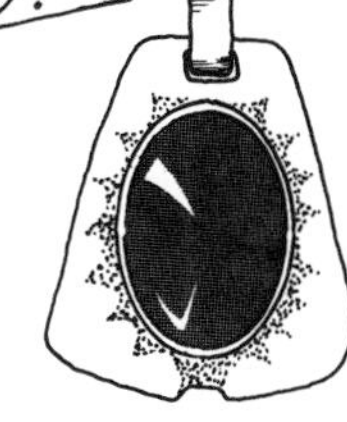

USES:

- AS SUBTLE DECORATION THAT CAN BE APPLIED LATE IN THE CONSTRUCTION.
- TO CAMOFLAGE SOLDER STAINS OR SIMILAR FLAWS.
- TO PROVIDE INTEREST ON AN OTHERWISE PLAIN SURFACE.
- TO MINIMIZE THE UNPLEASANT LOOK OF FINGERPRINTS ON A SMOOTH SHINY SURFACE.

FOR DEEPER MARKS

A HAMMER HANDPIECE MAY BE USED. A WIDE RANGE OF TOOL FACES MAY BE MADE:

Engraving SURFACES

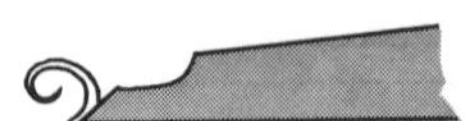

ENGRAVING IS A CUTTING PROCESS IN WHICH A STEEL TOOL (CALLED A GRAVER OR BURIN) SLICES SMALL BITS OF METAL AS IT IS PUSHED ALONG THE SURFACE OF A SHEET.

TOOLS

GRAVERS ARE MADE OF HIGH QUALITY TOOL STEEL AND ARE USUALLY SOLD IN THE HARDENED, UNTEMPERED STATE.

GRAVER	SHAPE	COMMON USES
SQUARE		LINES, SCRIPT, MOST CUTTING
FLAT		CARVING & TOUCH-UP, WIGGLE CUTS
KNIFE		FINE LINES
ROUND		WIDE LINES, DOTS
OVAL		CARVING, MEDIUM LINES
LINER		TEXTURES, FLORENTINE, WIGGLE CUTS

handles

GRAVER HANDLES ARE AVAILABLE IN SEVERAL STYLES; CHOICE IS A MATTER OF PERSONAL PREFERENCE. SINCE LARGE BULBOUS HANDLES CAN GET IN THE WAY WHEN MAKING SHALLOW CUTS, THOSE WITH A FLAT PORTION ARE GENERALLY PREFERRED. BECAUSE GRAVERS WILL GET SHORT WITH REPEATED SHARPENING, SOME ENGRAVERS START WITH A SHORT HANDLE AND LATER SWITCH TO A LONGER ONE TO PROLONG USE OF THE TOOL.

AN EFB ADJUSTABLE HANDLE IS OFTEN USED WITH A SQUARE GRAVER. THE TOOL IS HELD IN PLACE BY A METAL CONE SLID TIGHTLY ALONG ITS SHAFT. A NOTCHED PIECE OF BRASS PROVIDES FOR THE CHANGING LENGTH OF THE TOOL.

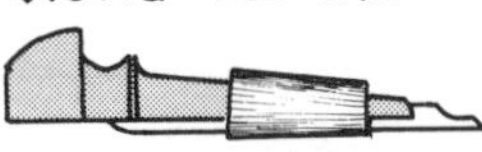

MOST GRAVERS ARE AVAILABLE WITH FLAT OR BENT SHANKS. THE CURVED SHAPE IS USUALLY PREFERRED FOR WORKING ON A CONCAVE SURFACE OR ON AREAS NOT EASILY ACCESSIBLE.

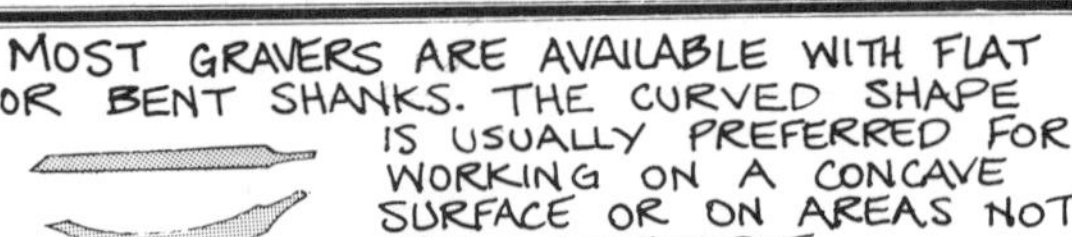

TO DETERMINE THE CORRECT LENGTH OF A GRAVER, HOLD A PENCIL AS SHOWN. MARK AND MEASURE AT THE ARROW.

Holding Devices

IT IS IMPOSSIBLE TO ENGRAVE WITH CONTROL UNLESS THE METAL IS SECURELY HELD. HERE ARE SEVERAL COMMON SOLUTIONS:

1. THE GRAVERS' BALL: A HEAVY STEEL SPHERE WITH VISE JAWS ON TOP. IT SITS ON A DONUT-SHAPED PAD TO PROVIDE FOR ANY ANGLE AND ROTATES ON A BEARING.

2. A SHELLAC STICK: THIS IS A PLATFORM AND HANDLE THAT IS HELD AGAINST THE BENCH PIN WHILE CUTTING. THE BEST STYLE HAS A FLARED NECK THAT FACILITATES ANGLING THE WORK. IT CAN ALSO BE MADE FROM A PIECE OF WOOD AND DOWEL AS SHOWN.

6-8"

THE PLATFORM IS COATED WITH A 3-5 MM LAYER OF FLAKE SHELLAC, SEALING WAX, OR A MIXTURE OF THE TWO. TO USE, GENTLY HEAT BOTH SHELLAC AND OBJECT AND PRESS TOGETHER.

3. VISE STICK: THIS IS MADE WHEN ENGRAVING SEVERAL OBJECTS OF THE SAME SHAPE. THE OUTLINE OF THE PIECE IS CARVED INTO THE ENDGRAIN OF THE WOOD.

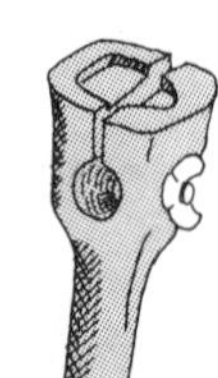

1. GRIP: THE GRAVER IS HELD BETWEEN THE FINGERTIPS AND THE LENGTH OF THE THUMB. THIS WILL FEEL AWKWARD AT FIRST BUT IS WORTH GETTING USED TO. THE HANDLE SHOULD REST IN THE FLESHY PART OF THE PALM: THIS IS WHERE THE PUSH COMES FROM.
2. POSTURE: WORK SHOULD BE AT MID-CHEST HEIGHT. WHEN USING A GRAVER'S BALL A TABLE LOWER THAN A JEWELER'S BENCH WILL BE NEEDED. THE ELBOWS SHOULD BE STABLE WHEN USING A SHELLAC STICK RESTED AGAINST THE BENCH PIN, SIT SIDEWAYS SO THE ARM THAT HOLDS THE GRAVER CAN BE ANCHORED.
3. MAGNIFICATION: MOST ENGRAVERS RELY ON A MAGNIFYING HEADSET OR LOUPE. IF YOU CAN'T SEE IT, YOU CAN'T CUT IT.

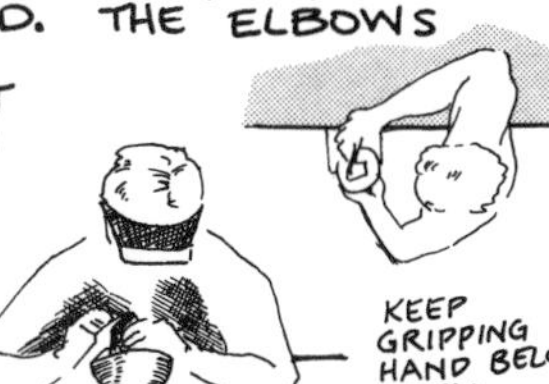

KEEP GRIPPING HAND BELOW METAL.

CUTTING

PROPER CUTTING INVOLVES A SLIDING RATHER THAN A SCOOPING STROKE. THE GRAVER IS LIGHTLY PRESSED DOWN INTO THE METAL AT THE BEGINNING OF THE CUT. THE TOOL IS THEN SLID FORWARD AT A CONSISTENT DEPTH. THE SCOOPING STROKE CAN BE RECOGNIZED BY A TELL-TALE SCAR LEFT BEHIND THE INTENDED BEGINNING OF THE LINE.

A LINE OF VARYING WIDTHS IS ACHIEVED NOT BY CHANGING THE DEPTH OF THE CUT, BUT BY 'ROLLING' THE GRAVER TO THE RIGHT AS IT IS PUSHED ALONG. BY ROLLING AND RETURNING TO A VERTICAL POSITION, A GRACEFUL LINE CAN BE CUT.

IN MOST CASES THE GRAVER IS HELD IN ONE POSITION AS THE WORK IS BROUGHT INTO IT. THIS IS ESPECIALLY TRUE OF CURVED LINES. CURVES AND CIRCLES ARE GENERALLY CUT COUNTER-CLOCKWISE.

Layout:

SINCE ENGRAVING IS A PRECISE AND DEMANDING PROCESS IT IS USUALLY UNWISE TO PLAN ON DESIGNING AS YOU CUT. CAREFUL LAYOUT WILL ALLOW YOU TO CONCENTRATE ON YOUR WORK. DRAWING DIRECTLY ON THE METAL WITH A PEN OR PENCIL WILL CREATE A WIDE LINE THAT CAN EASILY SMUDGE. A BETTER METHOD IS TO COAT THE METAL WITH A WHITE PAINT (CHINESE WHITE, TEMPERA, TYPEWRITER CORRECTION FLUID) AND DRAW ON THIS WITH A SHARP PENCIL. THE DESIGN IS THEN LIGHTLY TRACED WITH A SHARP SCRIBE OR SEWING NEEDLE HELD IN A PIN VISE. THE WHITE SURFACE CAN BE CUT THROUGH OR WASHED AWAY. THOUGH IT TAKES A LITTLE LONGER, THIS KIND OF PRECISION IS NEEDED FOR GOOD ENGRAVING.

WELL BEGUN IS HALF DONE.

-ARISTOTLE 350 B.C.

uses:

DECORATION

LETTERS

TEXTURE

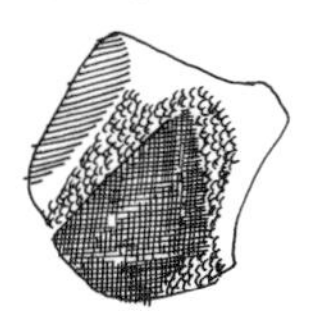

DETAIL

CARVING

STONE SETTING

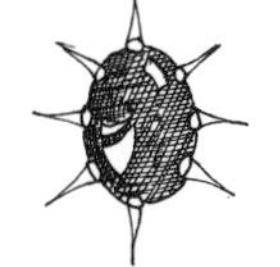

Wiggle Cut

A FLAT GRAVER OR LINER IS USED TO MAKE THIS SIMPLE AND VERSATILE CUT. THE TOOL IS HELD AT A STEEP ANGLE AND "WALKED" FORWARD, ROCKING FROM SIDE TO SIDE. ANY SIZE GRAVER MAY BE USED. THE AMOUNT OF SWING IN THE WRIST WILL ALTER THE CUT FROM BEING CLOSED TO OPEN.

GOOD FOR FILL IN

ALL ENGRAVING REQUIRES A KEEN PRECISE EDGE; REPEATED SHARPENING IS NEEDED TO KEEP THE CUTTING EDGE IN SHAPE. THOUGH SHARPENING CAN BE DONE BY HAND, AN INDEXING DEVICE IS RECOMMENDED SINCE IT KEEPS EACH SURFACE ABSOLUTELY FLAT.

1. BEFORE SHARPENING MOST TOOLS MUST BE SHORTENED. DETERMINE THE DESIRED LENGTH AS SHOWN ON THE PRECEDING PAGE, TAKING INTO ACCOUNT THE LENGTH OF THE HANDLE. TIGHTEN THE GRAVER IN A VISE AT THE RIGHT LENGTH AND HIT IT WITH A SHARP BLOW. FOR SAFETY, CATCH THE BROKEN PIECE IN A TOWEL.

2. THE TIP WILL BE SHARPENED MANY TIMES. TO SPEED THIS PROCESS THE TIP SURFACE IS REDUCED AS SHOWN. THIS IS DONE ON A GRINDING WHEEL (ANY SIZE; FLEX SHAFT IS OK). QUENCH OFTEN DURING GRINDING TO KEEP THE STEEL HARD. IF THE GRAVER GETS BLUE REFER TO THE PAGE ON HARDENING STEEL.

REMOVED

3. THE FACE ANGLE FOR MOST GRAVERS IS 45° – LESS FOR SOFT METALS AND SLIGHTLY MORE FOR HARD MATERIALS. USE A PROTRACTOR OR SIMILAR AID TO SET THIS ANGLE. THE FACE IS RUBBED AGAINST A SHARPENING STONE THAT HAS A COATING OF ANY KIND OF LIGHT OIL. SET BOTH STONE AND DEVICE ON A SMOOTH FLAT SURFACE LIKE A PIECE OF GLASS OR PLEXIGLAS. THE COARSE STONE IS FOLLOWED BY A SIMILAR STROKING ON A FINE STONE. CONTINUE THIS UNTIL ALL OBVIOUS LINES (SCRATCHES) ARE GONE.

4. BURS CREATED BY GRINDING ARE REMOVED BY JAMMING THE TOOL A COUPLE OF TIMES INTO A BLOCK OF HARDWOOD. THE GRAVER IS THEN CAREFULLY POLISHED BY RUBBING IT ALONG A PIECE OF 500 GRIT SANDPAPER HELD ON A HARD FLAT SURFACE. THE GRAVER IS USUALLY REMOVED FROM THE DEVICE AND HELD IN THE HAND FOR THIS. ONE OR TWO SLOW STEADY PASSES ARE USUALLY SUFFICIENT. THE GRAVER IS TESTED BY SETTING THE TIP AGAINST THE THUMBNAIL. IF PROPERLY SHARPENED IT WILL "BITE" RATHER THAN SLIP. TO FASTEN THE HANDLE, GRIP THE TOOL VERTICALLY IN A VISE AND POUND THE HANDLE ONTO THE TANG WITH A MALLET. IF THE HANDLE HAS A FLAT AREA IT SHOULD LAY ALONG THE UNDERSIDE (BELLY) OF THE TOOL.

Sharpening a Square Graver:

1. IF USING AN EFB HANDLE PREPARE THE BACK END BY FILING THE EDGES LIKE THIS. HEAT THE LAST HALF INCH TO BRIGHT RED WHILE HOLDING THE TOOL IN PLIERS. BEND A SLIGHT CURL AND QUENCH QUICKLY. BE SURE TO BEND UP FROM THE CORNER, NOT FROM A FLAT SIDE.
2. PREPARE AS IN STEPS 1 AND 2 ABOVE.

LIKE THIS   NOT THIS

3. GRIP THE TOOL IN THE SHARPENER AND SET THE BOTTOM EDGE OF THE GRAVER FLAT AGAINST THE STONE. USING A PROTRACTOR, RAISE THE TOOL TO AN ANGLE OF 8-12°.
4. THE TOOL IS NOW ROTATED SLIGHTLY BY TURNING THE BARREL OF THE SHARPENER; NOTE THE AMOUNT OF ROTATION BY COUNTING THE MARKINGS ON THE SHARPENER. GRIND THIS SURFACE (CALLED THE BELLY) FIRST ON A COARSE STONE AND THEN ON A FINE ONE. ROTATE THE BARREL BACK TO ITS ORIGINAL POSITION THEN CONTINUE TURNING TO THE SAME NUMBER OF NOTCHES USED ON THE FIRST SIDE. GRIND AS BEFORE ON THE OILSTONE. THE RESULT, FROM BELOW, SHOULD LOOK LIKE ONE OF THESE: NOTE THAT THE ANGLE CAN BE MODIFIED TO MAKE A TOOL THAT WILL CUT LINES OF VARIOUS WIDTHS. THE ANGLES DO NOT NEED TO BE IDENTICAL, BUT THEIR POINTS MUST MEET.
5. THE BARREL IS THEN TURNED AROUND SO THE FACE IS DOWNWARD. THE ANGLE IS SET AT 45° AND THE FACE IS GROUND ON THE COARSE AND FINE STONE.
6. THE TOOL IS THEN SET INTO ITS HANDLE AND ITS THREE CUTTING PLANES ARE POLISHED BY RUBBING ON 500 GRIT SANDPAPER HELD ON GLASS. TEST AGAINST A THUMBNAIL: TOOL SHOULD 'BITE' WITHOUT PRESSURE.

METHOD 1

SAW OUT ONE UNIT, EITHER THE POSITIVE OR THE NEGATIVE.

AFTER FILING THE SHAPE TO EXACTLY WHAT IS WANTED, TRACE AROUND IT ON THE OTHER PIECE OF METAL USING A SHARP SCRIBE. A SEWING NEEDLE IN A PIN VISE WORKS WELL.

SAW SECOND UNIT CAREFULLY AND THE TWO PIECES SHOULD MAKE A PERFECT FIT. FILE OR PLANISH IF NECESSARY. SOLDER PIECES TOGETHER.

METHOD 2

CLAMP OR GLUE TWO PIECES OF METAL TOGETHER. SAW THROUGH BOTH PIECES SIMULTANEOUSLY. SOLDER PIECES TOGETHER.

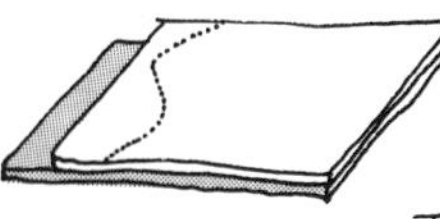

WHEN BOTH CONVEX AND CONCAVE CURVES ARE INVOLVED THE KERF OF THE SAW BECOMES A FACTOR. THIS SPACE WILL HAVE TO BE FILLED BY SOLDER SO MUST BE KEPT SMALL. USE A FINE BLADE: NO LARGER THAN 4/0

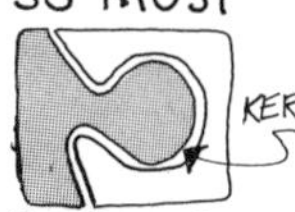

WHEN PIERCING SOME CUNNING IS NEEDED TO GET STARTED. DRILL AND SAW OUT TO PROPOSED LINE BEFORE GLUING ON THE SECOND SHEET.

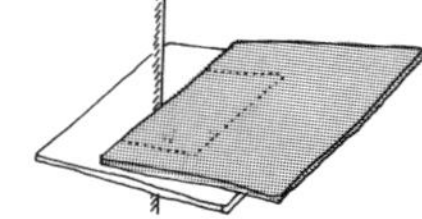

CLAMP, OR GLUE WITH EPOXY OR SUPER GLUE

Solder Inlay

THIS IS A FAST AND DIRECT METHOD OF ACHIEVING A FLUSH SURFACE MIXED METAL EFFECT. GENERALLY THE INLAY IS COMPLETED BEFORE THE OBJECT IS FABRICATED BUT SPECIFIC DESIGNS WILL REQUIRE DIFFERENT SEQUENCES.

SINCE THE INLAY MATERIAL IS SOLDER AND FLOWED INTO PLACE WITH HEAT, SUBSEQUENT HEATINGS MUST BE THOUGHT OUT CAREFULLY. IF THE INLAY IS DONE BEFORE FABRICATION, IT SOLDER (OR AT LEAST HARD) SHOULD BE USED. IF INLAY IS GOING TO BE A FINAL STEP, EASY OR EASY-FLO SOLDER MUST BE 'RESERVED' FOR THIS USE.

PREPARE A RECESS BY STAMPING CHASING, ENGRAVING, ROLL PRINTING OR ETCHING. IT SHOULD BE ABOUT ½ MM DEEP (.020").

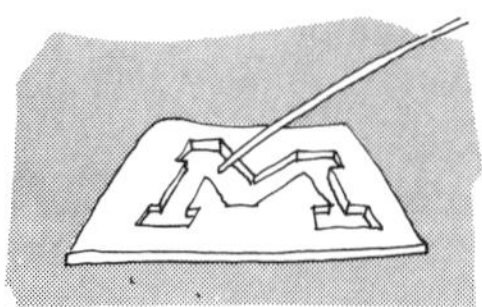

FLUX THE PIECE WELL AND FLOOD RECESS WITH SOLDER. WIRE SOLDER WORKS WELL BUT CHIPS MAY BE SET IN PLACE IF PREFERRED.

SOME SOLDER WILL SPILL OVER OUTSIDE THE RECESS. THIS IS UNAVOIDABLE BUT SHOULD BE KEPT TO A MINIMUM.

AFTER QUENCHING THE EXCESS SOLDER IS FILED OFF: BE CAREFUL NOT TO FILE AWAY TOO MUCH. THE DESIGN CAN BE LOST ALTOGETHER.

TO SHOW UP THE PATTERN WHILE FILING, DIP PIECE INTO LIVER OF SULPHUR SOLUTION. WORK IS SANDED, BUFFED AND COLORED AS USUAL. A HIGH POLISH DOES NOT SHOW THE EFFECT WELL AND IS NOT RECOMMENDED.

Solder:

SILVER SOLDER MAY BE USED ON:
- STEEL
- NICKEL SILVER
- COPPER
- BRASS / BRONZE
- 18K GOLD
- 14K COLORED GOLD
- 14K WHITE GOLD

GOLD SOLDER MAY BE USED ON:
- STEEL
- NICKEL SILVER
- COPPER
- BRASS/BRONZE
- STERLING: AVOID COLORED & WHITE SOLDERS.

SOFT SOLDERS LIKE STA-BRITE AND TIX WORK WELL FOR INLAY BUT OF COURSE CANNOT BE HEATED ABOVE 600°F (300°C)

BIG WORDS ARE ALWAYS PUNISHED.
-SOPHOCLES

SURFACES

LAMINATE INLAY

THIS IS A QUICK METHOD OF CREATING AN INLAID LOOK WHEN USING TWO METALS OF CONTRASTING COLOR. IT CANNOT BE USED WHERE SPECIFIC SHAPES ARE REQUIRED SINCE DISTORTION IS INHERENT IN THE PROCESS.

1. ONE PIECE OF METAL MUST BE THICKER THAN THE DESIRED GOAL AND THE OTHER MUST BE VERY THIN - AROUND 26 GA. THE TWO PIECES ARE CLEANED AND SOLDERED TOGETHER. THE BOND MUST BE COMPLETE, EXTENDING ALL THE WAY TO THE EDGES OF THE PIECE BEING SOLDERED. THE WAY TO ACHIEVE THIS IS WITH CAREFUL PREPARATION AND HEATING, NOT BY SUPPLYING SURPLUS SOLDER. EXCESS SOLDER WILL MAKE A VAGUE YELLOWISH GHOST IMAGE AROUND THE INLAY IN THE FINISHED PIECE.

2. AFTER PICKLING AND DRYING THE SHEET IS PASSED THROUGH THE ROLLING MILL UNTIL THE TWO SURFACES BECOME FLUSH. IF ROLLING IS TO TAKE PLACE IN BOTH DIRECTIONS ANNEAL BEFORE CHANGING THE DIRECTION OF THE STRETCH.

3. FINISH CONVENTIONALLY WITH FILES, PAPER, AND BUFFING IF DESIRED. SUBSEQUENT SOLDERING COULD SPOIL THE EFFECT. AS PRECAUTION USE A LOWER MELTING SOLDER AND PROTECT THE INLAY WITH YELLOW OCHER.

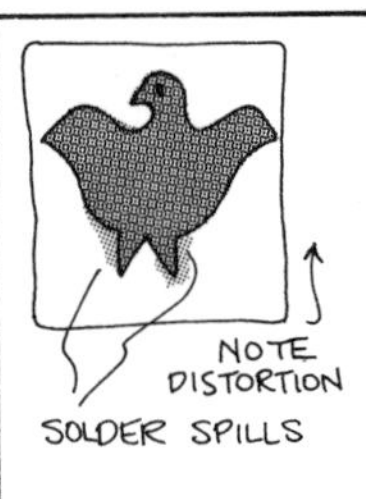

variations:

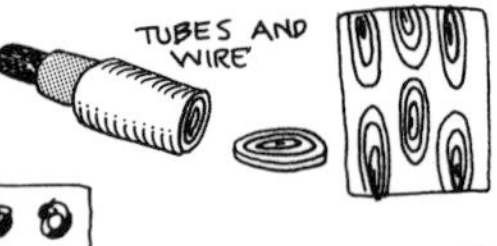

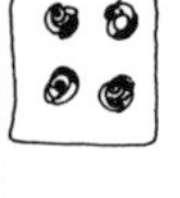

ART IS LONG, LIFE SHORT. -GOETHE

SAFETY ALERT

AMALGAMATION

THIS ANCIENT METHOD INVOLVES HIGH HEALTH RISK AND IS NOT TO BE TAKEN LIGHTLY. IF YOU TRY IT, WORK OUTSIDE WITH THE WIND AT YOUR BACK.

1. IN A MORTAR AND PESTLE KNEAD EQUAL WEIGHTS OF MERCURY AND FILINGS OF FINE GOLD.
2. PREPARE WORK PIECE (OF STERLING, COPPER, WHITE GOLD, ETC) WITH RECESSES MADE BY CHASING, ENGRAVING, OR ETCHING. THESE DO NOT NEED TO BE VERY DEEP; SAY .25 —.5 MM.
3. THE SURFACE OF THE THOROUGHLY CLEANED PIECE IS PRIMED BY RUBBING IT WITH A PASTE MADE OF MERCURY AND CHALK DUST. WEAR RUBBER GLOVES WHEN HANDLING MERCURY.
4. THE GOLD AMALGAM IS PRESSED LIKE PUTTY INTO THE GROOVES AND RECESSES OF THE PIECE. A SLIGHT EXCESS IS MOUNDED ABOVE EACH FILLED AREA.
5. LEAVE THE PIECE EXPOSED TO AIR - PREFERABLY OUTDOORS - FOR AT LEAST TWO DAYS. OCCASIONALLY PRESS THE AMALGAM FURTHER INTO THE RECESSES WITH A SMOOTH WOODEN STICK. THE MERCURY IS BEING DRIVEN OFF IN THE FORM OF A POISONOUS GAS.
6. PLACE WORK IN A WARM (NOT HOT) PLACE LIKE NEAR A RADIATOR OR A SUNNY WINDOW SILL TO DRIVE OFF FURTHER FUMES. AGAIN VENTILATION IS VERY IMPORTANT. LEAVE FOR SEVERAL HOURS.
7. SET WORK PIECE ON A SOLDERING BLOCK AND HEAT TO A DULL RED. FUMES ARE STILL BEING DRIVEN OFF. IF POSSIBLE SET A FAN BESIDE YOU TO BLOW THE GASES AWAY. THIS ASSUMES THERE IS NO ONE SITTING ACROSS FROM YOU.
8. AFTER AIR COOLING, BURNISH THE AMALGAM INTO THE RECESS, SCRAPE OFF EXCESS, RE-BURNISH, AND POLISH AS USUAL.

MOKUMÉ-GANE IS AN ANCIENT JAPANESE TECHNIQUE THAT USES LAMINATED METAL TO GIVE A MULTI-COLORED WOODGRAIN EFFECT. THERE ARE SEVERAL POSSIBLE METHODS FOR BUILDING UP THE BILLET OF LAMINATES, AND THEN SEVERAL OTHERS FOR MAKING THE PATTERN IN THE METAL. EACH METHOD HAS ITS VARIATIONS SO THE PROCESS IS DIFFICULT TO SUMMARIZE. THE TECHNIQUES DESCRIBED BELOW PROVIDE A STARTING POINT FOR PERSONAL INTERPRETATION.

DIFFUSION

THE BEST JOINING OF LAMINATES IS MADE BY DIFFUSION UNDER SLIGHT PRESSURE AT RED HEAT IN A REDUCING ATMOSPHERE. SUCH A BLOCK WILL BE PERFECTLY FUSED AND MAY BE TREATED LIKE ANY OTHER METAL MASS. THIS PROCESS IS EXPLAINED IN THE CHAPTER ON CONNECTING: SEE PAGE 58.

PATTERN DEVELOPMENT

bumping

THE SHEET OF LAMINATES IS SET ON A MEDIUM SOFT SURFACE LIKE PITCH OR SOFT WOOD AND WORKED WITH SMALL ROUND PUNCHES TO CREATE A BUMPY SHEET. THE BUMPS MAY BE RANDOM OR ARRANGED AND MAY BE MADE WITH A SINGLE OR SEVERAL TOOLS.

IF A BUMP IS MADE DEEPER THAN THE THICKNESS OF THE SHEET A HOLE WILL RESULT IN THE NEXT STEP WHEN THE TOPS OF EACH BUMP ARE FILED OFF. THE METAL IS FILED, SANDED, AND POLISHED USING CONVENTIONAL TECHNIQUES. THE RICHNESS OF THE PATTERN WILL NOT SHOW UNTIL THE MOKUME HAS BEEN COLORED.

carved

WHEN THE LAMINATE STACK IS COMPLETED THE STILL THICK BILLET IS DRILLED INTO WITH CARE TAKEN NOT TO GO ALL THE WAY THROUGH. THE POINTED RECESS OF THE DRILL TIP IS CONVERTED TO A ROUND-BOTTOMED HOLE WITH A SPHERICAL BUR. AN ALTERNATE METHOD IS TO CARVE A RECESS WITH A ROUND CHISEL.

FURTHER POSSIBILITIES ARE OPENED BY ENGRAVING OR MACHINING RECESSES. THE CARVED INGOT IS FORGED OR MILLED TO CAUSE THE PATTERN TO EMERGE. REPEATED CARVING AND FORGING STEPS ARE USED TO BRING OUT THE FULL PATTERN.

Soldering:

WHEN A SMALL PIECE OF MOKUME IS NEEDED AND WHEN VERY LITTLE FORMING IS TO BE DONE ON THE LAMINATE STACK, LAYERS MAY BE SOLDERED TOGETHER:

1. METAL SHEETS ARE FLATTENED, SCRUBBED WITH PUMICE, AND DEGREASED WITH AN ALCOHOL-TYPE SOLVENT. PIECES ABOUT 2 x 2 INCHES ARE A HANDY SIZE.
2. SHEET SOLDER IS FORGED OR MILLED AS THIN AS POSSIBLE AND SIMILARLY CLEANED. IT IS CUT INTO SHEETS SLIGHTLY SMALLER THAN THE LAMINATES.
3. LAMINATES ARE LIGHTLY FLUXED ON BOTH SIDES AND STACKED UP WITH A PIECE OF SOLDER BETWEEN EACH ONE. THIS PILE MAY HAVE 4-8 LAMINATE SHEETS, NOT COUNTING THE SOLDER.
4. HEAT THE WHOLE PILE WITH A LARGE BUSHY REDUCING FLAME. IF YOUR TORCH CANNOT BE ADJUSTED TO A REDUCING FLAME, WORK ON CHARCOAL PERHAPS, WITH ANOTHER CHARCOAL BLOCK SET BEHIND. THIS WILL HELP ABSORB OXIDES.
5. AIR COOL. DO NOT QUENCH, ESPECIALLY IN PICKLE.
6. FORGE OR MILL SHEET TO ABOUT HALF ITS ORIGINAL THICKNESS. CUT IN HALF, CLEAN AND FLUX TWO SURFACES. USING ANOTHER PIECE OF SOLDER, JOIN THESE TWO PIECES. THIS WILL DOUBLE THE NUMBER OF LAYERS IN THE STACK. AGAIN, AIR COOL.
7. THIN THE SHEET, CUT, AND REPEAT THE LAST STEP. DO THIS UNTIL THE DESIRED NUMBER OF LAYERS IS ACHIEVED. GENERALLY 10-30 LAYERS WILL YIELD PLEASANT RESULTS.
8. A DISADVANTAGE OF THIS METHOD IS THAT SUBSEQUENT SOLDERING CAN ENDANGER THE BOND BETWEEN THE LAYERS. USE HARD SOLDER FOR STACKING AND MINIMIZE SOLDERING AROUND THE MOKUME.

RECOMMENDED METALS:

COPPER (#110 RECOMMENDED)
FINE SILVER
GOLD
STERLING
BRASS (HIGH ZINC PREFERRED)
SHIBUCHI (APPROX. 75% Cu 25% Ag)
KUORMI-DO (99% Cu 1% ARSENIC)
SHAKUDO (2-8% Au, BAL Cu)

SURFACES

RETICULATION IS A PROCESS BY WHICH METAL IS MADE TO DRAW ITSELF INTO RIDGES AND VALLEYS, CREATING A UNIQUE RICH TEXTURE. THE BUCKLING OCCURS AS THE RESULT OF DIFFERENT COOLING RATES OF THE TWO STRATA SHOWN BELOW. THE COPPER OXIDE LAYER WILL REMAIN SOLID WHILE THE INTERIOR OF THE SHEET BECOMES MOLTEN. WHEN HEAT IS REMOVED THE INTERIOR CONTRACTS, PULLING THE COPPER OXIDE SKIN INTO RIDGES. THE EFFECT MAY BE ACHIEVED BY CAREFUL HEATING OF MOST NON-FERROUS METALS (HEAT SCARRING) BUT IS MUCH MORE DRAMATIC WHEN THE METAL IS PREPARED AS DESCRIBED BELOW.

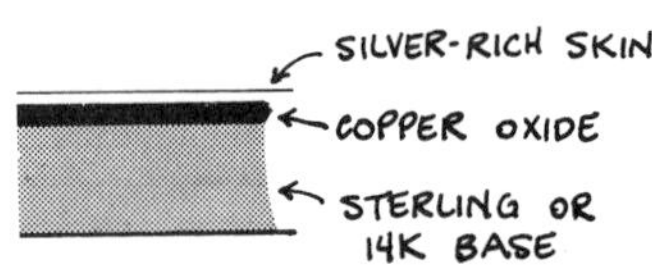

PROCESS

1. SINCE THE PROCESS IS SOMEWHAT ACCIDENTAL, WORK ON A PIECE OF METAL A LITTLE LARGER THAN YOUR ACTUAL NEED. 16-20 GAUGE SHEET WILL PRODUCE THE BEST RESULTS. HEAT THE PIECE OF STERLING OR 14K GOLD AT 650°C (1200°F) FOR FIVE MINUTES. THIS IS MOST EASILY DONE IN A KILN BUT CAN BE DONE WITH A TORCH – KEEP THE METAL AT A DULL RED. DO NOT USE FLUX, SINCE THE PURPOSE OF THIS STEP IS TO CREATE A LAYER OF COPPER OXIDE. AIR COOL. THE METAL WILL BE A UNIFORM DARK GRAY.

2. PICKLE IN HOT FRESH SPAREX OR A 10% SULFURIC ACID SOLUTION. THIS REMOVES COPPER OXIDE FROM THE SURFACE. LEAVING A SILVER-RICH SKIN AND "LOCKING IN" THE COPPER OXIDE LAYER BENEATH IT.

3. AFTER RINSING, HEAT AS BEFORE TO THE SAME TEMPERATURE THIS TIME FOR AT LEAST 10 MINUTES. OXYGEN CANNOT REACT MUCH WITH THE SILVER-RICH SKIN SO IT PENETRATES AND PROMOTES GROWTH OF THE COPPER OXIDE LAYER INTO THE SHEET (I.E. INTERIOR OXIDATION). AIR COOL. THE SHEET SHOULD BE ONLY SLIGHTLY GRAY. PICKLE AS BEFORE.

4. RETICULATION IS DONE WITH A TORCH. SINCE IT IS NECESSARY TO MAKE THE METAL MOLTEN THROUGHOUT ITS INTERIOR, IT IS WISE TO EITHER PREHEAT THE SOLDERING BLOCK AND THEN ALLOW THE HEAT TO RISE UP INTO THE SHEET OR TO WORK ON A WIRE MESH. THE SHEET IS BROUGHT TO RED WITH A SHARP HOT FLAME. THE TORCH IS QUICKLY PASSED OVER AN AREA ALLOWING IT TO COOL INTERMITTENTLY, SINCE THE COOLING IS WHAT CAUSES THE METAL TO BUCKLE. THE SKIN MAY MELT AND CRAWL WITHOUT DAMAGING THE RESULTS, BUT THIS SHOULD BE MINIMIZED SINCE THIS FLOWING SOFTENS THE SHARPNESS OF THE RIDGES AND DIMINISHES THE EFFECT. ALLOW PIECE TO LOSE REDNESS BEFORE QUENCHING.

5. RETICULATED METAL MAY BE SOLDERED, COLORED, AND FINISHED LIKE ITS ORIGINAL STOCK. IT IS BRITTLE, AND EXTENSIVE FORMING IS NOT RECOMMENDED. THE COPPER OXIDE LAYER SEEMS TO BE POROUS AND "SOAKS UP" SOLDER WHEN JOINING AN EXPOSED EDGE. TO MAKE A NEAT AND STRONG JOINT, BURNISH EDGES BEFORE SOLDERING.

SINCE COPPER PLAYS AN IMPORTANT ROLE IN RETICULATION, HIGHER COPPER CONTENT GENERALLY ENHANCES RESULTS. 14K YELLOW OR ROSE GOLD WILL WORK BETTER THAN 14K GREEN OR WHITE, OR ANY COLOR OF 18K. AN ALLOY OF 820 PARTS SILVER, BALANCE COPPER, IS ESPECIALLY RESPONSIVE TO THE TREATMENT DESCRIBED ABOVE. IT MAY BE MADE BY ADDING 10% (WEIGHT) COPPER TO STERLING OR CAN BE PURCHASED FROM HAUSER & MILLER, 10950 LIN-VALLE, ST. LOUIS, MO 63123.

GRANULATION IS AN ANCIENT DECORATIVE PROCESS IN WHICH A METAL IS ORNAMENTED BY THE APPLICATION OF MANY TINY BEADS (GRANULES). THE TECHNIQUES INVOLVED WILL WORK FOR OTHER SHAPES OF APPLIED ORNAMENT AND RELY ON A DIFFUSION PROCESS RELATED TO EUTECTIC BONDING. THE TERM GRANULATION IS OFTEN EXPANDED TO INCLUDE ALL THESE ASPECTS OF THE PROCEDURE.

THE STUNNING QUALITY OF PROPER GRANULATION IS ACHIEVED BY HAVING TINY GRANULES ADHERED TO THE SURFACE BY AN ALMOST IMPERCEPTIBLE BOND AT THE TANGENT POINT. BECAUSE THIS FINE PRECISION IS IMPOSSIBLE WITH CONVENTIONAL SOLDERING A DIFFUSION REACTION IS CREATED. A DETAILED DESCRIPTION OF THIS IS GIVEN ON PAGE 58. IN BRIEF - BY ONE OF SEVERAL METHODS A SMALL AMOUNT OF METAL (USUALLY COPPER) IS INTRODUCED TO THE CONTACT AREA. WHEN APPROPRIATE HEAT IS REACHED, THE METALS PRESENT FORM AN ALLOY OF A LOWER MELTING POINT THAN THE METALS BEING JOINED. THIS ALLOY FLOWS, CREATING A STRONG BOND.

PROCESS

WHEN WORKING ON FINE SILVER OR HIGH KARAT GOLD IT IS NECESSARY TO SUPPLY THE METAL THAT WILL MAKE THE BONDING ALLOY.

1. THE METAL AND GRANULES ARE COATED WITH A MIXTURE OF FLUX (CONTAINING A METALLIC SALT) AND GLUE WHICH CONTAINS CARBON. AT HIGH TEMPERATURES THE METAL BECOMES AN OXIDE (LIKE CuO_2). THE CARBON FROM THE GLUE THEN UNITES WITH THE OXYGEN AND PASSES OFF AS CARBON DIOXIDE GAS. THIS LEAVES A SMALL AMOUNT OF METAL TO ALLOY AT THE JOINT.

FLUXES USED INCLUDE PRIP'S, ANTIMONY TRIOXIDE, COPPER CHLORIDE, OR A COPPER NITRATE MADE BY DISSOLVING COPPER INTO AMMONIA (ALLOW SCRAPS OF COPPER TO SIT IN A CLOSED JAR OF AMMONIA UNTIL THE SOLUTION TURNS BLUE; ABOUT 24 HOURS).

ANY ORGANIC GLUE CAN BE USED: GUM TRAGACANTH, ELMER'S, MUCILAGE, HIDE GLUE, ETC. THIN THESE WITH WATER TO A PALE SOUPY CONSISTENCY.

2. COPPER CAN BE SUPPLIED BY PLATING THE GRANULES BEFORE APPLYING THEM. SET GRANULES IN A STEEL CONTAINER LIKE A JAR LID AND POUR IN OLD (COPPER-SATURATED) PICKLE. THE PLATING SHOULD BE THIN - A ROSY COLOR - SINCE TOO MUCH COPPER WILL CAUSE A FLOODING OF THE BONDING ALLOY. WHEN USING COATED GRANULES ANY FLUX MAY BE USED AND THE GLUE MAY BE OMITTED.

WHEN GRANULATING ON STERLING, THE COPPER CONTENT OF THE ALLOY ITSELF PROVIDES THE METAL NEEDED TO CREATE A LOW MELTING SOLUTION AT THE POINT OF CONTACT. USE ANY FLUX AND A TRACE AMOUNT OF GLUE. A DISADVANTAGE IS THE PROXIMITY OF THE FUSION POINT (1500°F, 815°C) AND THE MELTING POINT OF STERLING (1640°F, 960°C). GRANULES OF FINE SILVER ARE RECOMMENDED.

Making Granules:

A LINE A COFFEE CAN WITH ABOUT 3/4" OF POWDERED CHARCOAL. THIS CAN BE MADE BY FILING BRIQUETS. SPRINKLE ON THIS TINY CHIPS OF METAL KEEPING THE NUMBER SMALL ENOUGH THAT THE CHIPS WON'T TOUCH EACH OTHER. METAL CHIPS ARE MADE BY COLLECTING FILINGS OR BY CLIPPING THIN SHEETS IN THE WAY THAT SOLDER IS CUT. BUILD UP ALTERNATE LAYERS OF CHARCOAL AND METAL. SET INTO A KILN UNTIL IT GLOWS RED HOT. HOLD IT AT THIS HEAT FOR ABOUT 15 MINUTES. TO TEST FOR GRANULE FORMATION, REMOVE A SPOONFUL OF THE MIXTURE AND RINSE AWAY THE CHARCOAL. IF SHOT IS NOT COMPLETELY SPHERICAL, CONTINUE HEATING. WHEN READY, AIR COOL THE CAN AND RINSE IN WATER. THE CHARCOAL WILL GO OFF, LEAVING THE GRANULES ON THE BOTTOM OF THE CAN.

B CUT METAL INTO CHIPS AND SPRINKLE ONTO A CLEAN FLAT CHARCOAL BLOCK. HOLD THE BLOCK IN YOUR HAND ABOUT 12" ABOVE A DISH OF WATER. USE A TORCH TO MELT THE METAL, HOLDING THE BLOCK AT AN ANGLE THAT ALLOWS THE SHOT TO ROLL OFF AS IT IS DRAWN INTO A SPHERE.

FIRING

GRANULES ARE SET ONTO CLEAN METAL WITH TWEEZERS OR A BRUSH. A SCRIBED LINE CAN BE USED TO HELP LOCATE SHOT. AVOID A SINGLE LINE OF GRANULES IF POSSIBLE SINCE IT IS WEAK. DIP GRANULES IN FLUX/GLUE MIX BEFORE APPLYING, BUT AVOID EXCESS LIQUID: SOAK IT UP WITH A TISSUE. ALLOW WORK TO DRY THOROUGHLY (HALF HOUR) BEFORE APPLYING THE TORCH. WITH A BROAD FLAME BRING THE WHOLE PIECE TO BRIGHT RED. WHEN JOINTS 'FLASH' WHICH LOOKS LIKE SOLDER FLOW, REMOVE HEAT. PICKLE AND FINISH, AVOIDING ROUGH HANDLING. SCRATCH BRUSHING IS RECOMMENDED.

Niello (nē·el´·ō) FROM THE LATIN 'NIGELLUM', BLACKISH. A METALLIC ALLOY OF SULFUR WITH SILVER, COPPER, AND LEAD. USED TO FILL AN INCISED DESIGN. THE PROCESS WAS USED IN ANCIENT TIMES BY GOLDSMITHS OF ALMOST EVERY CULTURE.

process:

1. GRIND THE PREPARED NIELLO TO A POWDER, BEING SURE IT CONTAINS NO IMPURITIES. IT MAY BE WASHED BY SWIRLING IN A SHALLOW DISH UNDER RUNNING WATER LIKE ENAMELS.
2. THE WORKPIECE SHOULD BE FINISHED THROUGH A MEDIUM SANDPAPER STAGE. ALL SOLDERING SHOULD BE DONE BUT STONES SHOULD NOT BE SET. THE NIELLO WILL FILL GROOVES AS DEEP AS ONE MILLIMETER. THESE MAY BE MADE BY ETCHING, STAMPING, CHASING, ROLL PRINTING, OR ENGRAVING. AFTER CLEANING THOROUGHLY, COAT THE METAL WITH DILUTED (MILKY) PASTE FLUX OR A THIN SOLUTION OF AMMONIUM CHLORIDE AND WATER.
3. THE NIELLO IS CAREFULLY LAID INTO PLACE WITH TWEEZERS, A BRUSH, OR A SMALL SPATULA-SHAPED TOOL. IT IS DRIED BY SETTING IT IN A WARM PLACE OR BY GENTLY PLAYING A TORCH OVER IT.
4. WHEN ALL MOISTURE HAS BEEN DRIVEN OFF, THE PIECE IS UNIFORMLY HEATED EITHER IN A KILN OR WITH A TORCH. KILN SHOULD BE SET FOR 700-1000 °F. WHEN USING A TORCH, HEAT FROM BELOW; AVOID TOUCHING NIELLO WITH FLAME. THE NIELLO WILL BEAD UP AND GLOW RED-ORANGE AS IT MELTS. IF IT DOES NOT FLOW THE NIELLO MAY BE SPREAD INTO POSITION WITH A STEEL OR CARBON ROD. BE CAREFUL NOT TO OVERHEAT. EVEN WITH HIGH-MELTING NIELLO THE METAL TO WHICH IT IS BEING FUSED SHOULD NEVER GO ABOVE A DULL RED. MOST NIELLO FUSES AROUND 700° F (380°C).
5. WHEN ALL RECESSES HAVE FILLED, REMOVE HEAT. TRY TO KEEP THE FUSING OPERATION BRIEF SINCE PROLONGED HEAT CAN CAUSE THE NIELLO TO PIT AND ATTACK SILVER AND GOLD. AIR COOL. THE WORK MAY THEN BE FINISHED WITH FILES, BURNISHER, AND SANDPAPER. SINCE MACHINE BUFFING WILL WEAR AWAY THE NIELLO FASTER THAN SILVER OR GOLD, HAND BUFFING IS PREFERRED.
6. IF THE FINISHED NIELLO CONTAINS PITS THE PROCESS MAY BE REPEATED ONCE, BUT NO MORE THAN THAT. FOR REPAIRS AND FILLING SMALL LINES NIELLO MAY BE LEFT AS AN INGOT AND RUBBED OVER THE FLUXED AND HEATED METAL.

NOTE: KEEP FILES, SANDPAPER, AND A SOLDERING BLOCK (IF TORCH FIRING) RESERVED FOR THIS WORK. TRACES OF NIELLO WILL CAUSE DAMAGE IF ACCIDENTALLY HEATED TO USUAL SOLDERING TEMPERATURES. BECAUSE OF LEAD CONTENT, WASH CAREFULLY AFTER WORKING WITH NIELLO.

SAFETY ALERT

Making Niello

THOUGH SOME RECIPES SEEM COMPLICATED NIELLO IS REALLY A SIMPLE MIXTURE OF THREE METALS AND ALL THE SULFUR THEY CAN HOLD. IN GENERAL TERMS:

- SULFUR CAUSES BLACKNESS.
- COPPER DEEPENS BLACKNESS.
- SILVER RAISES THE MELTING POINT.
- LEAD PROVIDES FOR FUSION AND 'SPEADABILITY.'

1. MELT METALS IN BORAX-LINED CRUCIBLE: STIR WITH CARBON ROD.
2. ADD SULFUR AND CONTINUE TO STIR. RESULTING SMOKE IS DRAMATIC AND ACRID. PROVIDE VERY GOOD VENTILATION.
3. WHEN NO MORE SULFUR CAN BE ABSORBED POUR INTO WATER OR WARM INGOT MOLD.
4. GRIND IN MORTAR TO A FINE POWDER. MIX WITH SULFUR AND REMELT.
5. POUR AND REGRIND. NIELLO IS NOW READY TO USE.

RECIPES:

HEINRICH	SILVER OR STERLING	1 OZ.
	COPPER	2
	LEAD	3
	SULFUR	6
AUGSBERG #1	SILVER	1
	COPPER	1
	LEAD	2
	SULFUR	8
RUKLIN #2	SILVER	1
	COPPER	2
	LEAD	4
	SULFUR	5
PERSIAN	SILVER	1
	COPPER	2.5
	LEAD	7
	SULFUR	25
	AMMONIUM CHLORIDE	2.5

PREPARATION

IT IS ALWAYS IMPORTANT THAT THE METAL TO BE COLORED IS CLEAN. THE BEST WAY TO ACHIEVE THIS IS TO AVOID GREASY MATERIALS LIKE STEEL WOOL AND BUFFING COMPOUNDS IN THE FIRST PLACE. ALTERNATE FINISHING MATERIALS WOULD INCLUDE PUMICE, SANDPAPER, SCOTCHBRITE, AND A SCRATCHBRUSH.

WHEN GREASE IS PRESENT CLEAN THE WORK IN AN ULTRASONIC MACHINE OR SCRUB IT IN A STRONG SOLUTION OF AMMONIA, SOAP, AND WATER. WHEN METAL IS THOROUGHLY CLEAN, WATER WILL "SHEET" OR COVER THE WHOLE SURFACE RATHER THAN BEAD UP. WHEN THE WORK PASSES THIS TEST, DRY IT WITH A SOFT CLOTH OR TISSUE, OR DROP IT IN A BOX OF ABSORBENT MATERIAL LIKE SAWDUST. FROM HERE ON HANDLE THE WORK ONLY BY THE EDGES.

IF THE PIECE CANNOT BE COLORED RIGHT AWAY PROTECT IT FROM OXYGEN BY KEEPING IT COVERED WITH WATER. JUST BEFORE COLORING, CLEAN THE METAL BY WIPING WITH AN ALCOHOL-TYPE SOLVENT.

Simple Copper Plating

BRASS, GOLD, AND PLATINUM ARE NOTORIOUSLY HARD TO DARKEN. ONE SOLUTION IS TO PLATE THE WORK WITH A THIN LAYER OF COPPER, WHICH CAN BE READILY COLORED. BECAUSE IT IS THIN THIS LAYER WILL NOT WITHSTAND WEAR BUT IN CASES WHERE A RECESS IS TO BE DARKENED THIS METHOD OFFERS A USEFUL ANSWER.

USED PICKLE (BLUE-GREEN IN COLOR) IS ACTUALLY A COPPER PLATING SOLUTION IN THAT IT IS AN ACID CHARGED WITH FREE COPPER IONS. THESE HAVE A TENDENCY (ESPECIALLY WHEN ELECTRICITY IS INTRODUCED) TO ATTACH TO A METAL OBJECT. AN EASY WAY TO CREATE A SLIGHT ELECTRICAL CHARGE IS TO PUT INTO THE SAME ACID SOLUTION A FERROUS AND A NON-FERROUS METAL. THE ACTION IS INCREASED WITH HEAT.

① IF USED PICKLE IS NOT AVAILABLE, SET COPPER SCRAPS IN PICKLE OVERNIGHT OR UNTIL IT SHOWS COLOR.

② WRAP OBJECT LIGHTLY WITH IRON OR STEEL WIRE. BINDING WIRE WORKS WELL.

③ SET OBJECT INTO THE PICKLE; HEAT IT FIRST TO HASTEN THE ACTION. EVERYTHING IN THE PICKLE AT THIS TIME WILL BE PLATED. AFTER REMOVING THE STEEL THE PICKLE CAN BE USED AS NORMAL: IT HAS NOT BEEN DAMAGED.

④ RINSE AND COLOR OBJECT AND BUFF COPPER FROM RAISED AREAS.

Preservation:

IT IS THE NATURE OF MOST METALS TO REACT WITH THEIR ENVIRONMENT. THIS PRODUCES (AMONG OTHER THINGS) THEIR COLORS. IN CHOOSING ONE PARTICULAR PATINA WE ARE SINGLING OUT ONE POINT ON A CONTINUUM AND TRYING TO PRESERVE IT. THERE ARE SOME METALS, LIKE TIN, THAT OXIDIZE TO A STABLE FILM BUT MOST ARE LIKELY TO CONTINUE TO CHANGE.

IF THIS CHANGE IS UNDESIRABLE THE METAL MUST BE EITHER RETURNED TO ITS ORIGINAL FINISH (AS MOST SILVER HOLLOWWARE IS HAND POLISHED) OR SEALED OFF FROM THE ENVIRONMENT. A HARD FILM LIKE LACQUER WILL RESIST MARRING BUT CAN EVENTUALLY BE CHIPPED AWAY. A SOFT FILM LIKE WAX IS MORE LIKELY TO BE VULNERABLE TO WEAR BUT WILL PROBABLY JUST SMUDGE ACROSS THE PROTECTED SURFACE, KEEPING THE FILM MORE OR LESS INTACT. IN ARTICLES TO BE WORN, WAX CAN RUB OFF ON CLOTHING.

Lacquer:

USE ONLY TOP QUALITY LACQUER. THIS MAY BE BOUGHT AS A LIQUID OR A SPRAY AT AN ART SUPPLY STORE. APPLY A THIN COAT, TAKING CARE TO AVOID BUBBLES AND TRAPPED DUST. A COUPLE OF THIN COATS ARE PREFERRED TO A SINGLE THICK ONE. POROUS MATERIALS (IVORY, WOOD, CLAY, ETC) ARE LIKELY TO CONTAIN OILS THAT CAN NEVER BE COMPLETELY CLEANED AWAY. SINCE THIS CAN CAUSE LACQUER TO BEAD OR DISCOLOR, A TEST SAMPLE IS A GOOD IDEA. IF THE LACQUER DOES NOT "TAKE" ON THE WORK, CLEAN IT OFF RIGHT AWAY WITH THINNER. FOR SMALL AREAS, CLEAR NAIL POLISH IS A GOOD LACQUER. IT MIGHT NEED TO BE THINNED TO SPREAD WELL.

Wax

BEESWAX OR PARAFFIN ARE COMMONLY USED TO PROTECT METAL OBJECTS. COMMERCIAL PRODUCTS ARE CALLED "MUSEUM WAX" OR "RENAISSANCE WAX."

ONE METHOD OF APPLICATION IS TO WARM THE OBJECT AND RUB THE WAX OVER IT. ANOTHER METHOD IS TO REDUCE THE WAX TO A PASTE BY MELTING IT AND POURING IT INTO TURPENTINE. USE ABOUT ONE-THIRD AS MUCH TURPENTINE AS WAX. THIS IS THEN RUBBED ONTO THE PIECE. FURNITURE WAX (LIKE BUTCHER'S WAX) CAN ALSO BE USED, BUT AVOID POLISHES WHICH OFTEN CONTAIN OTHER ADDITIVES, LIKE SILICONE.

Recipes SURFACES

SURFACES

THESE RECIPES HAVE BEEN CULLED FROM THE SCORES AVAILABLE IN THE LITERATURE OF METALSMITHING. IN SELECTING THEM, AN EFFORT HAS BEEN MADE TO COLLECT A FEW SIMPLE MIXTURES THAT PROVIDE A WIDE RANGE OF COLOR POTENTIAL. RECIPES THAT USE DANGEROUS CHEMICALS HAVE BEEN OMITTED, AS HAVE THOSE THAT DUPLICATE EFFECTS POSSIBLE WITH THE SOLUTIONS LISTED HERE.

MANY VARIABLES WILL AFFECT THE RESULTS YOU WILL GET. FOR INSTANCE, DIFFERENT BRASSES WILL REACT IN DIFFERENT WAYS TO A SINGLE SOLUTION. THE TEMPERATURE OF THE MIXTURE AND THE METAL WILL ALTER RESULTS, AS WILL THE FINISH AND CLEANLINESS OF THE PIECE. IN EACH CASE, EXPERIMENTATION IS REQUIRED.

SINCE SOME PROJECTS WILL REQUIRE A LARGE QUANTITY OF SOLUTION AND SOME WILL NEED ONLY A SPOONFUL, I HAVE USED THE UNSCIENTIFIC SYSTEM OF PERCENTAGES IN THESE RECIPES. MEASUREMENTS DO NOT NEED TO BE PRECISE; EITHER WEIGHT OR VOLUME UNITS MAY BE USED. INGREDIENTS USUALLY DISSOLVE FASTER WHEN WARM. SOLUTIONS ARE GENERALLY BEST WHEN FRESH, AND OFTEN DO NOT REMAIN POTENT WHEN STORED.

TO HEAT-COLOR A PIECE I RECOMMEND USING A HOT PLATE OR KITCHEN STOVE. SET THE PIECE ON A BURNER AND WATCH IT OXIDIZE TO A DESIRED SHADE. THE WORK IS THEN QUICKLY LIFTED OFF WITH TWEEZERS AND QUENCHED, USUALLY IN WATER. THIS PROCESS IS MOST SUCCESSFUL ON PIECES OF A CONSISTENT CROSS-SECTION.

MANY OF THE SOLUTIONS WORK BEST ON SLIGHTLY WARM METAL. HOLD THE PIECE UNDER RUNNING HOT TAP WATER, DIP IT INTO A SOLUTION, AND FLUSH AGAIN WITH HOT WATER WHILE SCRATCH BRUSHING. REPEAT THIS WARM-DIP-RINSE CYCLE UNTIL THE CORRECT COLOR APPEARS.

SMALL QUANTITIES OF CHEMICALS ARE SOMETIMES AVAILABLE THROUGH THE CHEMISTRY DEPARTMENT OF A LOCAL HIGH SCHOOL OR COLLEGE.

THE ARROWS USED HERE INDICATE A PROGRESSION OF COLOR.

COLORING GUIDE

THE NUMBERS GIVEN HERE REFER TO THE RECIPES ON THE FOLLOWING PAGE.

Gold
PURPLE → DARK GRAY **6**
GRAY → BLACK **9**

Sterling
GOLDEN → SCARLET → BLUE → PLUM → GRAY → BLACK **1**
MULTICOLORED IRIDESCENCE (LIKE AN OIL SLICK) **5**
DEEP PURPLE **8**

Brass
VARIEGATED GREEN FILM **3,7**
FLAT PALE GREEN **5**
DEEP BLUE → PURPLE → GRAY **8**

Bronze
VARIEGATED GREEN FILM **3,7**
SCARLET → BLUE → BROWN → GRAY **2,8**
RICH BROWN → BLUE **4**
DARK GRAY **5**

Copper
DEEP PURPLE → BLUE → BROWN → BLACK **1**
BRILLIANT SCARLET → BLUE → VIOLET **2,8**
GREEN FILM **7**
RICH BROWN → BLUE **4**
FLAT SEA GREEN **5**
RICH BLUE → PURPLE → GRAY **8**

Pewter
GRAY → BLACK **3,5**

SURFACES *Recipes*

1. DISSOLVE A SMALL AMOUNT OF LIVER OF SULFUR IN WARM WATER. A PEA-SIZE PIECE TO A CUP OF WATER IS USUAL. A TRACE AMOUNT OF AMMONIA ADDED TO THE SOLUTION WILL BRIGHTEN THE COLORS. IF THE SOLUTION IS TOO STRONG THE RESULTING SULFIDE LAYER IS BRITTLE AND WILL BE EASILY CHIPPED OFF. SOLUTION MAY BE WARMED, BUT DO NOT BOIL.

SOLUTION MAY BE BRUSHED ON, OR WORK MAY BE IMMERSED. I RECOMMEND THE DIP-RINSE-BRUSH METHOD TO SLOWLY CREATE A WIDE RANGE OF COLORS.

2. DISSOLVE A PINCH OF BARIUM SULFIDE IN A CUP OF WATER.

BRUSH ON OR IMMERSE WORK.

3.

1 PART	AMMONIUM CHLORIDE	(1 %)
6 PARTS	COPPER SULFATE	(11)
60 "	WATER	(88)

APPLY TO METAL AND ALLOW TO DRY. REPEAT SEVERAL LAYERS, ALLOWING EACH TO DRY. SCRATCH BRUSHING WILL ALTER COLORS.

4.

1 PART	LEAD ACETATE	(10 %)
1 "	ACETIC ACID	(10)
2 PARTS	SODIUM THIOSULFATE	(20)
6 "	WATER	(60)

APPLY AS IN #3 ABOVE.

5.

1 PART	ZINC CHLORIDE	(2 %)
2 PARTS	ACETIC ACID	(3)
4 "	AMMONIUM CHLORIDE	(5)
4 "	TABLE SALT	(5)
8 "	COPPER SULFATE	(10)
60 "	WATER	(75)

DISSOLVE ALL INGREDIENTS TOGETHER AND BRUSH ONTO OR IMMERSE WORK.

6. DISSOLVE A FEW IODINE CRYSTALS IN ALCOHOL UNTIL THE SOLUTION IS SATURATED. THIS IS A HIGH-STRENGTH MERCUROCHROME, AND WILL STAIN SKIN.

BRUSH ON OR IMMERSE WORK.

7. SUSPEND WORK IN A CALCIUM CARBONATE ENVIRONMENT.

UNDER A BOX, SET A BOWL OF WATER AND A SECOND BOWL OF VERY DILUTED HYDROCHLORIC ACID OR VINEGAR. DROP BITS OF CHALK INTO THE ACID (VINEGAR) PERIODICALLY. THIS METHOD CAN TAKE SEVERAL DAYS TO PRODUCE A COLOR.

8. <u>BIRCHWOOD CASEY GUN BLUE</u>

THIS COMMERCIAL PREPARATION IS AVAILABLE AT MOST SPORTING GOODS STORES, OR WRITE TO:
BIRCHWOOD CASEY CO.
EDEN PRAIRIE, MINN. 55344

USE SOLUTION FULL STRENGTH BY BRUSHING ONTO OR IMMERSING WORK.

9. COMMERCIAL GOLD OXIDIZER

AVAILABLE THROUGH MOST JEWELRY SUPPLY COMPANIES.

THIS IS USED FULL STRENGTH AND, ON GOLD, MUST BE IN CONTACT WITH STEEL TO CAUSE THE COLORING REACTION. USE A NAIL OR PIECE OF WIRE (PAPER CLIP) TO COLOR SPECIFIC AREAS. FOR BROADER APPLICATIONS, GRIP A BIT OF STEEL WOOL IN TWEEZERS AND USE AS A DAUBER. TO COLOR SMALL PIECES OR CHAINS, DISSOLVE STEEL WOOL IN SOLUTION AND IMMERSE.

SAFETY FIRST

- WORK IN A WELL-VENTILATED SPACE: ACID FUMES ARE HARMFUL.
- WEAR PROTECTIVE CLOTHING SUCH AS A PLASTIC APRON (SOLD SPECIFICALLY FOR THIS), HEAVY RUBBER GLOVES AND GOGGLES.
- WORK NEAR A SOURCE OF RUNNING WATER. IF THIS IS IMPOSSIBLE OBTAIN A PORTABLE EYE WASH UNIT.
- STORE ACIDS IN NARROW-NECKED TOUGH PLASTIC BOTTLES, KEPT IN A COOL DARK PLACE WITH A FREE FLOW OF AIR. LABEL ALL CONTAINERS CLEARLY AND BOLDLY. NEVER STORE ACIDS UP HIGH OR IN A BOTTLE NOT DESIGNED FOR THIS PURPOSE.
- TO DISPOSE SMALL QUANTITIES OF ACID, PUT INTO A LARGE CONTAINER (BUCKET) AND SLOWLY ADD BAKING SODA. THE MIXTURE WILL BUBBLE AND SWELL. WHEN THE ACTION HAS SUBSIDED, REPEAT WITH MORE SODA, CONTINUING UNTIL THE SOLUTION NO LONGER BUBBLES. THEN POUR INTO THE GROUND AWAY FROM ANY SOURCE OF WATER.

Important: ALWAYS ADD ACID TO WATER.

TO INSURE AN IMMEDIATE MIXING, POUR THE DENSER LIQUID (ACID) INTO THE WATER. STIR WITH A GLASS ROD OR SWIRL CONTAINER AS YOU POUR. MIX ACIDS AT ARM'S LENGTH.

ACIDS

CHEMICALLY PURE (C.P.)		100 %
REAGENT	APPROX.	70 %
COMMERCIAL	APPROX.	50 %

MOST ACIDS ARE AVAILABLE IN THESE THREE GRADES. THE FIRST IS VERY EXPENSIVE AND NOT NECESSARY. MOST PEOPLE USE REAGENT GRADE AND MOST FORMULAS, INCLUDING THOSE ON THE NEXT PAGE, ARE WRITTEN FOR THIS. IF YOU HAVE A MORE DILUTE ACID, MODIFY THE FORMULAS ACCORDINGLY.

BY ITS NATURE ACID IS A TEMPERMENTAL COMMODITY. UP TO A POINT, DILUTING IT MAKES IT STRONGER. IN SOME CASES IT GETS STRONGER AS IT ABSORBS OTHER CHEMICALS, SO "OLD" ACID IS BETTER THAN A FRESH MIX. INCREASED TEMPERATURE WILL ACCELERATE THE ACTION OF ACID BUT ONCE IT GETS GOING IT PRODUCES ITS OWN HEAT AND SO WILL CONTINUE A STRONG BITE. THE ONLY RULE IS THAT THERE ARE NO RULES AND EACH TIME YOU ETCH YOU MUST PAY ATTENTION TO WHAT IS HAPPENING.

Process:

MANY VARIATIONS ARE POSSIBLE BUT USUALLY EITHER A DESIGN IS PAINTED ON:

OR THE WHOLE PIECE IS PAINTED AND, WHEN DRY, THE PAINT IS SELECTIVELY SCRATCHED AWAY.

TO ACHIEVE DIFFERENT HEIGHTS MAKE THE DESIGN AS ABOVE AND AFTER ETCHING A WHILE, PULL IT OUT, RINSE, AND STOP OUT THE AREAS THAT HAVE SUFFICIENT DEPTH WITH ANY RESIST. ALLOW THE FRESH RESIST TO DRY AND RE-SUBMERGE. THE OPPOSITE APPROACH CAN YIELD THE SAME EFFECT: I.E. SCRATCH OUT ONLY THAT PART OF THE DESIGN YOU WANT DEEPEST AND BEGIN ETCHING. AFTER A WHILE PULL THE PIECE OUT, RINSE, AND SCRATCH AWAY MORE OF THE DESIGN. THE FIRST DESIGN WILL CONTINUE ETCHING DEEPER AS THE NEW DESIGN IS CUT. THIS MAY BE DONE REPEATEDLY TO ACHIEVE SEVERAL DISTINCT LAYERS. THIS TECHNIQUE IS ESPECIALLY APPROPRIATE TO SUBSEQUENT BASSE TAILLE ENAMELING.

LEARNING WITHOUT THOUGHT IS LABOR LOST; THOUGHT WITHOUT LEARNING IS PERILOUS. CONFUCIOUS

RESISTS	SOLVENTS
ASPHALTUM	TURPENTINE
OIL BASE PAINT (SPRAY OR PAINT ON)	TURPENTINE
SHELLAC (COLOR WITH DYE)	ALCOHOL
LACQUER (E.G. NAIL POLISH)	LACQUER THINNER
PRESS TYPE GREASE PENCIL	TURPENTINE OR THINNER

NONE OF THESE RESISTS WILL BOND WELL TO UNCLEAN METAL. SINCE A SLIGHTLY ROUGHENED SURFACE IS ALSO HELPFUL, CLEAN BY SCRUBBING WITH PUMICE. RINSE, PAT DRY, AND AVOID TOUCHING THE SURFACE.

During etching, GASES ARE RELEASED IN THE FORM OF BUBBLES. IF THESE REMAIN ON THE METAL THEY WILL PREVENT THE ACID FROM REACHING IT, CAUSING AN UNEVEN BITE. TO REMOVE BUBBLES, BRUSH THE WORK LIGHTLY WITH A FEATHER OR A MOP MADE FROM STRING. TO SAFELY LOWER WORK INTO THE ACID BATH, USE A LOOP OF STRING. TWEEZERS MAY ALSO BE USED BUT THEY OFTEN SCRATCH THE RESIST.

SAFETY ALERT

WE CAN'T ALL AND SOME OF US DON'T. THAT'S ALL THERE IS TO IT.

EEYORE (A.A. MILNE)

MORDANTS

UNLESS SPECIFIED THESE ARE FLUID MEASURES. THE FIGURE IN PARENTHESIS IS %. FOR BREVITY THE WORD 'ACID' HAS BEEN OMITTED.

gold	1 PART NITRIC (25) 3 PARTS HYDROCHLORIC OR SULFURIC (75) MAY BE DILUTED. THIS IS AQUA REGIA	8 PARTS HYDROCHLORIC (14) 4 PARTS NITRIC (7) 1 PART IRON PERCHLORIDE (2) 40-50 PARTS WATER (77)
silver, sterling	1 PART NITRIC (25) 3 PARTS WATER (75)	5 PARTS NITRIC (50) 4 PARTS WARM WATER (40) 1 PART ISOPROPYL ALCOHOL
copper, brass, nickel	1 PART NITRIC (50) 1 PART WATER (50)	10 PARTS HYDROCHLORIC (10) 2 " POTASSIUM CHLORIDE (2) 90 PARTS WATER (88)
pewter, lead, tin	1 PART NITRIC (20) 4 PARTS WATER (80)	
alum-inum	3/4 OZ. AMMONIA (2) 2.5 GRAM COPPER SULPHATE (2) 7 OZ. SODIUM HYDROXIDE (6) 1-2 DROPS 75% PHOSPHORIC ACID 1 GAL. WATER (92)	ETCH FOR 3 MINUTES, RINSE, AND SET IN 40% NITRIC SOLUTION (I.E. 60% WATER).
iron, steel	2 PARTS HYDROCHLORIC (67) 1 PART WATER (33)	1 PART NITRIC (9) 10 PARTS WATER (91) 1 DROP 75% PHOSPHORIC ACID

A VERY CLEAN ETCH CAN BE ACHIEVED ON COPPER AND BRASS WITH A FERRIC CHLORIDE SOLUTION. THIS IS A SALT RATHER THAN AN ACID AND BEHAVES A LITTLE DIFFERENTLY. WORK MUST BE LEVEL AND SHOULD BE SUSPENDED JUST BELOW THE SURFACE OF THE FLUID. ELECTRICAL TAPE OR CONTACT PAPER CAN BE USED TO MASK THE BACK. IN ADDITION TO THE RESISTS LISTED ABOVE FELT TIP PEN CAN BE USED. ETCH WILL TAKE ONE TO FOUR HOURS. RINSING IN WATER WILL NOT STOP THE CORROSIVE ACTION: SCRUB WITH AMMONIA OR BAKING SODA. SOLUTION CAN BE BOUGHT AT ELECTRONICS SUPPLIES LIKE RADIO SHACK.

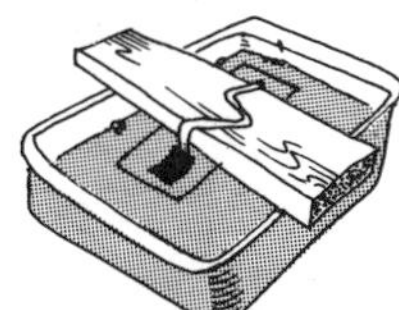

THE ETCHING IN THIS PROCESS IS THE SAME AS THAT DESCRIBED IN THE TWO PRECEDING PAGES. PHOTOGRAPHIC TECHNIQUES ARE USED TO CREATE AN IMAGE OF RESIST AND EXPOSED AREAS. IN SIMPLE TERMS, A PIECE OF METAL IS COATED WITH A LIGHT-SENSITIVE MATERIAL SO IT WILL REACT JUST LIKE PHOTOGRAPHIC PAPER. IT IS THEN EXPOSED TO LIGHT PASSING THROUGH A PIECE OF FILM. A CHEMICAL DEVELOPER IS THEN USED TO REMOVE THOSE AREAS EXPOSED TO LIGHT. WHEN THE RESIST HAS DRIED, THE METAL IS ETCHED.

THIS PROCESS IS FUNDAMENTAL TO THE MICRO-ELECTRONICS INDUSTRY AND IS GROWING AND IMPROVING AT AN AMAZING PACE. METALSMITHS PURSUING THIS TECHNIQUE ARE ENCOURAGED TO CONSULT TECHNICAL MAGAZINES, INDUSTRY, AND TECHNICAL COLLEGES FOR MORE INFORMATION.

The Image:

MUST BE A HIGH-CONTRAST NEGATIVE; THAT IS, ONE WITHOUT GRAY AREAS. THIS CAN BE MADE WITH KODALITH FILM OR BY DRAWING ON ACETATE WITH BLACK INK. SOME COPYING MACHINES WILL PRINT AN IMAGE ON ACETATE, ALLOWING THE USE OF DRAWINGS OR PRINTED IMAGES FROM BOOKS. THE FILM IMAGE MUST BE THE ACTUAL SIZE YOU WANT ON THE METAL.

PROCESS

CUT METAL SLIGHTLY OVERSIZE, DRILL HOLE IN ONE CORNER AND HOOK A 6" PIECE OF WIRE BY WHICH TO HANDLE AND HANG THE METAL.

CHECK METAL FOR FLATNESS AND CLEAN THOROUGHLY WITH PUMICE UNTIL WATER RUNS OFF IN SHEETS. DO NOT SHORTCUT ON CLEANLINESS. DRY WITH A LINT-FREE CLOTH.

UNDER A SAFE LIGHT COAT METAL WITH RESIST BY DIPPING. FUMES FROM THE RESIST ARE DANGEROUS: WEAR A PROPER RESPIRATOR. VENTILATE AND AVOID UNNECESSARY EXPOSURE. REMEMBER TO PUT RESIST AWAY BEFORE TURNING ON ROOM LIGHTS SINCE EXPOSURE WILL RUIN IT.

HANG METAL TO DRY IN A LIGHT-TIGHT BOX, PREFERABLY COVERED WITH A TIGHTLY WOVEN DARK FABRIC. WARM AIR WILL SPEED UP DRYING. THIS CAN BE SUPPLIED WITH A HAIR DRYER. SINCE THIS WILL KICK UP DUST, BE SURE THE BOX IS CLEAN. DO NOT HEAT ABOVE 100° C (200°F). DRYING WILL TAKE ABOUT 45 MINUTES WITH HEAT, 2-3 HOURS WITHOUT.

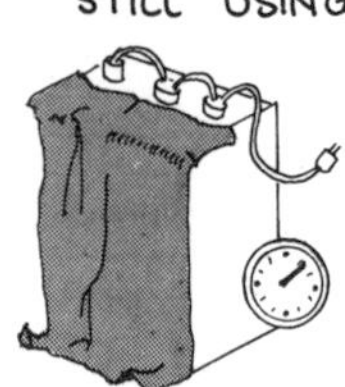

STILL USING SAFE-LIGHT CONDITIONS, CONTACT PRINT THE NEGATIVE ONTO THE METAL. THE EXPOSURE TIME WILL DEPEND ON THE BRAND OF RESIST, FRESHNESS OF LIGHT BULBS, AND DISTANCE FROM LIGHT SOURCE TO METAL. IT WILL PROBABLY BE BETWEEN 5 AND 20 MINUTES. A STANDARD STEP TEST WITH 2 MINUTE INTERVALS IS RECOMMENDED.

DEVELOP IN NORMAL LIGHT FOR 1-2 MINUTES BY SETTING THE METAL IN A TRAY OF DEVELOPER MADE FOR THE RESIST BEING USED. AGITATE SLIGHTLY TO PROVIDE A FRESH WASH OF CHEMICALS TO ALL AREAS. THOUGH HARD TO SEE, THE IMAGE SHOULD BE VISIBLE AS A PATTERN OF COATED AND BARE METAL. RINSE IN COOL WATER.

DRY IN HEATED BOX FOR 30 MINUTES OR IN WARM AREA FOR ABOUT 2 HOURS.

COAT BACK AND EDGES WITH CONVENTIONAL RESIST LIKE ASPHALTUM OR LACQUER.

ETCH AS DESCRIBED ON PRECEDING TWO PAGES WITH MORDANTS LISTED. A DETAIL-PRESERVING SLOW ETCH IS RECOMMENDED.

AFTER ETCHING REMOVE RESIST WITH ACETONE OR LACQUER THINNER.

CHEMICALS

SAFETY ALERT

CMR 5000 PHOTO RESIST
CMR 5000 THINNER
CMR 5000 DEVELOPER

FOR NEAREST DISTRIBUTOR:
TIKOL/DYNACHEM
P.O. BOX 2047
SANTA ANNA, CA 92711

THERE ARE OTHER BRANDS; CONSULT A LOCAL DIRECTORY OR INDUSTRIAL USER. ONLY SMALL QUANTITIES ARE USED BUT SUPPLIERS WILL HAVE A MINIMUM ORDER.

Materials

1. A DARKROOM OR LIGHT-PROOFED CLOSET OR BATHROOM.
2. A DRYING BOX (CARDBOARD)

3. A RESPIRATOR
4. EITHER LIGHT BOX. TOP ONE IS CHEAPER BUT GETS VERY HOT AND GIVES UNEVEN LIGHT AS BULBS BURN OUT.

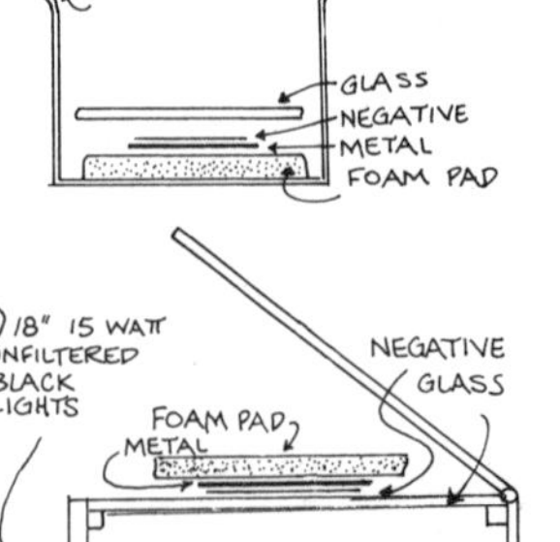

SHAPING

THE DRAWPLATE IS A BAR OF HARDENED STEEL PERFORATED WITH A SERIES OF TAPERED HOLES IN A SEQUENCE OF DECREASING DIAMETER. WIRE DRAWN THROUGH THE PLATE WILL CONFORM TO THE SIZE AND CROSS SECTION OF THE HOLE.

USES:
- TO MAKE WIRE THINNER
- TO CHANGE SECTION, E.G. TO MAKE ROUND WIRE SQUARE
- TO MAKE TUBING
- TO MAKE WIRE HARDER
- TO STRAIGHTEN BENT WIRE.

to lubricate:

A PIECE OF RAG OR SPONGE IS CLAMPED ONTO THE PLATE AFTER HAVING BEEN MOISTENED WITH A LIGHT OIL LIKE WINTERGREEN OR OLIVE OIL. THE WIRE MAY ALSO BE RUBBED WITH WAX BUT THIS CAN CLOG SMALL HOLES.

CLOTHES PIN

ANNEALING WIRE:

LIKELY TO MELT HERE

COIL WIRE CLOSELY TO AVOID MELTING IT. IF WRAPPED WITH COPPER OR BRASS THE COIL CAN BE QUENCHED IN PICKLE IMMEDIATELY AFTER ANNEALING

DRAWPLATES CAN BE MADE OF A HARD WOOD LIKE MAPLE. THESE ARE BEST FOR TUBEMAKING AND CHAIN DRAWING BUT IN A PINCH CAN BE USED TO REDUCE WIRE. USE A PIECE OF HALF INCH WOOD, DRILL A SERIES OF HOLES AND CARVE A FUNNEL SHAPE WITH A BUD BUR OR TAPERED REAMER.

FOR EXTRA STRENGTH AND GREATER PRECISION A THIN SHEET OF STEEL MAY BE BOLTED ONTO THE FRONT OF THE PLATE. THE HOLES ARE THEN DRILLED AND ENLARGED AS NEEDED WITH BURS OR FILES. THE STEEL DOES NOT HAVE TO BE HARDENED.

PROCESS:

① THE PLATE IS HELD IN A VISE LIKE THIS → NOT LIKE THIS.

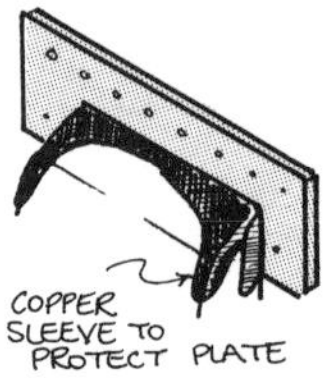

COPPER SLEEVE TO PROTECT PLATE

② A GRADUAL TAPER IS FILED ON THE TIP OF THE WIRE TO BE DRAWN. A NOTCH FILED IN THE BENCH PIN MAKES THIS EASIER.

③ THE TIP OF THE WIRE IS FED THROUGH THE UNNUMBERED SIDE OF THE PLATE, INTO THE FIRST HOLE IT FITS SNUGLY. HEAVY-DUTY GRIPPING PLIERS, CALLED DRAW TONGS, ARE USED TO PULL THE WIRE THROUGH IN A SLOW SMOOTH MOTION.

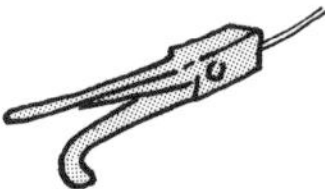

④ THE WIRE IS PULLED THROUGH SUCCESSIVE HOLES UNTIL IT FEELS TOUGH AND SPRINGY. IT IS ANNEALED, DRIED, AND DRAWING CAN CONTINUE. OFTEN THE POINT WILL NEED TO BE RE-FILED AS DRAWING PROGRESSES.

IF A VISE IS NOT AVAILABLE HOLD DRAWPLATE ON A PAIR OF BOARDS ACROSS A DOOR JAMB. YOU'LL NEED HELP TO HOLD THESE IN PLACE.

AMERICAN INDIAN SILVERSMITHS USED TO ANCHOR THEIR PLATES AGAINST PEGS IN THE GROUND.

TO DRAW HALF-ROUND WIRE IN A ROUND DRAWPLATE SLIGHTLY FLATTEN TWO STRIPS OF WIRE AND SOLDER THEM TOGETHER FOR ABOUT AN INCH AT ONE END. TAPER THIS AND DRAW AS USUAL, ANNEALING AS NECESSARY. IN THE SAME WAY, RECTANGULAR WIRE CAN BE MADE IN A SQUARE PLATE.

THIS DO-IT-YOURSELF DRAW BENCH IS MADE FROM HARDWARE STORE MATERIALS AND AN OLD BELT. IT REQUIRES THAT BOTH TIPS OF THE TONGS BE BENT: HEAT BEND-AREA TO RED AND GRIP WITH PLIERS. THERE IS NO NEED TO HARDEN OR TEMPER.

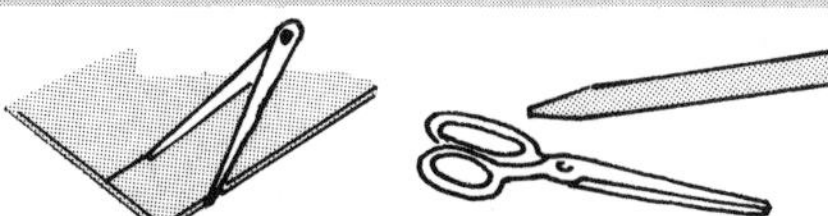

CUT A STRIP OF METAL HAVING PARALLEL SIDES; DIVIDERS ARE HANDY FOR MARKING THIS. CUT A POINT ON THE STRIP.

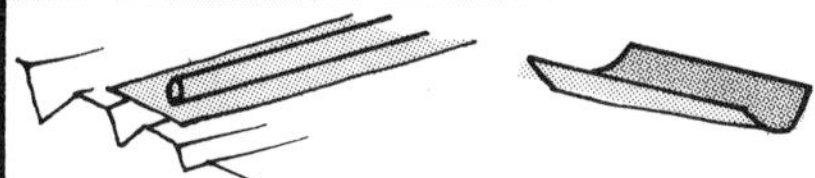

LAY ANNEALED STRIP IN A V-BLOCK, ACROSS THE OPEN JAWS OF A VISE, OR ON A LEAD BLOCK. SET A STEEL ROD ALONG ITS CENTER LINE AND STRIKE IT WITH A MALLET.

WITH PLIERS OR A MALLET CONTINUE BENDING THIS TROUGH INTO A TUBE NEAR THE POINT. ONCE THIS IS CURLED ENOUGH TO FIT IN THE DRAWPLATE THE REST OF THE STRIP WILL CURL EVENLY.

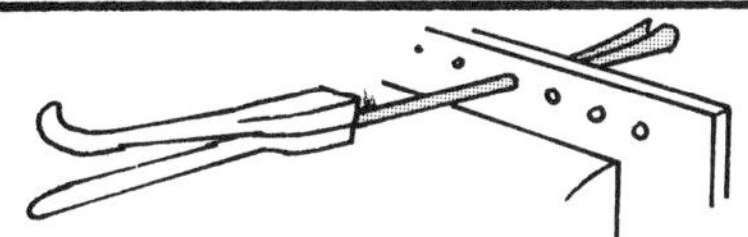

PULL THROUGH THE DRAWPLATE JUST AS FOR WIRE. TRY TO PULL THE TUBE STRAIGHT OUT; I.E PERPENDICULAR TO THE DRAWPLATE.

NOT LIKE THIS

CONTINUE PULLING UNTIL THE EDGES JUST MEET. IF THE SEAM LOOKS ROUGH, PAUSE BEFORE CLOSING TO EVEN THE EDGES WITH A NEEDLE FILE. DO NOT OVERLAP THE SEAM. IF TUBE RIPPLES, ANNEAL.

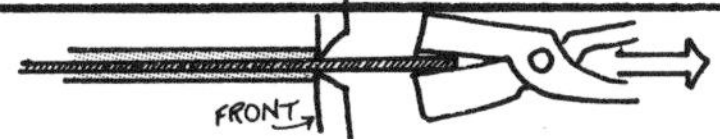

TO ACHIEVE A DESIRED INSIDE DIAMETER, SET A STEEL OR BRASS WIRE OF THIS DIAMETER INTO THE TUBE MIDWAY IN THE PROCESS. OIL THE WIRE LIGHTLY AND BE SURE IT IS LONGER THAN THE TUBE. TO REMOVE IT PUT THE WIRE THROUGH THE PLATE FROM THE FRONT AND PULL IT OUT WITH TONGS.

Calculations:

FOR A GIVEN O.D. :

O.D. + THICKNESS* X π (3.14)

FOR A GIVEN I.D. :

I.D. − THICKNESS* X π (3.14)

* THICKNESS OF SHEET BEING USED.

MOST DEALERS OF METALS SELL SOMETHING CALLED EXTRUDED TUBE. THIS HAS BEEN MADE IN A CONTINUOUS CASTING TECHNIQUE AND SO HAS NO SEAM. THOUGH IT IS NEVER ESSENTIAL, IN SOME APPLICATIONS THIS IS A HANDY FEATURE.

TUBEMAKING WORKS BEST WITH METAL 24 GAUGE OR THINNER. TO MAKE THICK-WALLED TUBING FOLLOW THE DIRECTIONS ABOVE AND SOLDER THE SEAM. AFTER PICKLING AND DRYING, THE TUBE MAY BE DRAWN LIKE WIRE, MAKING IT LONGER AND THICKER-WALLED.

twisting—

TWISTED WIRES MAY BE COMBINED IN MANY WAYS FOR STRIKING DECORATIVE EFFECTS. THEY MAY BE USED AS BORDERS, FILIGREE, BEZELS, OR TO INDICATE ROPE.

TWIST USING A HAND VISE, VISE GRIP PLIERS, OR A VARIABLE SPEED DRILL.

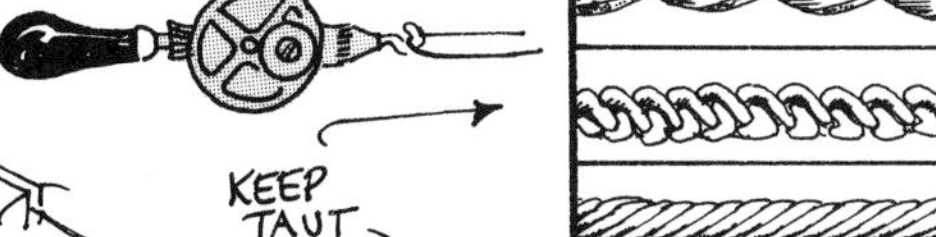

	A SQUARE WIRE, TWISTED.
	SQUARE WIRE, TWISTED LEFT AND RIGHT, SPACED
	A ROUND WIRE, DOUBLED OVER AND TWISTED.
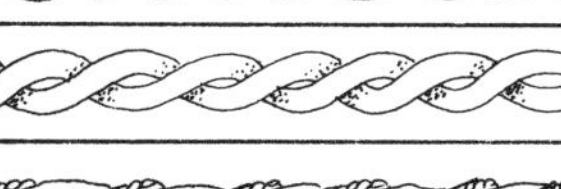	AS ABOVE, FLATTENED. (ON EDGE, USED FOR FILIGREE)
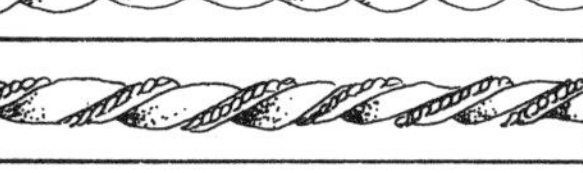	LARGE SQUARE WIRE TWISTED. SMALL TWISTED WIRE THEN WOUND AROUND IT.
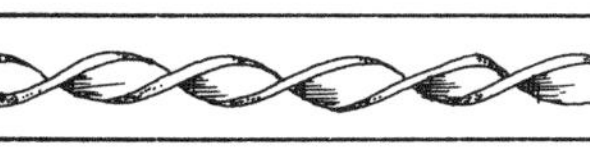	RECTANGULAR WIRE, TWISTED.
	ROUND WIRE TWISTED RIGHT, CUT IN HALF, LAID SIDE BY SIDE AND TWISTED LEFT.
	3-7 STRANDS TWISTED TOGETHER.
	TWO ROUND WIRES TWISTED TOGETHER; ONE IS THEN REMOVED
	SQUARE WIRE AND TWIST WOUND ON THICK WIRE.

SHAPING

Definition & Effect

FORGING MAY BE DEFINED AS THE CONTROLLED SHAPING OF METAL BY THE FORCE OF A HAMMER. THE TECHNIQUE LENDS ITSELF TO GRACEFUL TRANSITIONS FROM PLANE TO PLANE AND APPEALING CONTRASTS OF THICK AND THIN SECTIONS.

IT IS EQUALLY APPROPRIATE FOR LARGE AND SMALL WORK. GOLD, STERLING, AND COPPER FORGE VERY WELL. LOW-ZINC BRASSES CAN ALSO BE FORGED BUT WILL REQUIRE FREQUENT ANNEALING.

IT IS A SIGN OF GOOD FORGING TO REQUIRE VERY LITTLE FILING. FORCE AND CONTROL MUST WORK TOGETHER.

CONTROL IN FORGING COMES FROM THE CROSS PEEN. ITS WEDGE SHAPE CAN PUSH THE METAL IN ONLY TWO DIRECTIONS. THIS 'PUSH' CAN BE DIRECTED ALONG THE AXIS TO INCREASE LENGTH OR OUTWARD FROM THE AXIS TO INCREASE BREADTH.

TIPS

- SIT OR STAND CLOSE TO THE WORK IN A POSTURE YOU CAN COMFORTABLY MAINTAIN.
- WORK ON A SMOOTH, HARD, STABLE SURFACE
- KEEP FINGERS AND THUMB WRAPPED AROUND HANDLE, NOT POINTING ALONG IT.
- ANNEAL AS NEEDED: DON'T PRESS YOUR LUCK.
- KEEP HAMMER FACE POLISHED.
- DON'T HOLD WORK PIECE WHERE YOU INTEND TO HIT IT.
- HAMMER MUST MAKE SOLID SQUARE CONTACT WITH ANVIL.

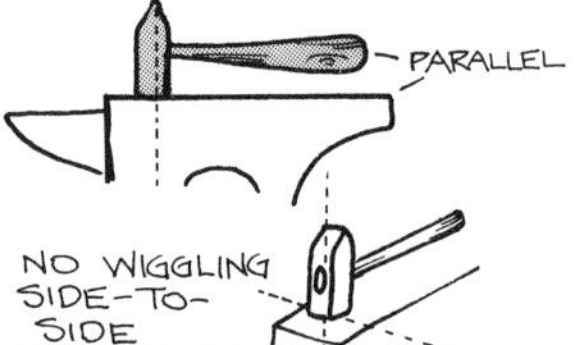

forging a taper

WORK ON SQUARE STOCK STRIKING ALL SIDES EQUALLY.

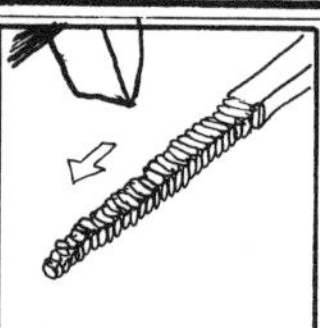

PLANISH OUT BUMPS BY ROTATING.

HASTE IN EVERY BUSINESS BRINGS FAILURES.

HERODOTUS, 450 BC

the rhombus

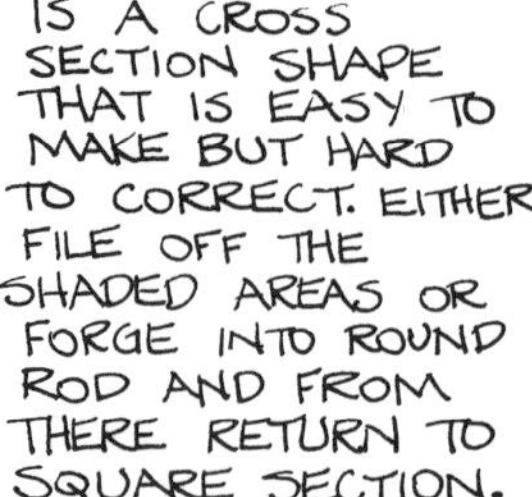

IS A CROSS SECTION SHAPE THAT IS EASY TO MAKE BUT HARD TO CORRECT. EITHER FILE OFF THE SHADED AREAS OR FORGE INTO ROUND ROD AND FROM THERE RETURN TO SQUARE SECTION.

AN ALTERNATIVE FORGING METHOD USES THE CURVE OF AN ANVIL HORN OR STAKE TO FORCE THE METAL TO FLOW OUTWARD FROM THE POINT OF CONTACT.

ANY WILL WORK

Bending

PERHAPS BECAUSE IT IS SUCH A BASIC TECHNIQUE MANY SMITHS FORGET HOW VERSATILE AND EFFECTIVE BENDING CAN BE.

HERE ARE SOME GENERAL RULES:

USE YOUR FINGERS AS MUCH AS POSSIBLE. WOOD OR RAWHIDE TOOLS ARE USED NEXT, AND STEEL TOOLS (HAMMERS, PLIERS, ETC) ONLY WHEN ABSOLUTELY NEEDED.

WHENEVER POSSIBLE, ANNEAL THE METAL. THIS STEP TAKES LESS TIME THAN REMOVING THE MARKS THAT MIGHT BE THE RESULT OF WORKING ON HARD MATERIAL.

TO ACHIEVE A SHARP BEND, SCORE METAL AT LEAST 3/4 OF THE WAY THROUGH. AFTER BENDING, THE CREASE SHOULD BE STRENGTHENED WITH SOLDER.

scoring

IS THE REMOVING OF METAL ALONG THE LINE OF A PROPOSED FOLD. IT MAY BE DONE WITH A SHARP SCRIBE ON THIN SHEET AND WITH A GRAVER, NEEDLE FILE, OR AN OLD FILE TANG CONVERTED LIKE THIS:

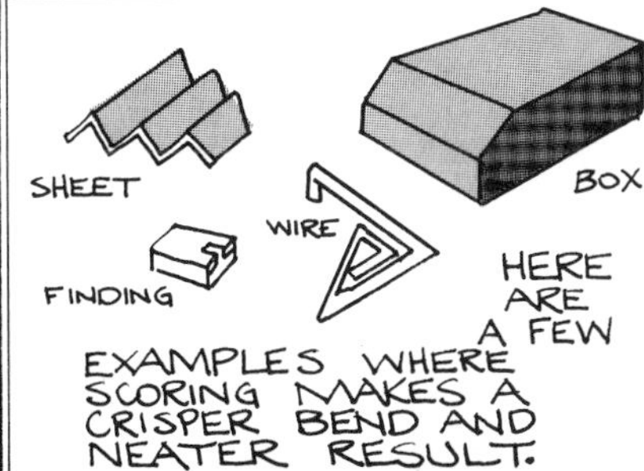

HERE ARE A FEW EXAMPLES WHERE SCORING MAKES A CRISPER BEND AND NEATER RESULT.

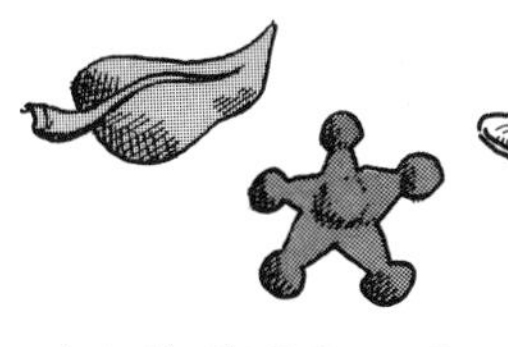

shallow forming

SHALLOW FORMING (ALSO CALLED 'BOSSING') IS A METHOD OF GIVING A MINOR CURVATURE OR DOMING TO SHEET METAL. IT USUALLY MAKES A PIECE LOOK THICKER AND SINCE CURVED SURFACES SHOW MORE REFLECTIONS THAN FLAT SHEETS, THE RESULT IS OFTEN BRIGHTER AND MORE DYNAMIC.

1. SAW OUT THE SHAPE.
2. IF PIECE IS TO BE STAMPED, CHASED, TEXTURED OR HAVE MARRIED METALS THESE ARE DONE NEXT.

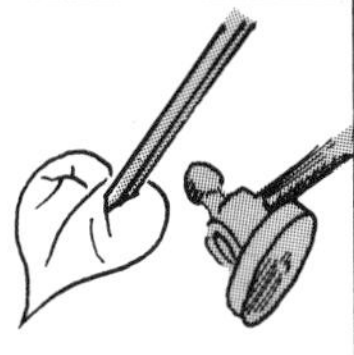

3. AFTER ANNEALING THE METAL IS SET ON A SEMI-SOFT SURFACE AND WORKED WITH MALLET, HAMMER, OR PUNCH.
4. IF EDGES DO NOT BLEND INTO THE PIECE THEY MAY BE FORMED OVER A DAPPING PUNCH OR HAMMER SIMILARLY HELD.

working surfaces

PITCH

LEAD – SCRUB HANDS & METAL AFTER USING.

MICRO-CRYSTALINE WAX

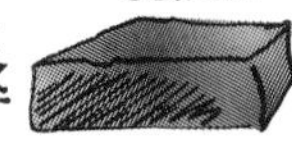

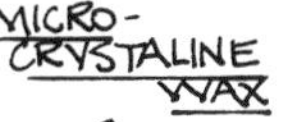

LEATHER

SOFT WOOD (PINE)

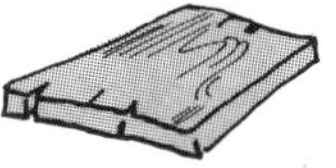

GENERAL

REPOUSSÉ IS ONE OF THE OLDEST METALSMITHING TECHNIQUES AROUND; VIRTUALLY EVERY ANCIENT CULTURE HAS LEFT EXAMPLES. IT IS A VERSATILE PROCESS APPROPRIATE TO ANY SCALE AND ALL MALLEABLE METALS, FROM ALUMINUM TO STEEL. MANY APPROACHES ARE USED SO THE OUTLINE BELOW MUST BE TAKEN ONLY AS AN INTRODUCTION AND POINT OF DEPARTURE.

THE WORD COMES FROM THE FRENCH VERB MEANING "TO PUSH BACK." SIMPLY STATED REPOUSSÉ IS THE PROCESS OF CREATING VOLUMETRIC FORMS BY PUSHING METAL. DESPITE THE NAME THIS PUSHING IS USUALLY DONE ON BOTH THE FRONT AND THE BACK.

PROCESS

1. DRAW DESIGN ON ANNEALED METAL.
2. WARM PITCH WITH GENTLE TORCH FLAME AND SET METAL RIGHT SIDE UP ONTO SMOOTH AREA. PITCH MAY BE PULLED ONTO METAL TO ACHIEVE BETTER GRIP.
3. GO OVER DESIGN LIGHTLY WITH TRACER PUNCH.
4. LIFT METAL OUT OF PITCH BY PRYING, RAPPING POT, OR WARMING IT AND THEN LIFTING WITH TWEEZERS. REMOVE EXCESS PITCH BY BURNING OR (BETTER) BY DISSOLVING IT IN BABY OIL OR TURPENTINE. IF BURNED, DO NOT ALLOW IGNITED PITCH TO DRIP BACK INTO POT. IT IS BRITTLE AND MUST BE DISCARDED.
5. TURN METAL OVER AND SET BACK INTO PITCH. BOSS UP FORM WITH WHATEVER ROUND TIPPED TOOLS WILL FIT. WHEN METAL FEELS STIFF AND CORNERS ARE CURLING OUT OF THE PITCH, REMOVE THE WORK, CLEAN OFF PITCH AND ANNEAL.
6. DRY METAL AND RETURN IT TO THE PITCH FOR FURTHER WORK ON EITHER THE FRONT OR BACK AS NEEDED.

THE MATERIAL USED TO SUPPORT THE METAL IS VERY IMPORTANT; MOST COMMONLY PREFERRED SUPPORT IS PITCH.

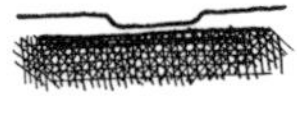

TOO HARD
METAL IS THINNED.

TOO SOFT
NO CONTROL

GOOD PITCH
HARD ENOUGH TO HOLD SHAPE
SOFT ENOUGH TO YIELD

NOTE: TEXTURES MAY BE ADDED AS THE METAL IS BEING FORMED BY USING SHARP-EDGED PUNCHES. FOR A SMOOTH SURFACE USE MIRROR-FINISHED PLANISHING TOOLS.

TRACERS
FOR MAKING LINES

MODELING AND PLANISHING

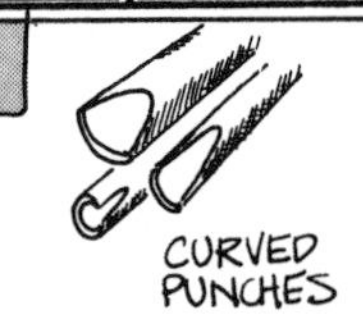

CURVED PUNCHES

MATTING TOOLS

TOOLS

THOUGH TOOLS CAN BE BOUGHT MANY PEOPLE PREFER TO MAKE THEIR OWN. ONLY A FEW ARE NEEDED TO BEGIN BUT WITH TIME A COLLECTION OF 40-50 IS OFTEN FORMED. THESE TOOLS, ESPECIALLY THE MODELING PUNCHES, DO NOT HAVE TO BE HARDENED AND TEMPERED BUT MOST PEOPLE PREFER TO DO THIS. TOOL STEEL MAY BE BOUGHT OR SALVAGED FROM BROKEN TOOLS. SEE TOOLMAKING ON PAGE 122.

TO MAKE MATTING TOOL FILE A LINE AROUND STEEL ROD AND SNAP IT OFF. A FINE GRAIN PATTERN WILL RESULT.

7 PARTS BURGUNDY OR SWEDISH PITCH
10 PARTS PLASTER OR PUMICE
1 PART LINSEED OIL OR TALLOW

PITCH MAY BE BOUGHT FROM MOST TOOL SUPPLY COMPANIES (SEE APPENDIX) OR:

NORTHWEST PITCHWORKS
5705 26TH AVE N.E.
SEATTLE, WASH 98105
(206) 525-4136

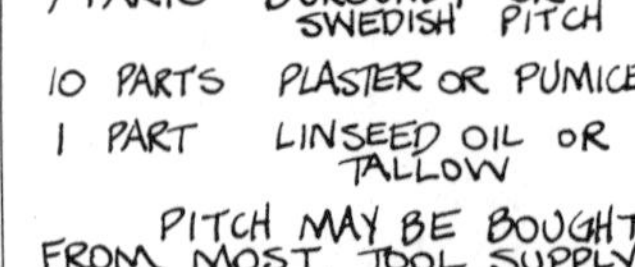

SHAPING *Stretching & Sinking*

STRETCHING

STRETCHING IS A TECHNIQUE THAT CAUSES SHEET METAL TO DOME BY FORGING IT AGAINST A SOLID FLAT SURFACE LIKE AN ANVIL. AS TENSION IS CREATED BETWEEN EXPANDED (STRETCHED) AND UNHAMMERED AREAS THE METAL IS PULLED TO A DOMICAL SHAPE.

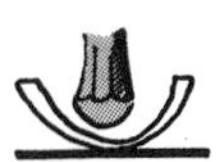

THE ADVANTAGES OF STRETCHING ARE THE THICK EDGE THAT RESULTS, ITS RAPID PROGRESS, AND THE FACT THAT THE SIZE DOES NOT CHANGE. THIS MIGHT BE IMPORTANT WHEN FITTING, AS IN THE CASE OF A LID.

DISADVANTAGES ARE THAT USUALLY A THICK STARTING BLANK IS NEEDED AND SINCE ALL THE HAMMERING IS DONE FROM THE INSIDE NOT MUCH DEPTH IS POSSIBLE.

SINKING

SINKING IS A VERSATILE TECHNIQUE USED TO CREATE DOMICAL FORMS IN SHEET METAL BY POUNDING THE METAL INTO A HEMISPHERICAL DIE. ON A SMALL SCALE IT IS CALLED DAPPING. ON A LARGER SCALE SINKING IS USED BY ITSELF AND AS A STEP PRELIMINARY TO RAISING.

1. CUT A DISK USING THE FORMULA

DIAMETER + ½ HEIGHT

TO CALCULATE THE DIAMETER OF THE STARTING BLANK. 16 TO 20 GAUGE STOCK IS USUALLY USED.

2.

YOU MAY WANT TO DRAW PENCIL GUIDELINES, USING A COMPASS, ON THE INSIDE OF THE FORM. WITH A LITTLE PRACTICE THESE BECOME UNNECESSARY.

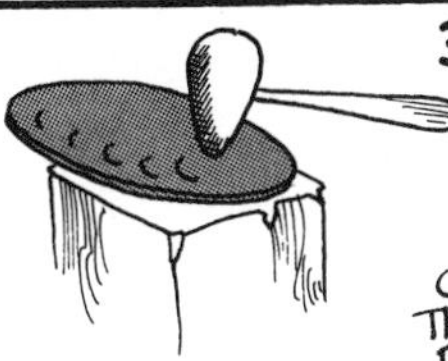

3. WITH THE EDGE OF THE DISK HELD ACROSS THE CENTER OF THE CARVED DEPRESSION THE METAL IS SUNK WITH A BALL-FACED HAMMER OR MALLET. WORK PROGRESSES FROM CIRCUMFERENCE INWARD TOWARD CENTER.

4. WHEN THE DESIRED DEPTH HAS BEEN REACHED—AND THIS COULD TAKE SEVERAL ANNEALINGS AND COURSES FROM EDGE TO CENTER—THE FORM IS MADE MORE EVEN (LESS LUMPY) IN A STEP CALLED BOUGING, OVER A MUSHROOM STAKE.

← WOODEN OR RAWHIDE MALLET

SINKING BLOCKS ARE USUALLY MADE OF WOOD, PREFERABLY USING THE END GRAIN. CARVE DEPRESSIONS WITH GOUGES, OR

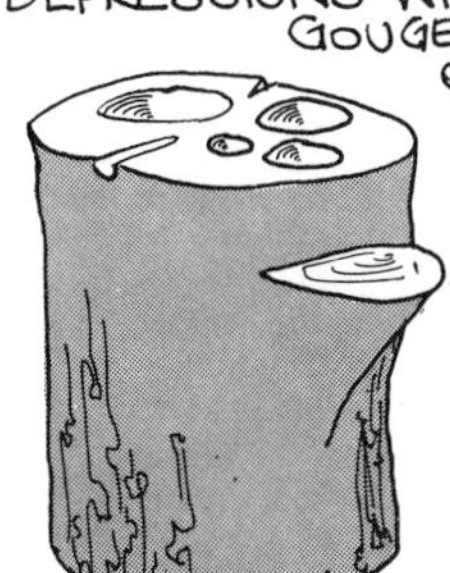

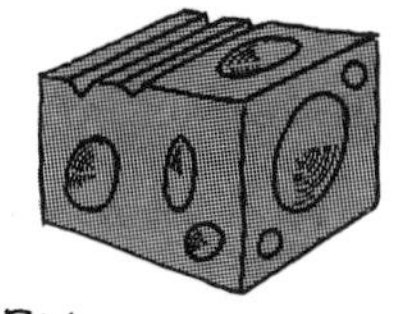

TURN ON A LATHE.

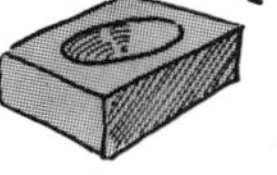

THE DIE SHAPES ARE GENERAL FORMING AIDS AND DO NOT EXACTLY FIT THE FINAL SHAPE.

HE WHO KNOWS OTHERS IS WISE.

HE WHO KNOWS HIMSELF IS ENLIGHTENED.

LAO TZU 500 BC

A DISK CUTTING DIE MAY BE USED FOR SINKING AS SHOWN. FIRST FORM DOME, THEN CUT WITH PUNCH.

THERE IS ONLY ONE RIGHT WAY TO RAISE: THE WAY THAT WORKS. METHODS WILL DIFFER DEPENDING ON THE SIZE AND SHAPE OF THE PIECE, THE METAL AND THICKNESS BEING RAISED, THE TOOLS AVAILABLE, AND THE PREFERENCES OF THE SMITH. THE ILLUSTRATIONS BELOW SHOW IN GENERAL TERMS THE STEPS IN RAISING A VESSEL. THIS INFORMATION, ALONG WITH THE ADJACENT PAGES, WILL GET YOU STARTED. ONLY BY ACTUALLY DOING THE WORK WILL THESE TIPS SYNTHESIZE INTO A METHOD.

1. MAKE AN ACTUAL SIZE DRAWING AND A TEMPLATE FROM IT.

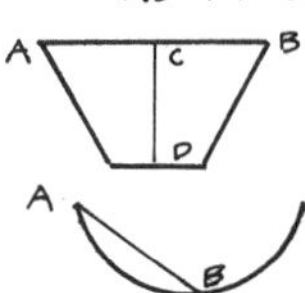

THE DIAMETER OF THE STARTING DISK IS THE SUM OF THE WIDEST AND TALLEST MEASURES (AB + CD) OR, FOR A DOME, TWICE THE LENGTH OF LINE AB.

2. THE CENTER IS PUNCHED AND A CIRCLE DRAWN AND CUT. FILE AND BURNISH THE EDGE, AND ANNEAL THE DISK. THE FLAT SHEET IS BENT UPWARDS BY STRETCHING, SINKING, OR CRIMPING. AFTER ANNEALING ANY OF THESE MAY BE REPEATED TO QUICKLY BRING THE SIDES UP. MOST PEOPLE HAVE A FAVORITE METHOD AND STICK WITH IT, BUT THESE MAY BE USED IN COMBINATION. IF THE VESSEL IS TO HAVE A FLAT BOTTOM, THIS AREA IS LEFT UNTOUCHED.

3. WHEN THE METAL IS TOO THICK TO CRIMP OR ITS CURVE RESTRICTS ACCESS, THE VESSEL IS WORKED FROM THE OUTSIDE OVER A STAKE, EITHER OF WOOD OR STEEL. A CROSS PEEN HAMMER OR MALLET IS USED. STARTING AT THE BASE-LINE WITH THE SHEET HELD AT ABOUT A 30° ANGLE, THE HAMMER IS BROUGHT DOWN SO THE HANDLE IS PARALLEL TO THE FLOOR. THE DISK IS ROTATED AS THE HAMMER FALLS IN EVEN BLOWS. AFTER GOING AROUND THE POT ONCE, THE DISK IS SLID BACK (OFF THE STAKE) ABOUT A HALF INCH AND RAISING CONTINUES. FOR A TALL VESSEL, REVERSE THE DIRECTION OF ROTATION. THE EFFECT IS TO MAKE A BULGE AND THEN WORK IT TO THE EDGE, AS YOU WOULD IN SMOOTHING OUT A CARPET.

4. THE PROGRESS FROM BASE LINE TO TOP EDGE IS CALLED A COURSE AND CONSISTS OF CONCENTRIC CIRCLES ABOUT A HALF INCH APART. THESE MAY BE DRAWN IN PENCIL WITH A COMPASS AFTER EACH ANNEALING TO SERVE AS GUIDELINES. KEEP IN MIND THAT THE EDGE HAS TO MOVE A LOT MORE (I.E. COVER A GREATER DISTANCE) THAN THE METAL NEAR THE BASE LINE. THAT MAKES IT EASY TO CREATE THIS SHAPE: TO PREVENT IT, STRIKE MORE GENTLY NEAR THE BASE AND TAKE SPECIAL CARE NEAR THE TOP EDGE. IF THE BOWL IS TOO FLARED OUT, RAISE A COURSE OR TWO STARTING AT MID-HEIGHT (ARROW).

5. AS RAISING CONTINUES THE TOP EDGE WILL THICKEN. THIS MAY BE EXAGGERATED BY TAPPING THE EDGE WITH A CROSS-PEEN AT THE END OF EACH COURSE. SUPPORT THE WORK ON A SANDBAG OR IN THE HAND WHILE DOING THIS.

6. IN AN OPEN VESSEL PLANISHING IS DONE AFTER THE FORM IS COMPLETE. ON A NECKED-IN SHAPE PLANISHING MUST BE DONE MIDWAY IN THE PROCESS WHILE THE STAKES STILL FIT INSIDE.

7. CURVED FORMS ARE USUALLY RAISED AS A SERIES OF FLAT SECTIONS LIKE THIS. THESE ARE THEN GIVEN ROUNDNESS WITH BLOCKING HAMMERS ON A WOODEN BLOCK OR SAND BAG.

8. THE STRAIGHTNESS OF THE POT IS CHECKED WITH A SURFACE GAUGE OR BY DRAWING PENCIL LINES AS SHOWN.

THE TOP IS CUT AS NEEDED AND FILED SMOOTH. PLANISHING REQUIRES THAT HAMMERS AND STAKES BE MIRROR-LIKE. BEGIN WITH A MEDIUM WEIGHT HAMMER AND FINISH WITH SEVERAL COURSES WITH A LIGHTWEIGHT HAMMER. OVERLAP BLOWS AND DON'T HURRY.

Crimping

IS A TECHNIQUE USED IN THE EARLY STAGES OF RAISING TO QUICKLY CHANGE A FLAT SHEET INTO A VOLUMETRIC FORM.

SOME PEOPLE PREFER TO BEGIN RAISING BY SINKING BUT THE ADVOCATES OF CRIMPING HOLD THAT IT IS A FASTER METHOD.

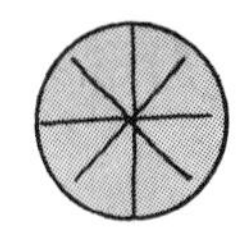

THE DISK IS MARKED INTO SEGMENTS AND HELD ACROSS A NOTCHED STAKE SO THE LINE IS OVER THE CENTER OF THE NOTCH. A CROSS PEEN MALLET OR HAMMER IS USED TO MAKE A FLUTED BOWL SHAPE. THESE FLUTES ARE THEN SMOOTHED OUT OVER A T-STAKE IN A USUAL RAISING OPERATION.

Planishing

THIS WORD COMES FROM THE LATIN PLANUS, WHICH MEANS FLATTEN OR LEVEL. IT IS THE SMOOTHING, TOUGHENING OR POLISHING OF METAL BY ROLLING OR HAMMERING.

THE EFFECT OF PLANISHING CAN BE ONLY AS GOOD AS THE SURFACES BEING USED. HAMMER FACES AND STAKES OR ANVIL MUST BE MIRROR-FINISHED.

hammers:

ANY SMOOTH-FACED HAMMER MAY BE USED FOR PLANISHING BUT IT IS A GOOD IDEA TO HAVE THE ONE MADE JUST FOR THIS. IT WILL HAVE THE TWO FACES SHOWN. TO GET MAXIMUM CONTACT BUT AVOID LEAVING MARKS, USE THE FLAT FACE ON CURVED SURFACES AND THE DOMED FACE ON FLAT OR NEARLY FLAT SURFACES.

A HEAVY HAMMER (12-16 OZ) IS BEST FOR QUICK WORK AND FLATTENING WIRE BUT A LIGHTWEIGHT (3-6 OZ) IS NEEDED FOR FINAL FINISHING.

IT'S PRETTY, BUT IS IT ART?

RUDYARD KIPLING ·1890·

SOME COMMON SITUATIONS

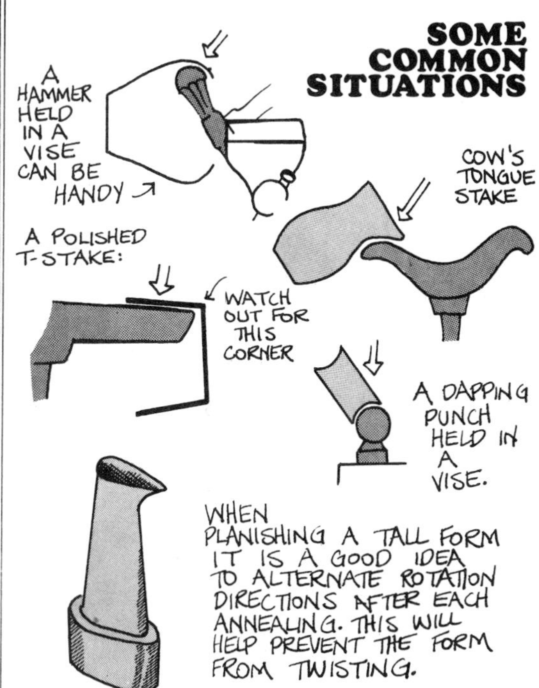

SHAPING

SHAPING

THOUGH ANY SHAPE CAN BE RAISED FROM A FLAT SHEET IT IS SOMETIMES MORE EFFICIENT TO FABRICATE A SHAPE THAT APPROACHES THE DESIRED END RESULT AND WORK MODIFICATIONS WITH FORMING TECHNIQUES FROM THERE. THE SOLDER SEAMS IN SUCH PIECES WILL RECEIVE A LOT OF STRESS SO SOME SPECIAL PROVISIONS MUST BE MADE.

preparing the joint

TO PROVIDE MORE SURFACE AREA FOR THE SOLDER TO GRAB, THE EDGES ARE FILED TO A BEVEL EQUAL TO ABOUT 5 THICKNESSES OF THE METAL. THE TWO ENDS OF THE STRIP ARE FILED IN OPPOSITE DIRECTIONS TO MAKE A SMOOTH JOINT.

TO PREVENT THE TWO EDGES FROM SLIDING OVER AND PAST EACH OTHER, TABS ARE CUT AND BENT AS SHOWN. THE EDGES ARE THEN BROUGHT TOGETHER AND THE FORM IS TIED WITH BINDING WIRE. THE TABS ARE MALLETED DOWN AS THE CYLINDER IS HELD ON A STAKE.

FLUX IS APPLIED BOTH INSIDE AND OUTSIDE ALONG THE SEAM. SOLDER IS USUALLY SET INSIDE.

examples:

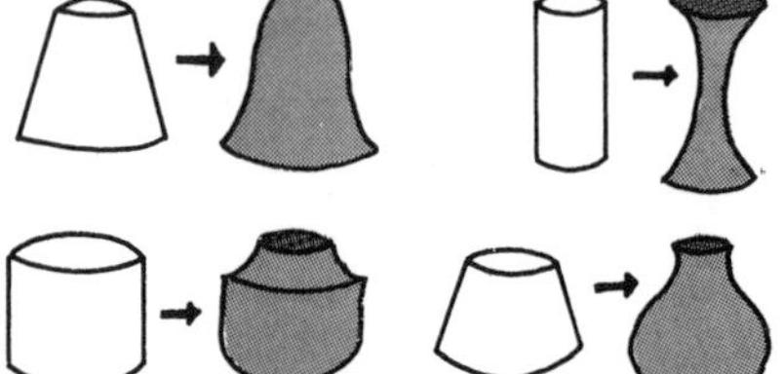

Interlocking Finger Joint

1. PLANISH THE TWO EDGES TO BE JOINED SO THEY THIN OUT EVENLY OVER ABOUT A HALF INCH AREA.

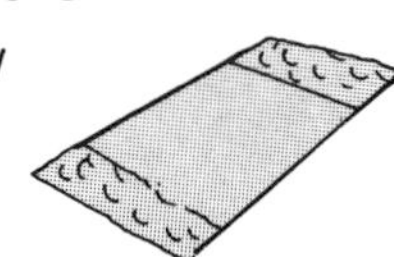

2. MARK A LINE ¼" IN FROM EACH EDGE AND LAY OUT THE SAME NUMBER OF TABS ON EACH EDGE. GENERALLY THE TABS ARE OF EQUAL SIZE, BUT ALL THAT MATTERS IS THAT THE TWO SIDES BE IDENTICAL.
3. SAW THESE LINES ON BOTH SIDES, STOPPING AT THE ¼" MARK. ON THIN STOCK (20 GA) A SINGLE CUT IS SUFFICIENT BUT FOR HEAVIER SHEET, CUT A SKINNY V.
4. ON ONE EDGE BEND THE EVEN-NUMBERED TABS UP ABOUT 30°. ON THE OTHER EDGE BEND THE ODD-NUMBERED TABS DOWN. PAINT THE WHOLE AREA WITH FLUX, SLIDE THE EDGES TOGETHER, AND TIE WITH BINDING WIRE.
5. SET OVER A STAKE AND MALLET THE TABS DOWN. A BURNISHER MAY BE USED TO PRESS THE END OF EACH TAB DOWN.
6. SOLDERING IS USUALLY DONE FROM OUTSIDE THE FORM.

RAISING AND PLANISHING ARE AS USUAL.

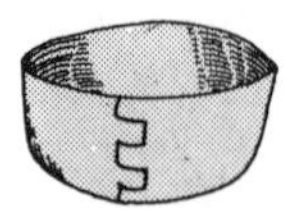

SHELL STRUCTURES

WHEN TWO OR MORE FORMED PIECES OF METAL ARE SOLDERED TOGETHER THE RESULTING HOLLOW FORM IS CALLED A SHELL STRUCTURE. FOR EXTENSIVE COVERAGE OF THIS TOPIC I RECOMMEND <u>FORM EMPHASIS FOR METALSMITHS</u> BY HEIKKI SËPPA.

IN MAKING HOLLOW FORMS, REMEMBER THESE RULES:

1. PIECES TO BE JOINED MUST BE CLEAN AND TIGHT-FITTING.
2. DO NOT TRAP AIR IN AN ENCLOSED SPACE — PROVIDE AN OUTLET.
3. WHEN USING BINDING WIRE ALLOW FOR EXPANSION BY PUTTING A ZIG-ZAG IN THE WIRE.

Some Typical Shell Structures

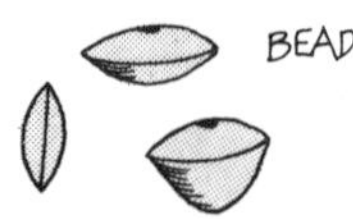

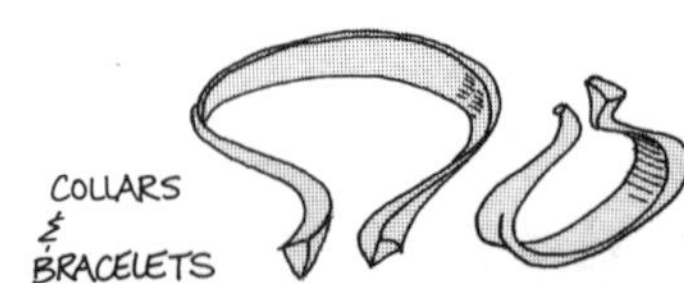

TOOLS

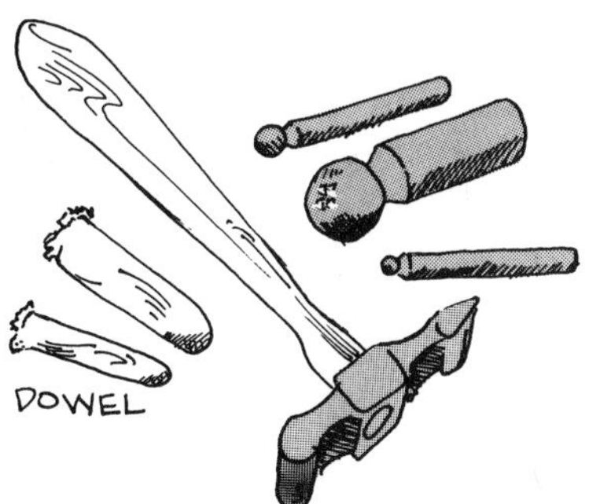

WHILE DIE FORMING CAN GET VERY SOPHISTICATED, IT ALSO LENDS ITSELF TO SIMPLE VARIATIONS. THESE TOOLS ARE COMMONLY USED AS RAMS.

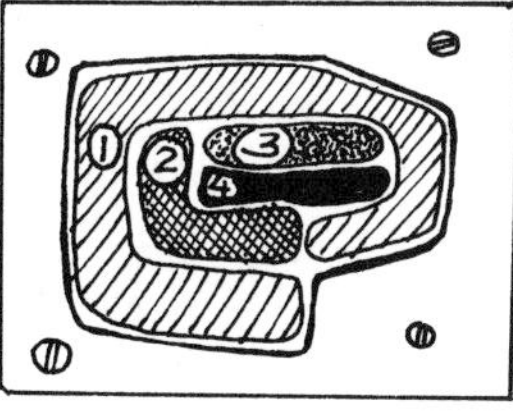

THIS SHOWS THE USUAL PROGRESSION OF TOOLING, MOVING FROM OUTSIDE EDGES INWARD. IF A PUNCH IS USED IT IS HELD AT A SLIGHT ANGLE SO THAT THE HAMMER BLOWS PUSH IT ALONG.

DO WHAT YOU CAN, WITH WHAT YOU HAVE, WHERE YOU ARE.

—THEODORE ROOSEVELT

THE FLANGE LEFT AROUND A DIE-FORMED SHAPE KEEPS THE FORM INTACT THROUGH MODERATE SURFACE DECORATING. IF MUCH DEFORMATION IS PLANNED, FILL PIECE WITH PITCH BY POURING FROM A PAN OR MELTING LUMPS RIGHT IN THE FORMED AREA. KEEP FLANGE INTACT FOR LATER REFITTING INTO DIE.

Description & Uses:

THE SILHOUETTE DIE DESCRIBED HERE IS A MODERN VERSION OF AN ANCIENT PROCESS. A RIGID MATERIAL IS PIERCED WITH AN OUTLINE (SILHOUETTE) OF A DESIRED SHAPE. THE WORK METAL IS THEN HELD AGAINST THE DIE THUS MADE AND PUSHED INTO THE OPEN AREA. THE PROCESS IS VERSATILE WITH VARIATIONS USUALLY INVOLVING THE DIE MATERIAL, THE RAM, THE FORCE BEHIND THE RAM AND THE METHOD OF HOLDING THE WORKPIECE TO THE DIE.

DIE FORMING HAS MANY APPLICATIONS BUT IS ESPECIALLY GOOD FOR:

- MATCHING HALVES LIKE SPOUTS & FABRICATED CONTAINERS.
- MATCHING PARTS AS WITH A BOX & LID.
- CERTAIN PRODUCTION METHODS.

the flange:

A UNIQUE FEATURE OF DIE FORMING IS THE FLANGE OR SKIRT THAT SURROUNDS THE FORM. BEFORE CUTTING IT OFF CONSIDER THESE USES:

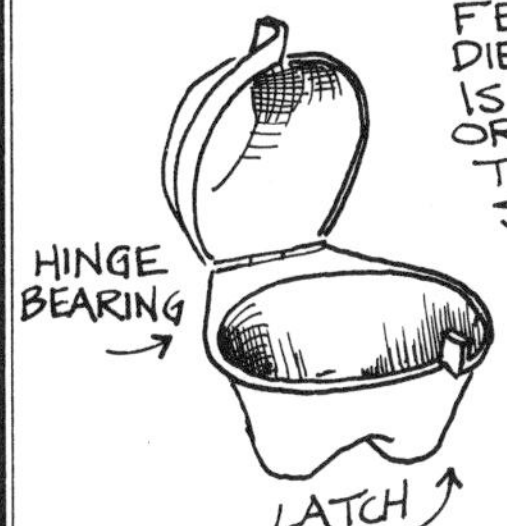

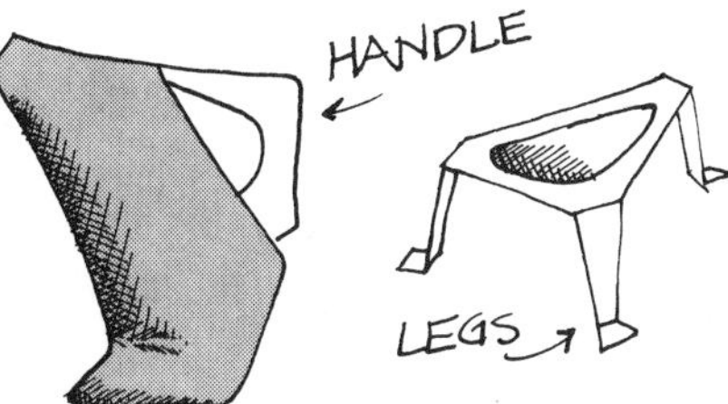

SHAPING

1. MAKE A DIE BLOCK BY GLUING TOGETHER PIECES OF PLYWOOD AND TEMPERED MASONITE AS SHOWN. USE A WHITE GLUE (ELMER'S) AND CLAMP OR WEIGHT.

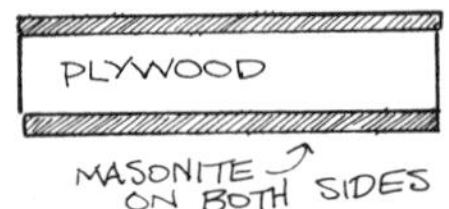

2. MARK DESIGN ON MASONITE AND CUT OUT DIE HOLE. USE A COPING SAW, SABRE SAW OR BAND SAW. IF A BAND SAW IS USED THE SAWN OPENING SHOULD BE GLUED CLOSED. INSERT STRIP OF WOOD (LIKE TONGUE DEPRESSOR) AND CLAMP.

NOTE: TAKE CARE IN SAWING THAT THE SIDES OF THE HOLE ARE VERTICAL. THE OPENING IN THE TOP MUST BE THE SAME SIZE AS THE OPENING IN THE BOTTOM.

3. PLAN LOCATION OF SCREWS AND DRILL HOLES: THESE SHOULD BE ABOUT 5MM (3/8") FROM THE DIE HOLE: SOME EXAMPLES:

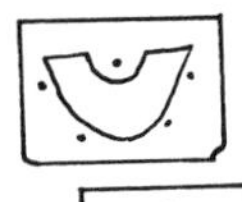

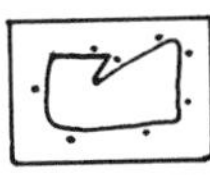

4. TO CUT THE RIGHT SIZE PIECE OF METAL MAKE A RUBBING OF THE DIE. USE THIS TO CUT METAL AND DRILL HOLES FOR SCREWS. THESE SHOULD BE OVERSIZE; ABOUT 1/4".

5. FASTEN THE METAL ONTO THE DIE WITH 3/4" SHEET METAL SCREWS. TO "FIND" THE OUTLINE OF THE FORM TAP THE METAL LIGHTLY WITH A MALLET OR HAMMER HANDLE. WHILE WORKING, THE DIE MAY SIT ON A BENCH, SANDBAG, OR VISE. TO ANNEAL, REMOVE THE SCREWS AND TAKE THE METAL OFF THE DIE.

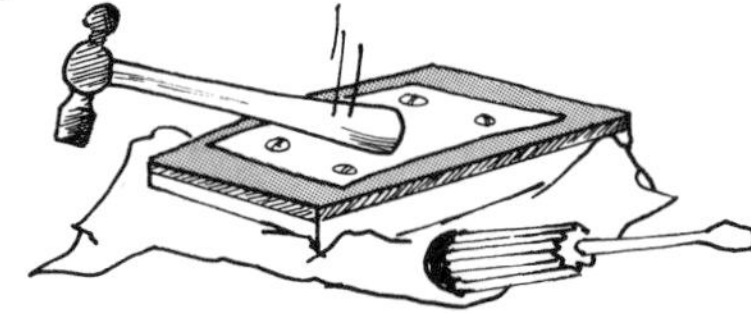

THIS IS A REVERSIBLE DIE: IT CAN BE USED TO MAKE TWO PIECES THAT FIT TOGETHER. TO MAKE THE OTHER PIECE, TURN THE DIE OVER AND START WITH #3.

A MASONITE-FACED DIE MAY BE USED SEVERAL TIMES BEFORE THE EDGE STARTS TO BREAK DOWN. FOR A MORE LASTING DIE, CUT A PIECE OF THIN STEEL SHEET WITH THE SAME HOLE AS THE REST OF THE DIE & FASTEN ONTO MASONITE WITH COUNTERSUNK FLAT HEAD SCREWS.

Variation:

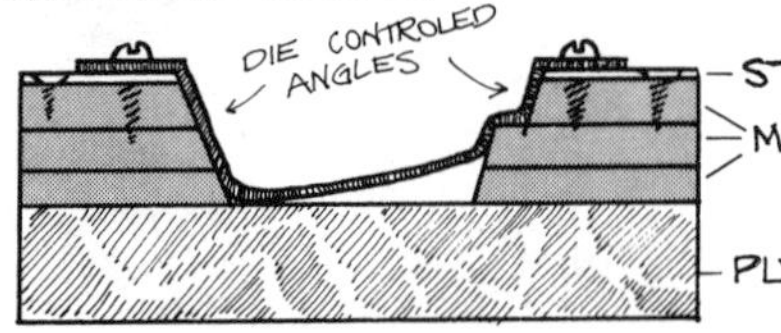

WHEN A SPECIFIC CONTOUR IS NEEDED, LIKE THE ANGLE SHOWN HERE, IT MAY BE BUILT INTO THE DIE BY FILING THE MASONITE TO THE CORRECT SHAPE. NOTE THAT THIS IS NOT A REVERSIBLE DIE; TWO DIES MUST BE MADE FOR MATCHING HALVES.

SHAPING

SHAPING *Steel Dies*

① DRAW DESIGN ON 1/8" STEEL SHEET WITH AN INDELIBLE MARKER. WHEN DRILLING A HOLE FOR PIERCING, COOL THE BIT WITH A GENEROUS BATH OF OIL. PIERCE WITH A JEWELERS' SAW USING A LARGE BLADE (E.G. 5). AFTER SAWING, SMOOTH OFF THE EDGES WITH AN OLD FILE AND SANDPAPER.

② PLAN LOCATION OF SCREW HOLES AND CENTER PUNCH FOR EACH, ABOUT 1/4" FROM THE DIE HOLE. AGAIN USE OIL WHEN DRILLING. BE SURE TO USE THE CORRECT SIZE BIT FOR THE TAP TO BE USED NEXT.

③ CUT THREADS WITH A TAP USING OIL TO LUBRICATE. A TAP AND HANDLE CAN BE BOUGHT AT A HARDWARE STORE FOR JUST A COUPLE DOLLARS. CUT SLOWLY WITH THE TAP, ADVANCING A QUARTER TURN THEN REVERSING TO CLEAR THE THREADS.

ANY SIZE TAP MAY BE USED: JUST BE SURE THAT THE BIT, TAP, AND SCREWS ALL COORDINATE IN BOTH DIAMETER AND NUMBER OF THREADS PER INCH. SHORT SCREWS WILL MAKE THE SCREWING AND UNSCREWING GO FASTER.

advantages

·BECAUSE STEEL IS SO MUCH STRONGER THAN WOOD A THINNER DIE WILL GIVE EQUAL SUPPORT. STEEL DIES ARE LESS CUMBERSOME AND MORE DURABLE THAN WOODEN DIES. BECAUSE THE DIE MATERIAL IS THINNER IT IS EASIER TO GUARANTEE THAT THE WALL OF THE DIE HOLE IS VERTICAL.

SHEET STEEL IS PROBABLY LOCALLY AVAILABLE: LOOK UNDER SHEET METAL OR STEEL. MANY SUPPLIERS WILL GIVE CHEAPER PRICES ON SCRAP PIECES. ANY KIND OF STEEL, INCLUDING STAINLESS, WILL DO.

④ MAKE A PENCIL RUBBING OF THE DIE TO DETERMINE HOW LARGE A SHEET OF METAL WILL BE NEEDED AND WHERE TO DRILL HOLES FOR THE HOLD-DOWN SCREWS. THESE HOLES SHOULD BE ABOUT 1/4" IN FROM THE EDGE AND HALF AGAIN AS LARGE AS THE SCREWS.

⑤ WITH THE ANNEALED METAL SCREWED ONTO THE DIE THE FORMING PROCEEDS AS SHOWN BEFORE. AFTER EACH ANNEALING, FORMING BEGINS FROM THE OUTSIDE EDGE PUSHING THE METAL DOWN INTO THE DIE HOLE.

Supports:

A STEEL DIE MUST REST ON SOMETHING TO KEEP IT UP OFF THE TABLE WHILE FORMING. SUGGESTIONS

STYROFOAM® SHEET

WOOD FRAME

PILLOW

SANDBAG

A Variation:

WHEN THE DIE HOLE IS FAIRLY SMALL – SAY LESS THAN 4 SQ. INCHES – STEEL THINNER THAN 1/8" MAY BE USED. THOUGH THIS MAY STILL BE THREADED (#3 ABOVE) AN ALTERNATE METHOD IS TO USE SHEET METAL (SELF-TAPPING) SCREWS. DRILL A HOLE ONLY AS LARGE AS THE SHANK OF THE SCREW AND FORCE IT IN. IT CAN BE UNSCREWED WITHOUT DAMAGE AND WILL WORK FROM BOTH SIDES.

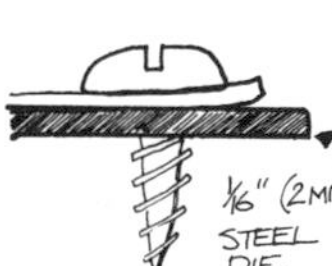

Definitions

A PRESS DIE CONSISTS OF A MATCHED PAIR OF COMPLEMENTARY SHAPES MADE OF ANY HARD MATERIAL. WHEN A SOFTER SUBSTANCE IS SET INTO POSITION, THE PARTS OF THE DIE ARE PRESSED TOGETHER, CAUSING A PLASTIC DEFORMATION OF THE SOFTER MATERIAL. BOTTLE CAPS, FOR INSTANCE, ARE MADE IN A KIND OF PRESS DIE.

THIS IS A CONFORMING DIE: THE TWO HALVES CONFORM TO ONE ANOTHER. THE SILHOUETTE DIES ON THE PRECEEDING PAGE ARE NON-CONFORMING: THEY HAVE ONLY THE OUTER (FEMALE) PART OF THE DIE. NON-CONFORMING DIES OFFER GREATER VERSATILITY BUT LESS CONSISTENCY FROM ONE PIECE TO THE NEXT.

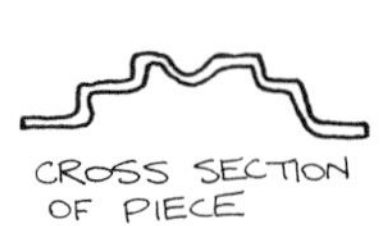

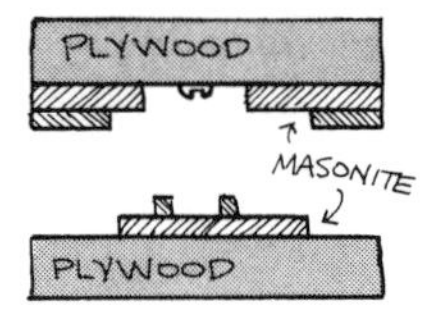

PLAN THE DESIGN AND DRAW A CROSS SECTION OF THE PIECE. FROM THIS, PLAN AND DRAW THE DIE.

1

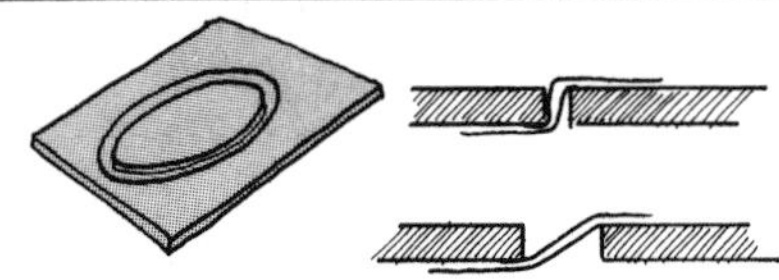

2

CUT PIECES FROM WOOD, MASONITE, BRASS AND/OR STEEL. AS YOU MEASURE ALLOW CLEARANCE; AS SHOWN, MORE CLEARANCE GIVES A SOFTER FORM.

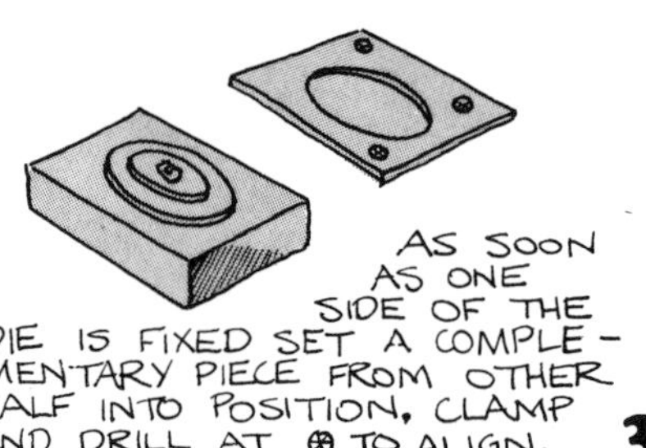

AS SOON AS ONE SIDE OF THE DIE IS FIXED SET A COMPLEMENTARY PIECE FROM OTHER HALF INTO POSITION, CLAMP AND DRILL AT ⊕ TO ALIGN.

3

USING HOLES IN TOP PIECE AS GUIDE, DRILL THE REST OF THE WAY.

4

COMPLETE THE DIE ASSEMBLY BY GLUING, SOLDERING &/OR SCREWING. REFER TO DESIGN DRAWING.

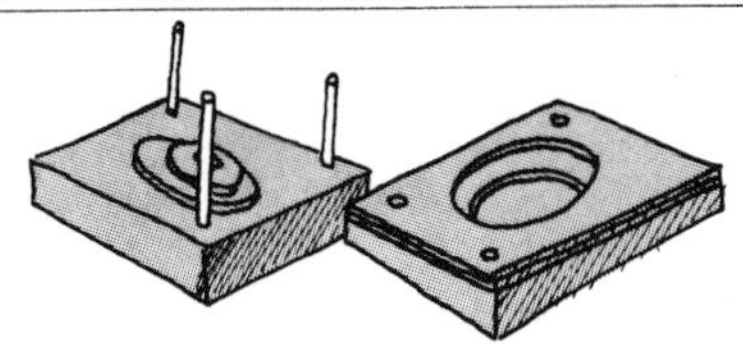

GLUE ALIGNMENT PINS (¼" DOWEL) INTO ONE HALF WITH ABOUT 1-2" EXTENDING OUT.

5

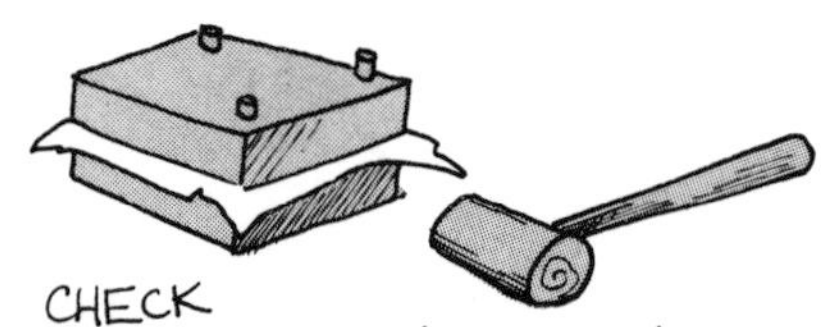

6

CHECK RESULTS ON A THIN METAL LIKE HEAVY DUTY ALUMINUM FOIL.

FILE, CARVE, OR BUILD UP WITH EPOXY OR AUTO BODY FILLER IF NEEDED.

7

PUT ANNEALED METAL INTO POSITION, SLIDE THE DIE INTO A LARGE VISE, AND SQUEEZE.

8

Some Generalities:

MACHINE TOOLS WORK SUBTRACTIVELY.

THEY ARE DESIGNED FOR ACCURACY, USUALLY TO .001 INCH. MOST HAVE PROVISION FOR MICROMETER READINGS.

THE EFFECT CREATED BY MACHINING IS CHARACTERIZED BY A CRISPNESS OF FORM. BECAUSE THE MACHINE IS CONSISTENT AS IT REPEATS AN OPERATION, THE PROCESS LENDS ITSELF TO GRIDS AND SIMILAR REPEATING PATTERNS.

"SOFT" METALS LIKE COPPER, SILVER ALLOYS, AND PURE ALUMINUM DO NOT MACHINE WELL. RECOMMENDED MATERIALS INCLUDE:

- BRASS (FREE TURNING)
- 6061 AND 2024 ALUMINUM
- 303 STAINLESS (FREE TURNING)
- MILD STEEL
- ACRYLIC, DELRIN, KYDEX, EPOXY RESINS
- IVORY AND BONE
- DENSE WOODS (LIKE EBONY)

SAFETY

- WEAR GOGGLES.

- KEEP LONG HAIR AND LOOSE CLOTHING TIED BACK.
- WHEN CUTTING PLASTIC, WOOD, OR GEM MATERIALS: USE A RESPIRATOR.

- KEEP YOUR MIND ON WHAT YOU'RE DOING.
- BEFORE TURNING ON THE MACHINE, DOUBLE CHECK FOR TIGHTNESS AND CLEARANCE.

SHAPING

MAN-MACHINE IDENTITY IS ACHIEVED NOT BY ATTRIBUTING HUMAN ATTRIBUTES TO THE MACHINE, BUT BY ATTRIBUTING MECHANICAL LIMITATIONS TO MAN.

—MORTIMER TAUBE

HERE ARE SOME EXAMPLES OF EFFECTS POSSIBLE WITH FOUR BASIC MACHINE TOOLS. ANY OF THE MACHINES MAY BE USED IN COMBINATION, CREATING UNLIMITED POSSIBILITIES.

Lathe

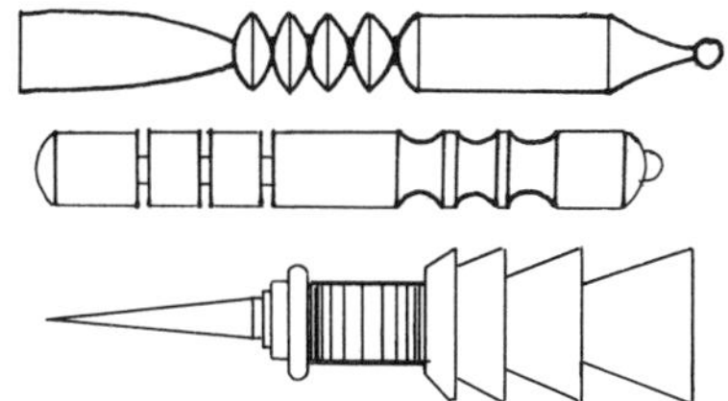

Shaper

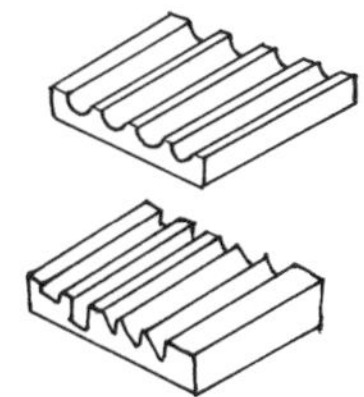

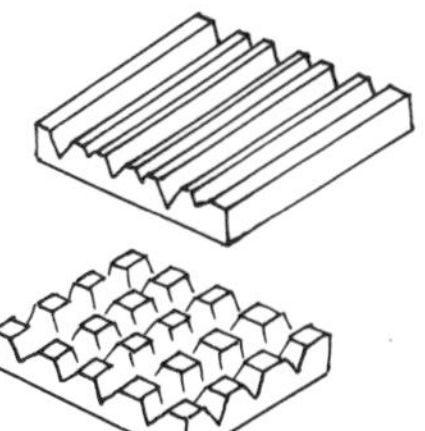

Horizontal Mill

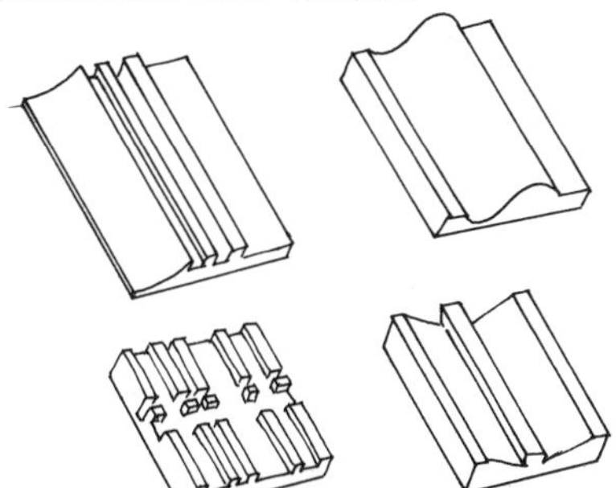

Vertical Mill

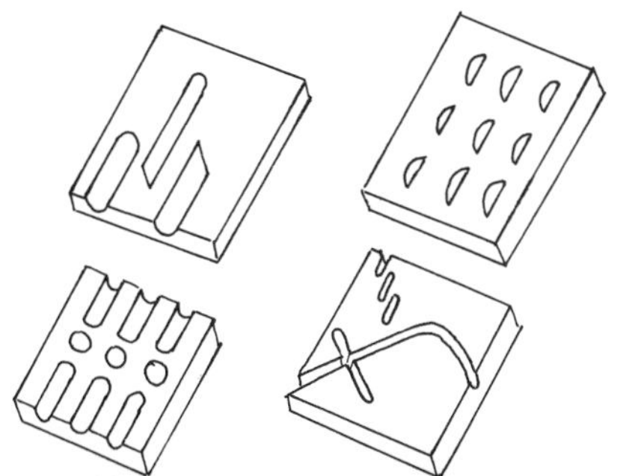

SAFETY ALERT

THE LATHE IS A VERSATILE PIECE OF EQUIPMENT IN WHICH A BAR OF MATERIAL IS ROTATED AGAINST A STATIONARY TOOL. METAL TURNING IS SIMILAR TO WOOD TURNING EXCEPT THAT THE TOOLS FOR METAL TURNING ARE SECURED ONTO THE MACHINE INSTEAD OF BEING HELD IN THE HAND. THE CARRIAGE ON WHICH THE TOOL IS MOUNTED MOVES ALONG THREADED RODS FOR PRECISE CONTROL. THIS MOVEMENT CAN BE MANUAL OR, IN MORE ELABORATE LATHES, AUTOMATED.

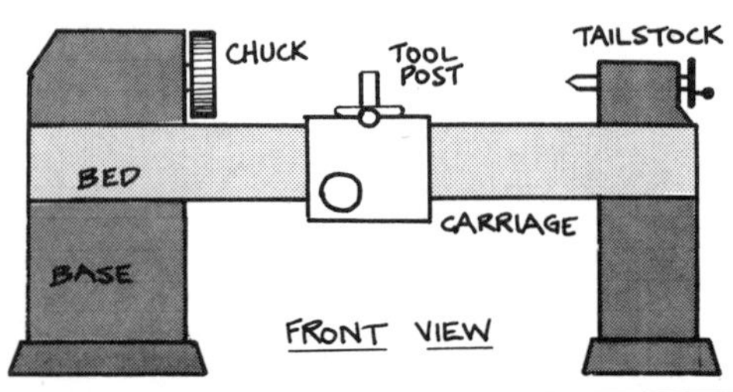

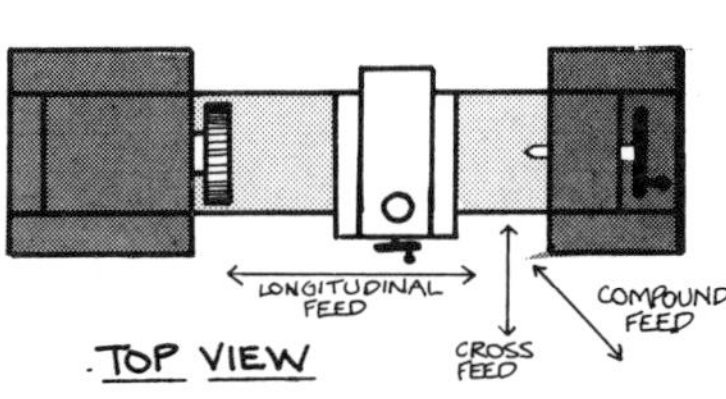

Tools

CUTTING BITS ARE AVAILABLE IN MANY SHAPES AND CAN ALSO BE MADE TO SUIT SPECIAL NEEDS. THE ANGLES OF THE TOOL ARE CRITICAL; MOST HAVE TO DO WITH CLEARING AWAY THE CHIP OF METAL THAT IS SLICED OFF AS THE TOOL CUTS.

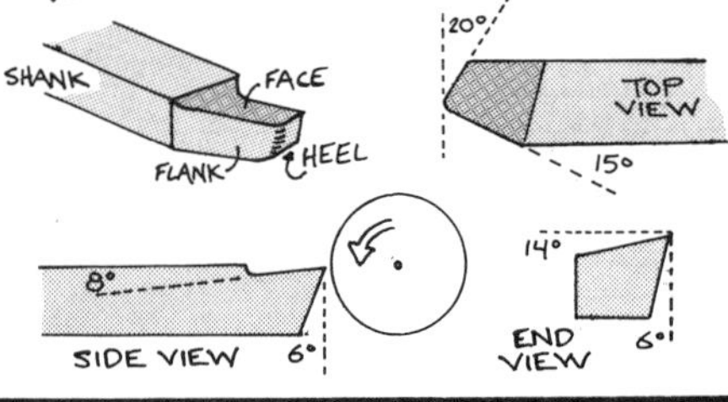

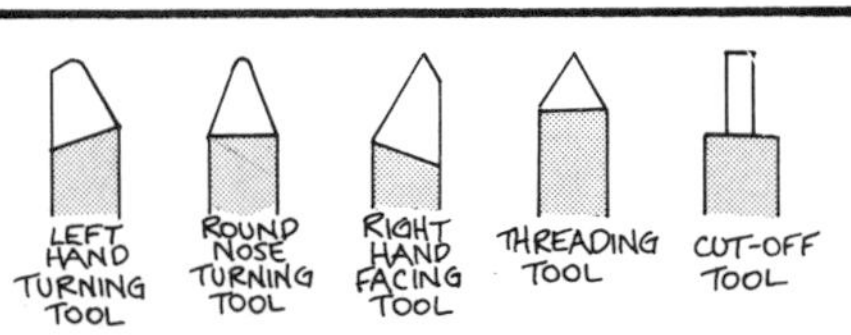

MATERIALS

THOUGH THE LATHE CAN BE USED WITH ALMOST ANY MATERIAL, SOME ARE BETTER SUITED THAN OTHERS. RECOMMENDED ARE:

- MOST BRASSES, 70/30 IS BEST
- 6061 AND 2024 ALUMINUM
- MILD STEEL
- * DELRIN, ACRYLIC, & EPOXY RESINS
- * DENSE WOODS (EBONY, ROSEWOOD, COCOBOLO, MAPLE, ETC.)
- * IVORY, HARD WAX
- * MARBLE, MALACHITE, JADE, ETC (OPAQUE, WAXY-LOOKING GEMS)
- 303 STAINLESS STEEL

* THESE MATERIALS GIVE OFF TOXIC DUST. A GOOD RESPIRATOR MUST BE WORN!

PROCESS:

THIS DESCRIPTION IS VERY GENERAL SINCE WORKING METHODS, EQUIPMENT, AND SCALE WILL ALTER THE STEPS NEEDED.

1. THE WORK IS SECURELY GRIPPED INTO THE HEADSTOCK. THE TAILSTOCK IS SLID INTO PLACE AND LOCKED THERE.
2. THE TOOL BIT IS SELECTED AND LOCKED INTO POSITION IN THE TOOL POST. ROTATE THE MACHINE BY HAND TO BE SURE THAT ALL IS IN ORDER.
3. GOGGLE UP.
4. TURN ON MACHINE AND BRING THE BIT INTO CONTACT WITH THE WORK — NEVER IN REVERSE ORDER!
5. WHEN BEING USED, ADD LUBRICATION, USUALLY THIN OIL.
6. TO REMOVE A LOT OF MATERIAL, SET THE BIT INTO THE BAR AND MOVE IT ALONG HORIZONTALLY (SIDEWAYS). THEN 'BITE' IN FURTHER AND GO BACK. THE POINT IS TO AVOID TAKING TOO MUCH OFF IN A SINGLE PASS. THIS WILL ABUSE THE BIT AND RISK TEARING THE MATERIAL AND JAMMING THE LATHE.

Finishing

WHEN PROPERLY USED THE BIT WILL LEAVE A PRETTY SMOOTH FINISH. IF IT LEAVES LINES SLOW DOWN THE RATE AT WHICH YOU MOVE IT SIDEWAYS (FEED). FOR A FINER POLISH, SANDPAPER AND/OR A CLOTH OR LEATHER DRESSED WITH BUFFING COMPOUND MAY BE HELD AGAINST THE WORK AS IT IS TURNING. BE GENTLE AND BRIEF OR YOU MAY WEAR AWAY THE ORIGINAL CRISPNESS OF THE FORM.

Shaper

THIS TOOL IS SIMILAR TO THE LATHE IN THAT ITS CUTTING EDGE SCRAPES OR SLICES AWAY A BIT OF MATERIAL AS IT IS PRESSED AGAINST IT. IN SMALLER APPLICATIONS (SHOWN) THE WORK IS HELD STATIONARY AS THE TOOL BIT IS PUSHED ACROSS THE SURFACE OF THE METAL. IN LARGER CASES THE WORK PIECE IS FASTENED TO A CARRIAGE THAT SLIDES PAST THE TOOL. IN EITHER CASE THE TOOL MAY PIVOT SO AS TO CUT ONLY ON THE FORWARD STROKE. DEPTH OF CUT IS CONTROLLED BY RAISING OR LOWERING THE TABLE.

SAFETY ALERT

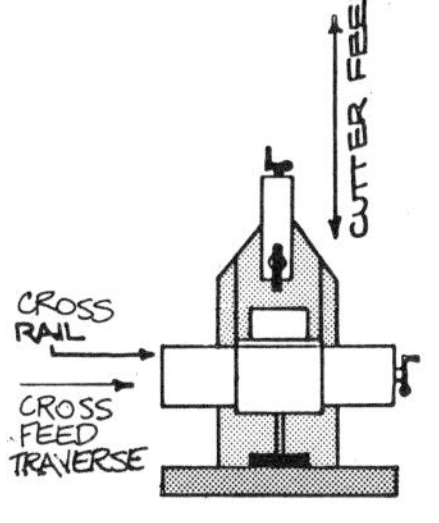

FRONT VIEW

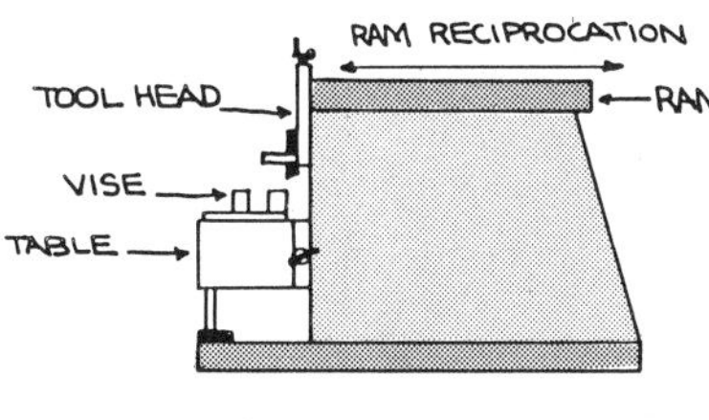

SIDE VIEW

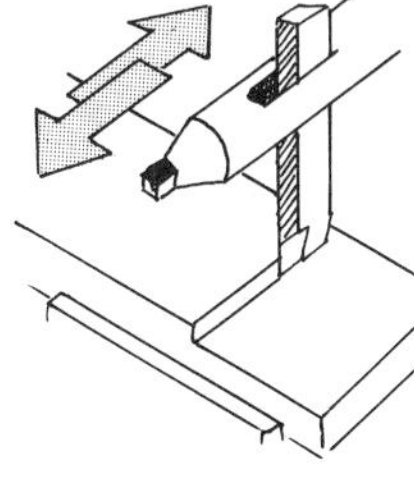

Horizontal Mill

A MULTI-TOOTHED CUTTER ROTATES IN A HORIZONTAL SPINDLE AS THE WORK PIECE IS FED INTO IT. THE HEIGHT OF THE TABLE IS ADJUSTED TO CONTROL DEPTH OF CUT. SINCE THE METAL MAY BE MOUNTED ONTO THE TABLE IN ANY ORIENTATION, COMPLEX LINEAR ARRANGEMENTS ARE POSSIBLE. CUTTING TOOLS ARE AVAILABLE IN MANY SHAPES AND SIZES.

FRONT VIEW

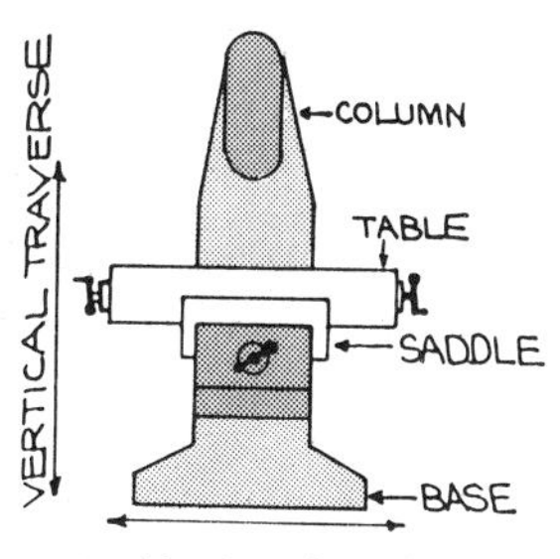

LONGITUDINAL FEED

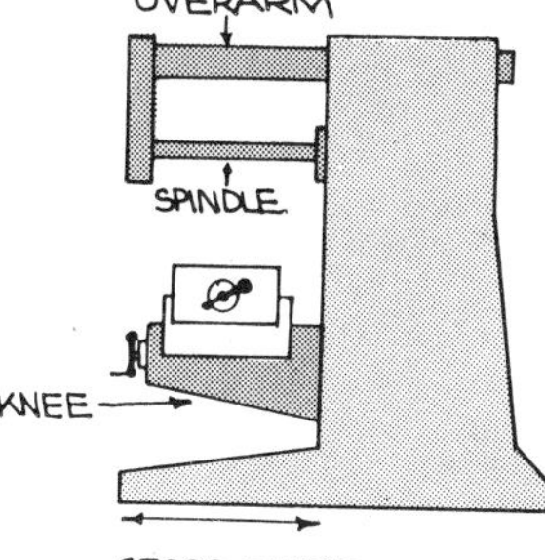

CROSS FEED

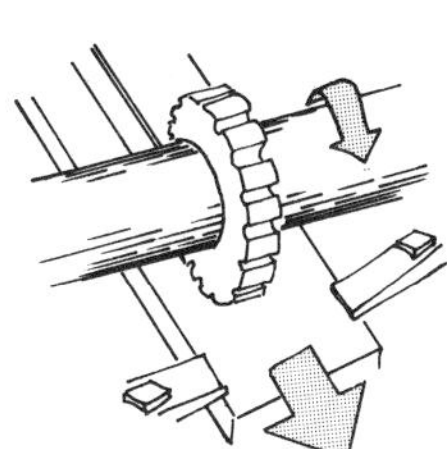

Vertical Mill

THIS IS A KIND OF DOUBLE-JOINTED DRILL PRESS. A CUTTING TOOL ROTATES IN A SPINDLE WHILE THE WORK IS FED INTO IT. A GREAT DEAL OF VERSATILITY IS POSSIBLE SINCE THE CUTTER MAY BE COCKED AT A COMPOUND ANGLE. THE DEPTH OF CUTTING AND ORIENTATION OF THE WORK AS IT IS FASTENED ONTO THE TABLE ALSO PROVIDE FOR GREAT FLEXIBILITY.

FRONT VIEW

VERTICAL TRAVERSE
HEAD
COLUMN
TABLE
SADDLE
BASE

LONGITUDINAL FEED

SIDE VIEW

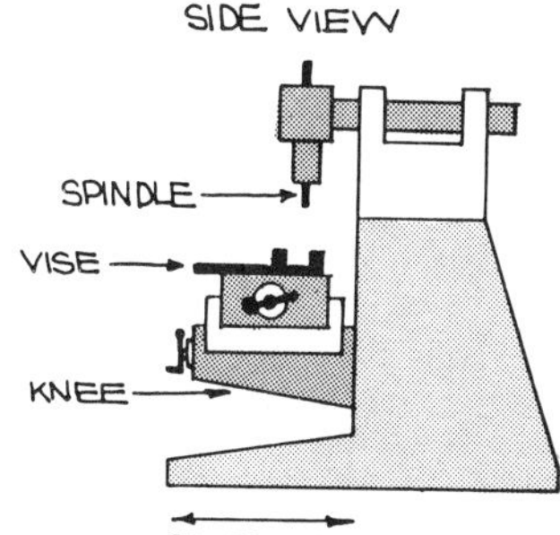

CROSS FEED

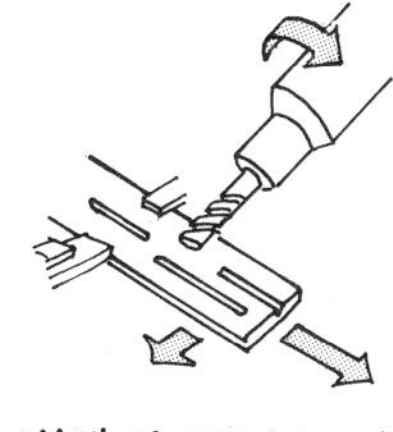

MACHINE DRAWINGS ARE COURTESY OF GARY GRIFFIN.

SHAPING

Description:

A FLAT METAL DISK MAY BE PRESSED AGAINST A WOODEN FORM AS BOTH REVOLVE ON A LATHE. THIS PROCESS, INVENTED BY THE CHINESE AND WIDELY USED BY THE GREEKS AND ROMANS, IS USED TODAY BOTH IN INDUSTRY AND HANDCRAFTING. THOUGH SPINNING IS USUALLY ASSOCIATED WITH COLD WORKING SOFT METALS LIKE ALUMINUM AND BRITANNIA, FLAME SPINNING OF TOUGH METALS LIKE MONEL AND STEEL IS DONE INDUSTRIALLY.

TOOLS

ANY LATHE CAN BE USED FOR SPINNING AS LONG AS IT HAS ENOUGH POWER TO PREVENT STALLING.

A LATHE MADE FOR TURNING WOOD WILL NEED TO BE FITTED WITH A TOOL REST AND BACK CENTER. THESE MAY BE MADE IN THE SHOP OR BOUGHT FROM A MANUFACTURER OF SPINNING EQUIPMENT.

THE FORMS AGAINST WHICH THE METAL IS PUSHED CAN BE MADE ON THE SAME LATHE. IF YOU ARE UNFAMILIAR WITH FACEPLATE TURNING, CONSULT A WOOD-WORKING TEXTBOOK.

FOR SPINNING ALUMINUM OR BRITANNIA A FLAT BACK TOOL MADE OF WOOD IS PREFERRED OVER THE STEEL TOOL. HICKORY IS COMMONLY USED: ONE SOURCE IS AN AXE HANDLE, CARVED TO SHAPE. ANY HARDWOOD MAY BE USED FOR THE BACKSTICK.

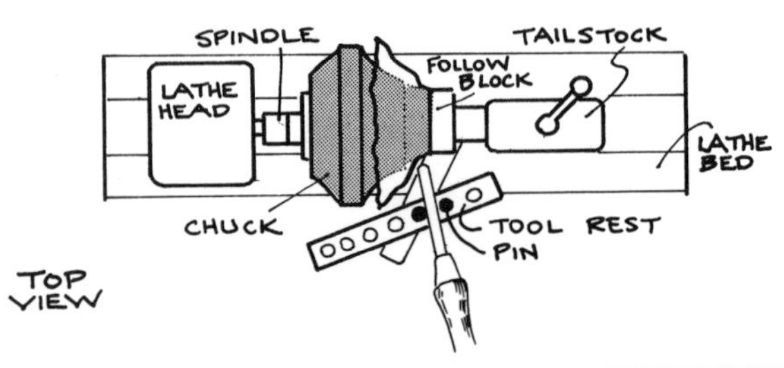

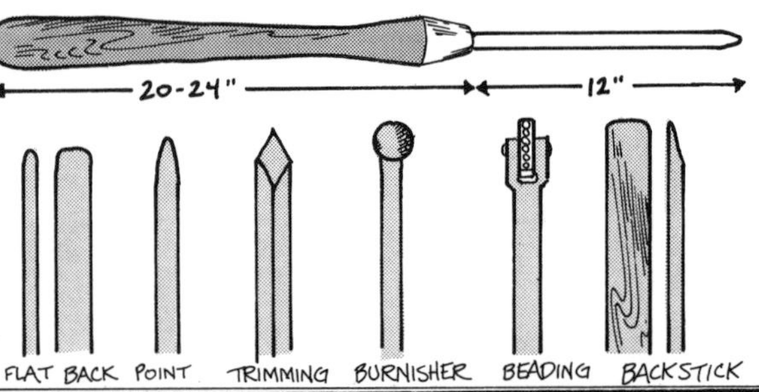

LUBRICANTS: TO REDUCE HEAT AND ALLOW EASY MOVEMENT, A WAX OR OIL IS APPLIED TO THE METAL DISK. COMMERCIAL SPINNING WAX IS AVAILABLE, OR TALLOW, LARD, OR YELLOW SOAP MAY BE USED.

SPEEDS: WHEN CONTROL IS OBTAINED FASTER SPEEDS ARE PREFERRED. A RULE OF THUMB IS 1200-1500 RPM FOR A MEDIUM SIZE DISK; 8-10". SLOWER SPEEDS FOR LARGER DISKS.

PROCESS

SAFETY: GOGGLES AND THICK LEATHER GLOVES.

1. MAKE A HARDWOOD FORM THE EXACT SHAPE OF THE INTERIOR OF THE POT.
2. CUT A DISK OF METAL. SIZE AND THICKNESS WILL BE THE SAME AS FOR RAISING. SEE PAGE 40.
3. LOCATE THE CENTER OF THE DISK AND WITH DIVIDERS MARK A CIRCLE JUST A BIT LARGER THAN THE BASE.
4. USING THIS CIRCLE AS A GUIDE HOLD THE METAL IN POSITION AND SLIDE THE FOLLOW BLOCK INTO PLACE. TIGHTEN. IF THIS SLIPS WHILE SPINNING, GRIP MAY BE IMPROVED BY PUTTING A LITTLE PUMICE OR DRY PITCH POWDER HERE.
5. SET SPINDLE SPEED. IT WILL PROBABLY BE AROUND 1200-1500 RPM. THIN METALS REQUIRE A SLIGHTLY HIGHER SPEED.
6. ADJUST THE TOOL REST TO BRING IT CLOSE TO THE CHUCK. THE TOOL SHOULD EXTEND ONLY A COUPLE OF INCHES BEYOND THE TOOL REST. THE TOOL SHOULD MAKE A RIGHT ANGLE WITH THE CHUCK AT THE POINT OF CONTACT.
7. TURN ON THE LATHE AND APPLY THE LUBRICANT. DO THIS REPEATEDLY THROUGHOUT THE PROCESS.
8. WITH THE HANDLE OF THE FLAT BACK TOOL HELD UNDER THE ARM, PRESS THE FLAT SIDE AGAINST THE METAL NEAR THE FOLLOW BLOCK, PUSHING IT AGAINST THE CHUCK.
9. WITH THE TRIMMING TOOL, CUT AWAY ANY EXCESS METAL AND TRUE THE DISK.
10. HOLD TOOL BELOW SPINDLE LEVEL AND PRESS METAL AGAINST THE CHUCK BY PIVOTING IN TOOL REST. ANNEAL AS NEEDED.

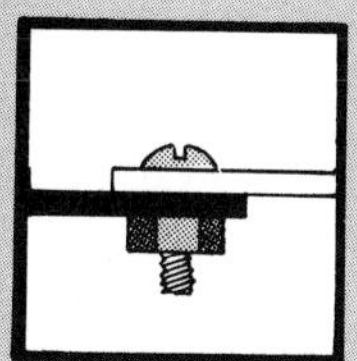

CONNECTING

Soldering CONNECTING

Description: WHEN METAL IS HEATED THE CRYSTALS OF WHICH IT IS MADE MOVE APART, CREATING MICROSCOPIC SPACES. THE IDEA OF SOLDERING IS TO USE AN ALLOY THAT WILL FLOW WHEN THERE ARE GAPS BETWEEN THE CRYSTALS. THIS ALLOY FLOODS INTO THE SPACES AND IS BONDED THERE AS THE TWO METALS CONTRACT.

RULES:

1. PIECES BEING JOINED SHOULD MAKE A TIGHT FIT. A NEATER AND STRONGER JOINT WILL RESULT FROM CARE ON THIS POINT.

2. JOINT AND SOLDER MUST BE CLEAN: NO GREASE, FINGER OILS, TAPE, PICKLE, BUFFING COMPOUND, GRAPHITE (PENCIL MARKS), ETC.

3. FLUX MUST BE USED TO PROTECT THE METAL FROM OXIDATION. EACH REHEATING USUALLY REQUIRES REFLUXING. SEE PAGE 55 FOR FLUXES.

4. ALL PIECES BEING SOLDERED SHOULD REACH SOLDERING TEMPERATURE SIMULTANEOUSLY. HEAT THE AREAS ADJACENT TO THE JOIN TO REDUCE THE CONDUCTION (FLOW OF HEAT) AWAY FROM THE JOINT. TAKE INTO ACCOUNT HEAT SINKS LIKE BINDING WIRE, STEEL MESH, AND LOCKING TWEEZERS.

5. SOLDER FLOWS TOWARD HEAT. WHEN POSSIBLE, USE THE TORCH LOCATION TO DRAW SOLDER THROUGH A JOINT.

GENERALLY, AVOID DIRECTING FLAME AT SOLDER. ALLOW HEAT TO TRAVEL THROUGH THE PIECE.

6. USE JUST ENOUGH SOLDER TO FILL THE JOINT; DON'T SETTLE FOR JUST WHAT IS HANDY. IT TAKES CONSIDERABLY LESS TIME TO CUT THE CORRECT SIZE PIECE OF SOLDER THAN TO REMOVE EXCESS LATER.

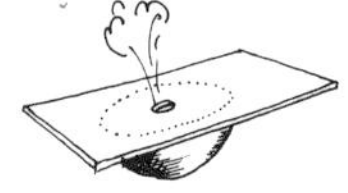

7. WHEN SOLDERING AN ENCLOSED AREA, PROVIDE AN ESCAPE FOR STEAM AND GASES TRAPPED INSIDE. THESE WILL EXPAND RAPIDLY AND, UNLESS VENTED, CAN CAUSE THE PIECE TO EXPLODE.

LIGHTING around the soldering area

METAL TEMPERATURES ARE JUDGED BY COLOR CHANGES AND THESE CAN BE SEEN BEST IN A DIMLY LIT AREA. WHATEVER YOUR LIGHTING, KEEP IT CONSISTENT.

SUCCESS IS A JOURNEY, NOT A DESTINATION.

some common problems:

Problem	Cause / Remedy
INCOMPLETE OR NON-EXISTANT JOINT.	NOT ENOUGH HEAT, METAL WAS DIRTY, NO FLUX, PROLONGED HEATING.
SOLDER BALLS UP.	HEAT MAY BE FLOWING AWAY FROM JOINT. METAL OR SOLDER MAY BE DIRTY. AVOID PLAYING FLAME DIRECTLY ON SOLDER.
SOLDER JUMPS TO ONE SIDE OF JOINT.	ONE SIDE IS HOTTER THAN THE OTHER. REFLUX AND TRY AGAIN.
PITS IN GOLD, SILVER, COPPER, OR BRASS.	LEAD/TIN ALLOY (E.G. SOFT SOLDER) HAS GOTTEN ONTO SURFACE AND BEEN HEATED ABOVE 500°F (260°C). WHERE SCRAPING OR FILING WON'T WORK, SOFT SOLDER CAN BE CHEMICALLY REMOVED: MIX 3 OZ. GLACIAL ACETIC ACID WITH 1 OZ. HYDROGEN PEROXIDE. HEAT BUT DO NOT BOIL. BRUSH ONTO AFFECTED AREA AND ALLOW SEVERAL DAYS TO WORK. TIN WILL BE LEFT AS A WHITE POWDER THAT CAN BE BRUSHED OFF.

What Happens:

WHEN A METAL IS HEATED TO TEMPERATURES APPROACHING ITS MELTING POINT THE CRYSTALS OF WHICH IT IS MADE MOVE APART, OPENING UP MICROSCOPIC SPACES. THE IDEA BEHIND HARD SOLDERING IS TO INTRODUCE AN ALLOY THAT IS FLUID JUST AT THE POINT OF MAXIMUM EXPANSION. THIS ALLOY, CALLED SOLDER, FLOWS INTO THE SPACES OF THE EXPANDED METAL AND JOINS WITH THE CRYSTALS THERE.

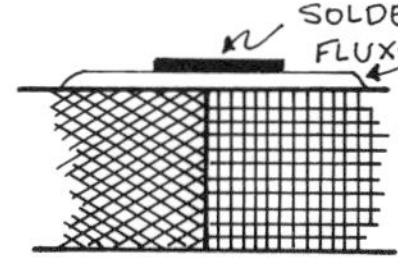

CRYSTALS PACKED TIGHT.

CRYSTALS MOVE APART; METAL HAS EXPANDED.

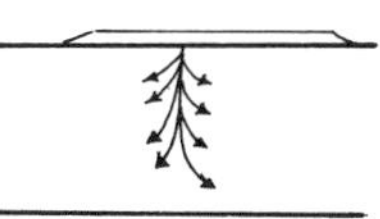

SOLDER MELTS AND IS DRAWN INTO SPACES BY CAPILLARY ACTION.

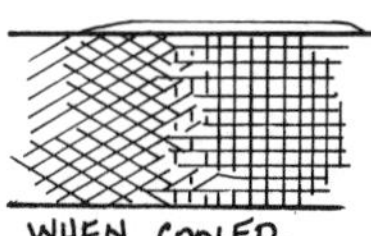

WHEN COOLED, SEAM IS INVISIBLE.

THIS SHOULD NOT BE CONFUSED WITH SOFT SOLDERING, WHICH USES AN ALLOY OF WHITE METALS LIKE TIN, LEAD, BISMUTH, ETC. SOFT SOLDER FLOWS AT TEMPERATURES ABOUT A THIRD OF THOSE NEEDED TO CAUSE THE CRYSTAL SPACES TO OPEN. THE HOLDING POWER OF SOFT SOLDER COMES FROM ITS ABILITY TO FUSE ONTO CLEAN METAL. SINCE THE GRIP IS ONLY SURFACE-TO-SURFACE, SOFT SOLDER CANNOT BE FILED FLUSH WITHOUT WEAKENING THE JOINT.

SOFT SOLDER JOINT

SILVER SOLDER

name	Ag	Cu	Zn	Cd	melt. pt.	
"IT"	80	16	4		809°	1490°
HARD	76	21	3		773°	1425°
MEDIUM	70	20	10		747°	1390°
EASY	60	25	15		711°	1325°
EASY-FLO	50	15	15	20	681°	1270°

IT IS THE AMOUNT OF ZINC IN SILVER SOLDER THAT CONTROLS ITS MELTING POINT. WHEN MAKING SOLDER CARE MUST BE TAKEN TO AVOID OVERHEATING SINCE THE ZINC WILL GO OFF IN A VAPOR, CHANGING THE PROPORTION. THIS VAPORIZATION IS ALSO A FACTOR WHEN SOLDERING. EACH TIME SOLDER BECOMES FLUID ITS MELTING POINT IS RAISED. OVERHEATING A PREVIOUSLY SOLDERED JOINT WILL BURN OUT THE ZINC AND CAN LEAVE A PITTED SEAM.

GOLD

GOLD MAY BE JOINED WITH SILVER SOLDER BUT TO ACHIEVE A COLOR MATCH A GOLD-BASED ALLOY IS USUALLY USED. GOLD SOLDERS ARE AVAILABLE IN MANY COLORS AND MELTING POINTS. WHEN BUYING, SPECIFY WHAT METAL YOU ARE JOINING: "14KARAT YELLOW SOLDER" REFERS NOT TO THE QUALITY OF THE ALLOY, BUT MEANS IT IS USED ON 14K GOLD. IN FACT SOLDER WILL BE A KARAT OR TWO LOWER THAN THE METAL IT WILL JOIN. ANY GOLD OF A LOWER KARAT CAN BE USED AS A SOLDER: 10K WILL BE A SOLDER FOR 14K; 14K WILL SOLDER 18K, ETC.

Spelter IS ANOTHER NAME FOR ZINC. TODAY THE WORD IS USED TO DESCRIBE BRASS WHEN IT IS USED AS A SOLDER FOR STEEL. THIS PROCESS IS CALLED BRAZING. FLUXES FOR BRAZING INCLUDE CREAM OF TARTAR, TABLE SALT, AND THE PASTE FLUXES USED IN SILVER SOLDERING. THE CHART ON PAGE 141 SHOWS THE MELTING POINTS OF COMMON BRASSES. THESE CAN BE USED LIKE THE DIFFERENT GRADES OF SOLDER.

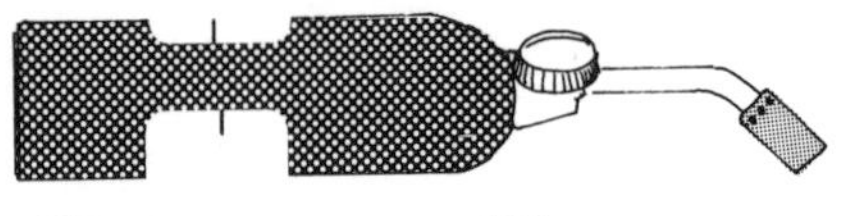
PROPANE CANISTER

ALSO CALLED: BERNZ-O-MATIC
COST: ABOUT $20

PROVIDES CLEAN HEAT IN A COMPACT, PORTABLE, INEXPENSIVE UNIT. CANISTER CAN BE AWKWARD TO HOLD, HAS LIMITED FLAME SIZE, AND CAN FLARE UP WHEN INVERTED.

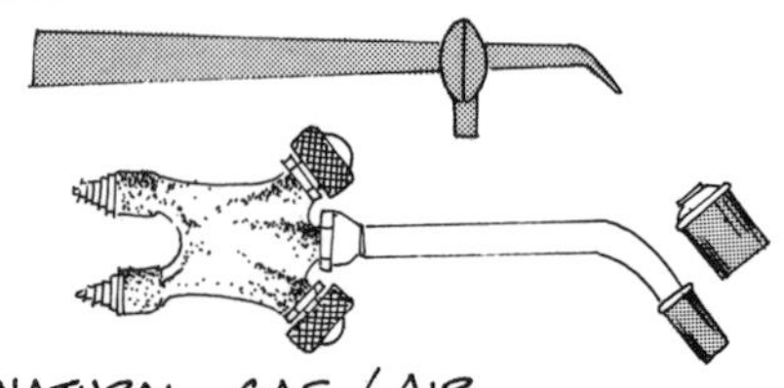
NATURAL GAS / AIR

THE GAS USED IS ALSO CALLED CITY GAS OR MANUFACTURED GAS.

AN INEXPENSIVE VERSATILE TORCH WHERE GAS IS AVAILABLE. AIR IS PROVIDED BY MOUTH, A FOOT BELLOWS, OR A COMPRESSOR. THE HI-HEAT TORCH SHOWN HAS NOZZLES THAT SUPPLY A WIDE RANGE OF FLAME SIZES.

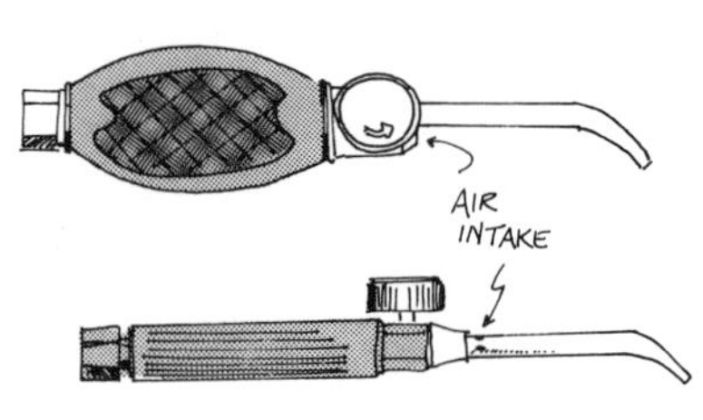

ACETYLENE / ATMOSPHERE

ALSO CALLED: PRESTO-LITE, ACETYLENE/AIR

THIS VERSATILE TORCH IS A COMMON FAVORITE AMONG SMITHS. IT PROVIDES A WIDE RANGE OF FLAME SIZES IN A CONVENIENT AND ECONOMICAL UNIT. AIR IS SUCKED INTO THE TIP AS THE PRESSURIZED GAS IS RELEASED. THE RATIO OF GAS-TO-AIR IS PRE-SET BY THE SIZE OF THE AIR INTAKE HOLES. COST WILL VARY ON TANK CHOSEN: CONSULT A LOCAL WELDING SUPPLY COMPANY.

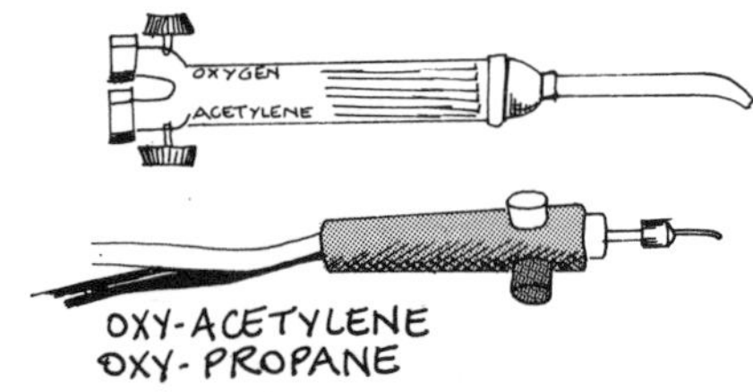

OXY-ACETYLENE
OXY-PROPANE

THESE TORCHES ARE SIMILAR IN THAT THEY BOTH USE PURE PRESSURIZED OXYGEN MIXED IN THE TIP WITH A FUEL GAS TO ACHIEVE VERY HIGH TEMPERATURES. THEY MAY BE USED FOR WELDING AND WILL REQUIRE A QUICK TOUCH WHEN SOLDERING GOLD OR SILVER. THE MINIATURE TORCH REQUIRES SPECIAL REGULATORS.

IRONS

THESE KINDS OF SOLDERING IRONS ARE USED ONLY FOR LOW-MELTING SOLDERS, USUALLY TIN-BASED. THEY GENERALLY REACH TEMPERATURES OF 600° F (280° C).

safety:

- NEVER LIGHT A TORCH UNTIL YOU UNDERSTAND ITS PROPER USE.
- KEEP MISCELLANEOUS COMBUSTIBLES AWAY FROM THE SOLDERING AREA.
- WHEN CHANGING TANKS, ALL FITTINGS SHOULD SCREW ON EASILY; DO NOT FORCE AND STRIP THREADS. TIGHTEN FIRMLY, BUT DO NOT USE BRUTE FORCE. CHECK FOR LEAKS BY BRUSHING LIQUID SOAP ONTO JOINTS. IF A TANK IS DEFECTIVE, RETURN IT TO THE DISTRIBUTOR AT ONCE.
- TANKS OF GAS SHOULD BE CHAINED TO A WALL OR STURDY BENCH. IF YOU MUST TRANSPORT THEM, BE SURE THE SCREW-ON CAP IS IN PLACE AND AGAIN SECURE VERTICALLY.
- WHEN DONE, SHUT OFF FUEL FIRST.
- TO EASE STRAIN ON REGULATORS, BLEED HOSES AT THE END OF THE WORK DAY. AFTER DRAINING THE HOSES, RE-CLOSE VALVES.

Flux:

FROM THE LATIN WORD FOR "FLOW". THE PURPOSE OF FLUX IS TO PROMOTE FUSION AND SOLDER FLOW BY PREVENTING THE FORMATION OF OXIDES. FLUXES ARE FORMULATED TO WORK ON CERTAIN METALS WITHIN A SPECIFIC TEMPERATURE RANGE. HERE ARE FLUXES COMMON TO METALSMITHING:

1. BORAX (BORIC ACID) - A MINERAL; POWDERED AND MIXED WITH WATER TO FACILITATE APPLICATION. FORMS A GLASSY RESIDUE BEST REMOVED BY PICKLING.
2. HANDY FLUX (PASTE FLUX) - A WHITE BORAX-BASE COMPOUND AVAILABLE FROM JEWELERS' AND WELDERS' SUPPLY COMPANIES. PROVIDES GOOD OXIDE PROTECTION; LEAVES FLUX GLASS. CLEAR AND FLUID AT 1100° F (591° C). FUMES CAN CAUSE DIZZINESS.
3. BATTERN'S (ALLCRA, HILLCO, ETC) - A FLUORIDE-BASE FLUX WITH THE CONSISTENCY OF WATER; USUALLY YELLOW OR GREEN. IT IS CALLED 'SELF-PICKLING' SINCE IT DOESN'T LEAVE MUCH FLUX GLASS. WILL BREAK DOWN (I.E. STOP ABSORBING OXIDES) AT SUSTAINED HIGH TEMPERATURES BEFORE THOSE ABOVE. BETTER FOR PRECIOUS METALS THAN COPPER ALLOYS.
4. BORIC ACID/ALCOHOL - ADD BORIC ACID TO DENATURED ALCOHOL UNTIL IT STOPS DISSOLVING AND MAKES A THIN PASTE. THIS WILL NEED TO BE SHAKEN OR STIRRED PERIODICALLY. WORK IS DIPPED INTO SOLUTION, SET ON SOLDERING BLOCK AND IGNITED. THE ALCOHOL BURNS OFF, LEAVING A WHITE SKIN OF BORAX. THIS IS BOTH A FLUX AND A FIRESCALE INHIBITER.
5. PRIPS FLUX - A HOME-MADE FLUX THAT PROVIDES GOOD OXIDE PROTECTION. THE RECIPE BELOW YIELDS A DOUBLE STRENGTH MIX; THIS IS THINNED WITH AN EQUAL PART OF WATER AS FLUX IS NEEDED. PIECE IS WARMED AND THEN FLUX IS BRUSHED OR SPRAYED ON. REPEAT SEVERAL TIMES UNTIL AN UNBROKEN COATING IS FORMED.

BORAX	2 OZ. TROY	OR 2 FL. OZ.	OR 75 ML
TRI-SODIUM PHOSPHATE*	2 " "	OR 2 " "	OR 75 "
BORIC ACID	3 " "	OR 3 " "	OR 90 "

BOIL THE ABOVE IN ONE QUART OF WATER. PROLONGED BOILING MAKES A BETTER SOLUTION.

* TSP IS SOLD IN HARDWARE STORES AS A CLEANING AGENT.

6. CUPRONIL - A COMMERCIAL FLUX SIMILAR TO #5 BUT BETTER AT PRESERVING FINISH THROUGH A HEATING OPERATION. ESPECIALLY HANDY FOR REPAIR WORK. AVAILABLE FROM: 45 LABS
BOX 11819
PHOENIX, AZ 85061

FLAME TYPES

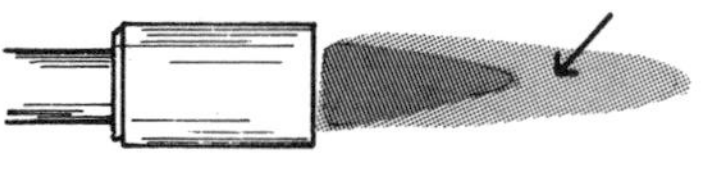

neutral

SHARP POINT, GENTLE HISS, MEDIUM BLUE COLOR, ALL GAS IS BEING IGNITED. HOTTEST POINT IS AT THE ARROW.

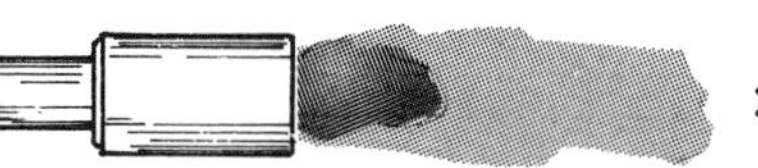

reducing

BUSHY, PULSING FLAME, DEEP BLUE COLOR FUEL-RICH HELPS ABSORB OXIDES; A "CLEANING" FLAME.

oxidizing

THIN CONE, ANGRY HISSING, PALE LAVENDER, FUEL-STARVED, HAS NO ADVANTAGES WHEN SOLDERING.

	METHOD	USES & ADVANTAGES
CHIP (PAILLON)	SOLDER FLOW IS DIRECTED BY LOCATION OF TORCH.	- MOST COMMONLY USED METHOD. - PUTS CORRECT AMOUNT OF SOLDER AT THE RIGHT PLACE. - SOLDER, ITSELF, SERVES AS TEMPERATURE INDICATOR.
SWEAT (TINNING)	FLUX SMALL PIECE IS PUT INTO PLACE ONLY AFTER SOLDER HAS FLOWED AND LARGER PIECE IS PRE-HEATED.	- KEEPS SOLDER OUT OF SIGHT WHEN DOING OVERLAY. - MORE CONTROL WHEN SOLDERING BIG AND LITTLE PIECES TOGETHER. - HELPS DIRECT SOLDER FLOW.
PROBE (PICK)	PROBE MAY BE COAT HANGER WIRE, OLD NEEDLE FILE, ETC.	- ESPECIALLY GOOD WHEN CONFIGURATION OF PIECE MAKES PLACEMENT OF SOLDER DIFFICULT. - FAST; GOOD FOR PRODUCTION WORK.
WIRE	MARK MARK SOLDER WIRE CAREFULLY SO IT ISN'T CONFUSED WITH STERLING.	- ADVANTAGES OF PROBE AND ELIMINATES THE SOLDER CUTTING STEP. — GOOD CONTROL IS IMPORTANT OR EXCESS SOLDER IS USED.
MUD	 SOLDER FILINGS MIXED WITH FLUX.	- COMMONLY USED IN COMMERCIAL ASSEMBLY LINE SOLDERING. - GOOD FOR DELICATE WORK LIKE FILIGREE. - FLUX OFTEN SPLATTERS MESSY JOINT ON SHEET.

Pickles

SAFETY ALERT

A PICKLE IS A CHEMICAL USED TO CLEAN OXIDES OFF A METAL SURFACE WITHOUT ABRASION. MOST ARE USED SIMULTANEOUSLY TO DISSOLVE FLUX GLASS. PICKLES WORK AT ROOM TEMPERATURE BUT ARE HASTENED WITH HEAT—EITHER BY WARMING SOLUTION OR BY QUENCHING METAL WHEN HOT.

SAFETY WARNING: WHEN MIXING, ALWAYS ADD ACID TO WATER. PROTECT YOURSELF FROM SPLASHES WHEN QUENCHING AND WASH HANDS BEFORE EATING OR SMOKING. KEEP BAKING SODA CLOSE AT HAND TO NEUTRALIZE SPILLS.

AS PICKLE IS USED IT ABSORBS COPPER IONS: THIS MAKES IT A PLATING SOLUTION. ELECTRICITY WILL CAUSE THE IONS TO ATTACH TO THE NEGATIVE POLE. THE INTRODUCTION OF STEEL CREATES AN ELECTRICAL CHARGE AND WILL CAUSE A PLATING REACTION. COPPER TONGS WILL NOT CREATE A CHARGE.

Formulas:

FERROUS METALS – SPAREX #1

ALL NON-FERROUS METALS – SPAREX #2 (SODIUM BISULPHATE)

STERLING – 1 PART SULPHURIC – 10 PARTS WATER

GOLD – 1 PART NITRIC – 20 PARTS WATER

KEEP PICKLE IN PYREX SAUCE PAN

OR CROCK POT THAT HAS HAD JOINTS SEALED WITH TUB CAULK.

CHARCOAL	CREATES REDUCING ATMOSPHERE, SOFT ENOUGH TO IMBED WORK, EXPENSIVE. A LITTLE MESSY. WILL REMAIN GLOWING FOR A LONG TIME: CAN BE A FIRE HAZARD.
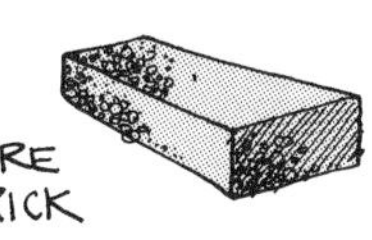 FIRE BRICK	SOFT ENOUGH, INEXPENSIVE, SAFE. CRUMBLES, ESPECIALLY WHEN A LOT OF FLUX IS USED. AVAILABLE THROUGH CERAMIC SUPPLIERS.
 COILED ASBESTOS	IS FLAT AND RELATIVELY SOFT. PARTICLES ARE GREAT RESPIRATORY DANGER; THIS IS BEST AVOIDED.
 CERAMIC & SYNTHETICS	GENERALLY GOOD BUT ARE HEAT SINKS AND GIVE OFF DISAGREEABLE FUMES WHEN HEATED. HEAT SOAK IN KILN TO REDUCE THESE.
WIRE NEST & PUMICE (CAKE PAN)	THE OLD STANDBY — VERY GOOD FOR ANNEALING OR CASES WHERE A FLAT SURFACE IS NOT NEEDED.

Maintenance:

LIKE ANY OTHER TOOL IN THE SHOP, SOLDERING SURFACES HAVE TO BE KEPT UP TO PROVIDE CONSISTENT SERVICE.

THE GREATEST PROBLEM IS THE BUILD UP OF FLUX GLASS, ESPECIALLY WHEN PASTE FLUX IS USED. TO AVOID THIS, HOLD WORK IN FINGERS TO APPLY FLUX RATHER THAN WHEN IT IS ON THE SOLDERING BLOCK.

SEVERAL MATERIALS, ESPECIALLY CHARCOAL, SHOULD BE QUENCHED IN WATER AT THE END OF THE WORK DAY TO PREVENT THEIR BURNING UP.

WITH USE, MOST SURFACES WILL BECOME IRREGULAR. A PIECE OF COARSE FLINT PAPER (THE KIND USED ON WOOD) MAY BE USED TO PERIODICALLY DRESS A SOLDERING BLOCK.

PAPER GLUED TO MASONITE

FIRESCALE

ALSO CALLED:
FIRE COAT
FIRE MARK
FIRE STAIN
#@!!@#"!

WHEN COPPER-BEARING ALLOYS ARE HEATED IN THE PRESENCE OF OXYGEN OXIDES ARE QUICKLY FORMED. CUPROUS OXIDE (CuO) IS A BLACK SURFACE LAYER THAT CAN USUALLY BE DISSOLVED IN PICKLE. CUPRIC OXIDE (CuO_2) IS A PURPLISH COMPOUND THAT FORMS SIMULTANEOUSLY WITHIN THE METAL. THIS IS FIRESCALE.

Prevention:

USE A 'HIT AND RUN' SOLDERING TECHNIQUE: AVOID PROLONGED HEATING. USE A BIG ENOUGH FLAME TO GET THE JOB DONE EFFICIENTLY. A SMALL FLAME CAN CAUSE, NOT PREVENT, FIRESCALE.

USE ENOUGH FLUX. FLUX ABSORBS OXYGEN, KEEPING IT FROM COMBINING WITH COPPER. JUST LIKE A SPONGE, FLUX WILL BECOME SATURATED, SO BE SURE YOU HAVE ENOUGH.

DO NOT OVERHEAT METAL WHEN SOLDERING. THERE IS NO ADVANTAGE TO KEEPING THE WORK HOT AFTER SOLDER HAS FLOWED. IN DIM LIGHT, METAL SHOULD NEVER NEED TO GO ABOVE A MEDIUM RED.

SEVERAL FLUXES ARE FORMULATED TO DEAL WITH FIRESCALE; I RECOMMEND HANDY FLUX, PRIP'S, OR
CUPRONIL — 4 S LABS
BOX 11819
PHOENIX, AZ 85061

Bright-dipping

IF FIRESCALE HAS FORMED IT CAN OFTEN BE REMOVED BY DIPPING THE WORK IN A STRONG NITRIC ACID SOLUTION. AFTER ALL SOLDERING AND ROUGH FINISHING ARE DONE, BUT BEFORE STONES ARE SET, ATTACH PIECE TO WIRE AND DUNK FOR ONLY A FEW SECONDS INTO A 50/50 SOLUTION OF NITRIC ACID AND WATER AT ROOM TEMPERATURE. FIRESCALE WILL TURN DARK GRAY. RINSE AND SCRATCHBRUSH.

REPEAT UNTIL SCALE IS GONE, NEUTRALIZE IN BAKING SODA AND WATER AND POLISH.

COMMERCIALLY A COMMON SOLUTION IS TO ELECTROPLATE OVER SCALE. THIS IS ESPECIALLY GOOD FOR WORK THAT WILL GET LITTLE WEAR TO RUB OFF THE PLATING. TO CREATE A SIMILAR COPPER-DEPLETED SKIN, HEAT WORK TO POINT WHERE GRAY OXIDE FORMS AND DROP IN CLEAN PICKLE; REPEAT TWO OR THREE TIMES.

FUSION

SIMILAR MATERIALS, WHEN FLUID, TEND TO MIX READILY. WHEN YOU POUR WATER INTO MILK, THE TWO MIX. A SIMILAR MIXING OCCURS WHEN TWO PIECES OF METAL ARE HEATED ABOVE THEIR MELTING POINT. MOLECULES FROM THE TWO PIECES INTERMINGLE. WHEN THE MATERIAL SOLIDIFIES (COOLS) THE INTERFACE BETWEEN THE TWO UNITS HAS DISAPPEARED. ANOTHER NAME FOR THIS, USUALLY REFERRING TO FERROUS METAL, IS WELDING.

THE DISADVANTAGE OF FUSION FOR METALS THAT ARE GOOD HEAT CONDUCTORS IS A LACK OF CONTROL OVER MELTING. THIS USUALLY MAKES FUSION INAPPROPRIATE FOR PRECISE WORK.

WHERE CONTROL IS NOT ESSENTIAL FUSING CAN BE USED TO GENERATE INTERESTING FORMS AND TEXTURES. PIECES OF METAL ARE COATED WITH FLUX AND HEATED TO THEIR MELTING POINT, WITH CARE TAKEN TO HEAT ALL UNITS SIMULTANEOUSLY. WITH SUFFICIENT TIME AND HEAT THE MASS WOULD DRAW UP INTO A LUMP, BUT IF THE TORCH IS LIGHTLY PLAYED OVER THE SURFACE IT CAN GUIDE THE MIRROR-LIKE FLASHES OF FUSION. PRECIOUS METALS WILL RESPOND BETTER TO THIS THAN COPPER ALLOYS LIKE BRASS.

DIFFUSION

SOMEWHAT RELATED TO FUSION, THIS TERM REFERS TO THE NATURE OF ELECTRONS IN METALS TO BE CONSTANTLY "WANDERING" (TECHNICALLY, "DELOCALIZED"). IF TWO PIECES OF METAL WERE CLAMPED TOGETHER AND LEFT AT ROOM TEMPERATURE THIS ELECTRON EXCHANGE WOULD EVENTUALLY BOND THE TWO PIECES TOGETHER. THE MOVEMENT OF ELECTRONS, AND CONSEQUENTLY THE RATE OF DIFFUSION, IS ACCELERATED BY HEAT. FOR CAREFUL DISCUSSION OF THIS TOPIC I RECOMMEND METALSMITH PAPERS (SEE PAGE 148). A SIMPLIFIED DESCRIPTION OF THE DIFFUSION PROCESS USED TO FORM A BILLET OF LAMINATES FOR MOKUMÉ-GANE FOLLOWS:

DIFFUSION IS BEST BETWEEN SIMILAR METALS: ALLOYS OF GOLD, SILVER, AND COPPER ARE COMMONLY USED. PUMICED-CLEAN SHEETS OF EQUAL SIZE AND SIMILAR THICKNESS ARE PILED UP AND BOUND TIGHTLY BETWEEN SHEETS OF STEEL. NO FLUX IS USED. AS THE STACK HEATS IT WILL EXPAND MORE THAN THE STEEL, HAVING THE EFFECT OF SQUEEZING THE LAMINATES TOGETHER. THIS TIGHTNESS PROHIBITS OXYGEN FROM ENTERING AND ALLOWS DIFFUSION (ELECTRON MIGRATION) BETWEEN SHEETS.

STEEL

THE UNIT IS HEATED IN A KILN OR FORGE UNTIL IT GLOWS RED AND SHOWS A LIQUID-LIKE FILM ("SWEATING"). THE PILE IS PULLED OUT, TAPPED LIGHTLY, AND QUICKLY REMOVED FROM ITS STEEL ENCLOSURE. IF STILL RED THE BILLET IS FORGED TO BE CERTAIN THAT ALL AREAS ARE IN CONTACT AND DIFFUSED. IF PROPERLY DONE DIFFUSION RESULTS IN A BOND THAT IS AS STRONG AS THE PARENT METALS; SUBSEQUENT FORMING AND SOLDERING WON'T DISTURB IT.

EUTECTIC BONDING

THE TERM EUTECTIC DEFINES THE SPECIFIC PROPORTION OF METALS IN AN ALLOY THAT HAS THE LOWEST MELTING POINT. THIS ALLOY WILL ALSO BE CHARACTERIZED BY PASSING DIRECTLY FROM A LIQUID TO A SOLID WITHOUT PASSING THROUGH A SLUSHY STATE. TECHNICALLY THIS IS A MEETING OF THE SOLIDUS AND LIQUIDUS TEMPERATURES. THE TWO WHITE AREAS AT THE SIDES OF THIS PHASE DIAGRAM INDICATE ALLOYS WHERE METALS ARE INTERMIXED BUT NOT YET IN A SOLUTION.

IF TWO PIECES OF METAL WERE COATED WITH THEIR EUTECTIC ALLOY, THEY WOULD EASILY JOIN AS SOON AS HEATED TO THE EUTECTIC TEMPERATURE. IN A WAY, THIS IS THE PRINCIPLE USED IN BONDING GRANULATION, DISCUSSED ON PAGE 24.

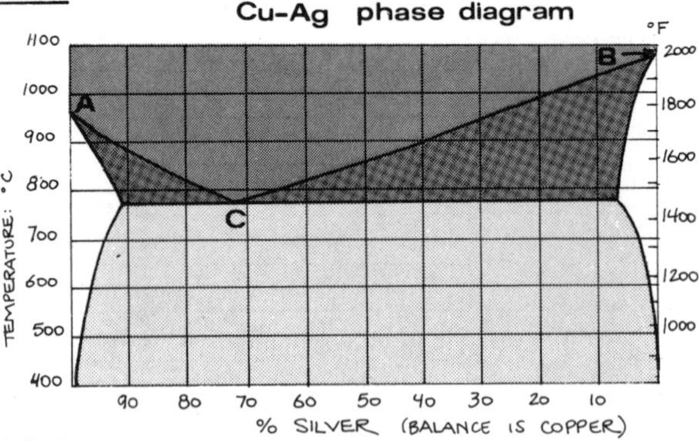

TOTALLY LIQUID

SOLID

SOLID-LIQUID MIX (SLUSHY)

A MELTING POINT OF SILVER: 962°C
B MELTING POINT OF COPPER: 1084°C
C EUTECTIC: 28.1 % SILVER, MELTS AT 780°C (1372°F)

making a standard rivet

① DRILL MATCHING HOLES IN PIECES TO BE JOINED. INSURE PROPER ALIGNMENT BY GLUING PIECES TOGETHER. IF MANY RIVETS ARE BEING MADE, SET TWO BEFORE DRILLING REMAINING HOLES.

② RIVET IS MADE OF <u>ANNEALED</u> WIRE THAT MAKES A SNUG FIT IN THE HOLES. CUT TO ½ DIAMETER EXTENSION ON EACH SIDE.

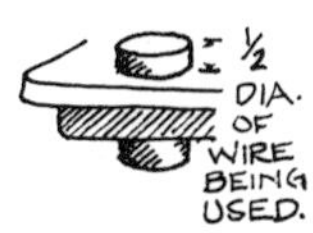

③ HOLDING PIECE SLIGHTLY ABOVE A STEEL SURFACE, TAP WIRE WITH SMALL CROSS PEEN IN TWO DIRECTIONS, MAKING AN **X**.

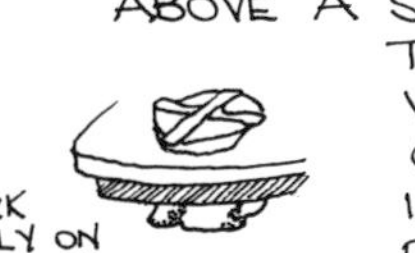

④ AS HEAD DEVELOPS AND RIVET IS HELD IN PLACE, USE FLAT FACE OF HAMMER TO SHAPE AND SMOOTH RIVET.

a variation:

(FORM ONE HEAD BEFORE INSERTING.)

① DRILL HOLES AND SELECT A TIGHT-FITTING WIRE.

② HOLD WIRE IN VISE OR IN PLIERS SUPPORTED AGAINST BENCH. FORM RIVET HEAD WITH CROSS PEEN AS IN #3 ABOVE.

PARALLEL JAW FLAT NOSE PLIERS PROVIDE A GOOD GRIP.

③ SLIDE PIN INTO WORKPIECE, TAP LIGHTLY TO SEAT, CUT OFF EXCESS WIRE AND FORM SECOND HEAD TO LOCK RIVET IN PLACE.

WHEN CONNECTING SOFT MATERIALS LIKE WOOD, LEATHER, ETC.

USE A WASHER TO KEEP THE RIVET HEAD FROM PULLING THROUGH.

<u>PLAN RIVETS SO AS TO PREVENT PIECES FROM ROTATING</u>:

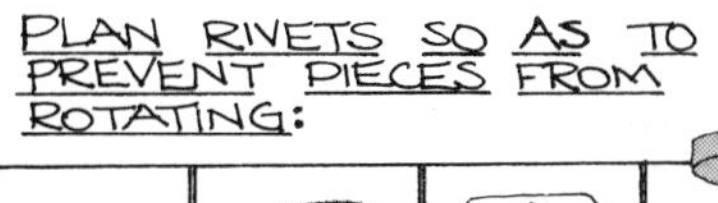

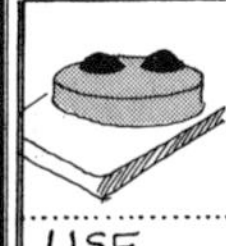

USE TWO	SOLDER WALL ON ONE SIDE	PIN WITH STITCHES	TACK ONTO RAISED BARBS

IF THE WIRE ON HAND IS A LITTLE TOO LARGE TO FIT A RIVET HOLE, SAND OR FILE A GRADUAL TAPER; IT'S FASTER THAN DRAWING THE WIRE DOWN.

TO ALLOW MOVEMENT,

WHEN A RIVETED PIECE IS TO SWING SIDE-TO-SIDE:

ADD A THIN PIECE OF CARDBOARD TO THE ASSEMBLY AND RIVET AS USUAL. REMOVE THE CARDBOARD BY BURNING OR SOAKING IN WATER TO PROVIDE CLEARANCE.

EVERY STYLE THAT IS NOT BORING IS A GOOD ONE.

—VOLTAIRE

WHEN FORMING A RIVET HEAD IN A TIGHT SPOT, A FLAT PUNCH HELD IN A VISE IS A USEFUL ANVIL.

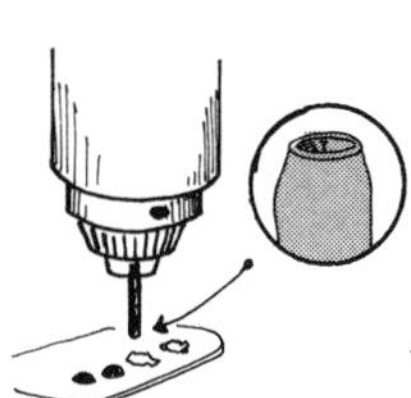

RIVET HEADS CAN BE SHAPED AND BURNISHED BY A BEADING TOOL SET INTO A DRILL PRESS. TOOLS MAY OF COURSE BE MADE FOR ANY SIZE HEAD THAT IS NEEDED.

process:

① DRAW A BEAD ON A WIRE THAT FITS TIGHTLY INTO RIVET HOLE. USE A HOT, SHARP-POINTED FLAME. GOLD AND SILVER WILL FORM BALLS MORE EASILY THAN COPPER OR BRASS.

THIS — NOT THIS

② SLIDE WIRE INTO TIGHT HOLE ON NUMBERED SIDE OF DRAWPLATE AS IT IS LAID ACROSS VISE JAWS OR ANVIL HOLE. HIT WITH PLANISHING HAMMER.

③ THE RESULTING NAIL**HEAD** MAY BE SHAPED WITH PUNCHES OR A NAIL SET WHILE STILL IN THE DRAWPLATE, OR REMOVED AND FILED TO A DESIRED SHAPE.

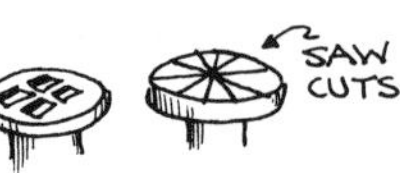

THESE WOULD PROBABLY BE MADE AFTER THE RIVET WAS IN PLACE.

④ INSERT WIRE INTO HOLE, TAP LIGHTLY TO SEAT, AND FORM STANDARD RIVET HEAD ON THE OTHER END.

NAILHEAD RIVETS ARE RECOMMENDED FOR **16-24** GAUGE WIRE. SMALLER WIRE DOESN'T LEAVE ENOUGH HEAD TO SHOW. HEAVIER WIRE DOES NOT EASILY FORM A BEAD.

TRADITIONAL ROSETTE

NAILHEADS ARE MADE WITH FOUR ANGLED BLOWS OF A PLANISHING HAMMER.

making flush rivets

THESE RIVETS ARE MADE AS DESCRIBED ON THE PREVIOUS PAGE (STANDARD RIVETS) EXCEPT THAT THE HOLE IS PREPARED BY **BEVELING** OUT ITS UPPER RIM. THIS ALLOWS FOR THE SWELL (BULGE) OF THE RIVET TO BE BELOW THE SURFACE OF THE PIECE BEING HELD.

SCRIBE; WIGGLE AND LEAN OUTWARD

ANY OF THESE TOOLS MAY BE USED. EITHER OR BOTH ENDS OF A RIVET MAY BE MADE FLUSH.

IF THE RIVET IS MADE OF THE SAME METAL AS THE PIECE IT IS HOLDING, THE RIVET WILL BLEND IN COMPLETELY. THIS IS CALLED A DISAPPEARING OR INVISIBLE RIVET.

AFTER FORMING RIVET HEAD WITH SMALL CROSS PEEN, PLANISH, FILE, SAND, AND FINISH AS USUAL.

① AS WITH OTHER RIVETS THE FIRST STEP IS TO DRILL A HOLE THAT MAKES A TIGHT FIT WITH THE CHOSEN TUBE THROUGH ALL THE PIECES BEING JOINED.

② THE TUBE SEAM SHOULD BE SOLDERED AND THE TUBE ANNEALED.

③ THE TUBE IS SLID INTO POSITION AND SAWN TO LEAVE HALF THE DIAMETER OF THE TUBE STICKING OUT ON EACH SIDE.

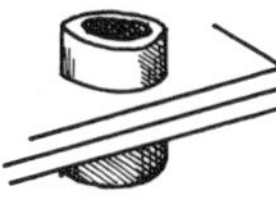

④ A SCRIBE IS SET INTO THE TUBE AND SWUNG AROUND TO FLARE OUT THE MOUTH. THIS IS REPEATED ON THE TUBE'S OTHER END.

⑤ TO FURTHER CURL OVER THE ENDS OF THE RIVET, IT IS SET ON A ROUND PUNCH OR HAMMER FACE AND TAPPED OR BURNISHED AS SHOWN.

uses:

THESE "GENTLE" RIVETS ARE RECOMMENDED WHEN THE HAMMERING NEEDED TO FORM A STANDARD RIVET HEAD MIGHT CAUSE DAMAGE. THIS WOULD INCLUDE ENAMELS, SHELLS, DELICATE MECHANISMS, & STONES.

PLASTIC RIVETS ALSO ADD COLOR TO A PIECE, ESPECIALLY WHERE LIGHT CAN BE SEEN THROUGH THE RIVET. TUBE RIVETS CAN BE USED TO PERFORATE A PIECE.

PLASTIC

SAFETY ALERT

PLASTIC RIVETS MAY BE MADE FROM ROD OR SHEET STOCK IN EITHER OPAQUE OR TRANSPARENT MATERIAL. SINCE EACH KIND OF PLASTIC HAS ITS OWN MELTING PROPERTIES IT IS A GOOD IDEA TO DO SOME TEST PIECES.

PLASTIC MAY BE BOUGHT FROM DISTRIBUTORS, WHO OFTEN HAVE SCRAP PIECES AVAILABLE. IT MAY ALSO BE SCAVENGED FROM HOUSEWARES AND TOYS.

① AS BEFORE, START WITH A HOLE (OR SLOT) THAT PIERCES ALL THE PIECES TO BE JOINED. PLASTIC MAY BE EASILY FILED OR SANDED TO FIT.

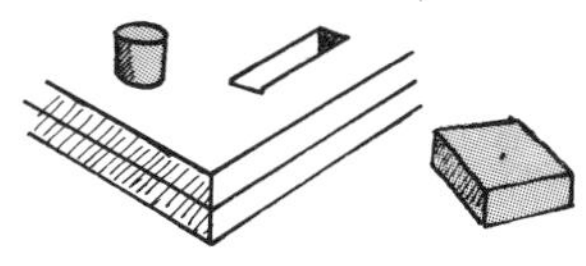

② SLIDE PLASTIC INTO PLACE AND SAW TO THE CORRECT LENGTH. THIS CAN BE FROM ½ TO ONE DIAMETER OR THICKNESS OF THE RIVET.

③ HEAT A STEEL TOOL IN A TORCH OR ALCOHOL FLAME AND PRESS FIRMLY ONTO THE PLASTIC WHILE SUPPORTING THE RIVET ON THE OTHER END.

THE TOOL SHOULD NOT BE TOO HOT TO HOLD IN THE HAND. A NEATER JOB WILL RESULT IF THE PLASTIC IS NOT HEATED TO THE POINT OF BUBBLING.

FOR ROUND RIVETS A CARPENTER'S NAIL SET WORKS WELL.

④ INVERT PIECE AND REPEAT TO FORM THE OTHER RIVET HEAD. PLASTIC MAY BE FINISHED WITH FILES, PAPER, AND ROUGE.

PROJECTING RIVETS

MAY BE MADE BY FILING BLANKS LIKE THESE:

THESE WORK IN METAL, TOO.

THE AIM OF ART IS TO REPRESENT NOT THE OUTWARD APPEARANCE OF THINGS, BUT THEIR INWARD SIGNIFICANCE.

-ARISTOTLE (344 BC)

LIQUID RESINS MAY BE USED TO JOIN PIECES BY POURING THE COMPOUND INTO A MOLD MADE BY PRESSING CLAY ONTO THE WORK PIECE.

DRIP SHAPED LATER

CLAY

CONNECTING

ADHESIVES USED AS A SUBSTITUTE FOR A PROPERLY MADE MECHANICAL CONNECTION ARE GENERALLY CONSIDERED A SIGN OF POOR CRAFTSMANSHIP. THERE ARE SITUATIONS, THOUGH, WHEN ADHESIVES ARE A LEGITIMATE AND IMPORTANT TECHNIQUE OF HEATLESS CONNECTING.

THERE ARE MANY KINDS OF GLUES: TO PURSUE THIS TOPIC I RECOMMEND A LOOK IN THE LIBRARY WHERE BESIDES TECHNICAL INFORMATION YOU WILL FIND AN ANNUAL DIRECTORY OF SUPPLIERS CALLED THE ADHESIVES RED BOOK.

1. Organic

THESE INCLUDE HOUSEHOLD GLUES LIKE ELMER'S, MUCILAGE, ETC. THEIR PRIMARY ADVANTAGES ARE LOW COST, EASY USE, AND BROAD APPLICATION. ON POROUS MATERIALS LIKE WOOD THESE GLUES PENETRATE AND HARDEN TO FORM A STRONG MECHANICAL BOND. MOST CAN BE THINNED WITH WATER BUT ARE WATERPROOF WHEN DRY.

2. Epoxies

THIS IS A CATEGORY OF THERMOSETTING PLASTIC KNOWN FOR ITS STRENGTH. IT IS AVAILABLE FROM MOST HARDWARE STORES IN BRANDS THAT OFFER DIFFERENT RATES OF CURE (I.E. SETTING TIME), COLOR, AND SPECIALIZED APPLICATION. ALL EPOXIES CONSIST OF TWO PARTS (RESIN AND HARDENER) THAT MUST BE MIXED TO START THE HARDENING REACTION. INCOMPLETE MIXING WILL GREATLY DECREASE BOND STRENGTH. USE A MIXING TOOL THAT PROVIDES A GOOD GRIP (NOT A BROKEN MATCH) AND STIR/FOLD THE TWO PARTS FOR AT LEAST A FULL MINUTE. THE HARDENING REACTION IS USUALLY HASTENED BY HEAT, UP TO ABOUT 40°C (100°F) - SPECIFIC GUIDELINES WILL BE ON THE PACKAGE. HARDENED EPOXY CAN BE DISSOLVED IN A COMMERCIAL SOLVENT. MOST EPOXIES WILL START TO BREAK DOWN AT ABOUT 200°C (400°F).

3. Cyanoacrylate

THIS RECENTLY DEVELOPED FAMILY OF ADHESIVES IS SOLD AS "SUPER GLUE," "EASTMAN 910," "KRAZY GLUE" AND OTHER BRAND NAMES. IT IS ANAEROBIC — THAT IS, IT HARDENS WHEN AIR IS EXCLUDED. FOR THIS REASON IT WILL NOT BOND POROUS MATERIALS AND CANNOT BE MADE TO FILL A GAP. ON TIGHT-FITTING NONPOROUS SURFACES IT IS QUICK AND STRONG, ITS DRAWBACKS BEING HIGH COST AND THE INABILITY TO RE-POSITION PIECES DURING GLUING.

RULES

MANY ADHESIVES ARE DELICATE COMPOUNDS THAT REQUIRE CERTAIN TEMPERATURES OR ENVIRONMENTS TO PROPERLY HARDEN. READ LABEL DIRECTIONS AND FOLLOW THEM METICULOUSLY FOR MAXIMUM HOLDING POWER.

SURFACES TO BE JOINED MUST BE ABSOLUTELY CLEAN. SINCE METAL IS NON-POROUS ONLY A THIN SURFACE FILM OF GLUE WILL BE DOING THE HOLDING. ON THIS SCALE, EVEN A TRACE AMOUNT OF OIL OR DEBRIS WILL AFFECT BOND STRENGTH. PROPER CLEANING BEFORE GLUING WILL USUALLY FOLLOW A SEQUENCE LIKE THIS:

1. ABRADE SURFACE.
2. PICKLE (IF METAL)
3. RINSE IN WATER
4. WIPE WITH SOLVENT
5. WARM SLIGHTLY TO DRIVE OFF SOLVENT.

THOUGH CLAMPING IS NOT ALWAYS NECESSARY, IT IS A GOOD WAY TO ASSURE PROPER FITTING AND WILL KEEP PIECES FROM BEING BUMPED APART.

TABS

A SIMPLE AND SECURE COLD CONNECTION CAN BE MADE BY BENDING OVER A FINGER OF METAL, AS SHOWN IN THE EXAMPLES HERE. BENDING IS USUALLY BEGUN WITH PLIERS AND FINISHED WITH A MALLET. FINISHING IS USUALLY DONE WHILE THE PIECES ARE SEPARATE.

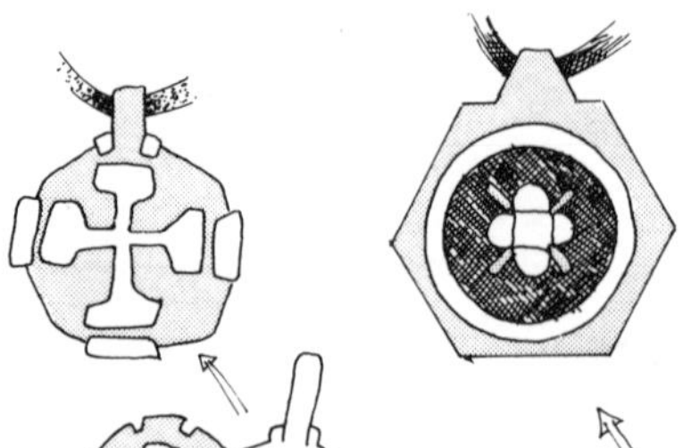

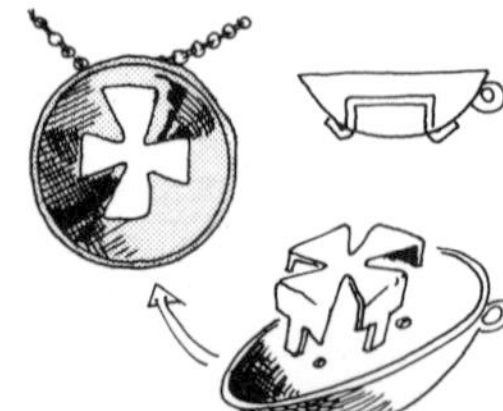

CASTING

method one

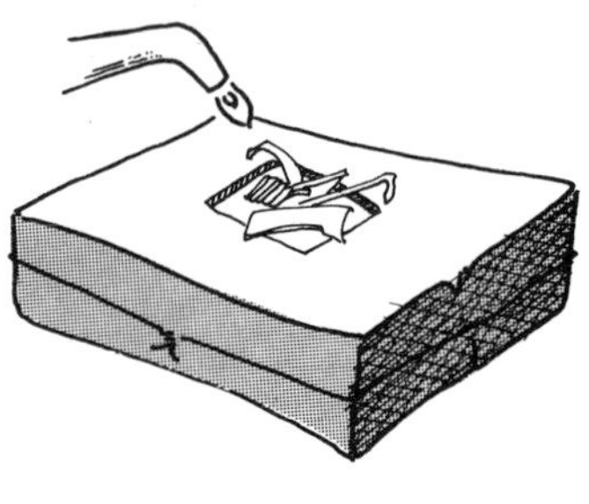

1. CARVE A RECESS TO THE THICKNESS AND SHAPE OF THE DESIRED PIECE.
2. MELT METAL DIRECTLY IN THE MOLD CAVITY. FLUX IS NOT USUALLY NEEDED BECAUSE OF THE PURIFYING ATMOSPHERE CREATED BY THE CHARCOAL.

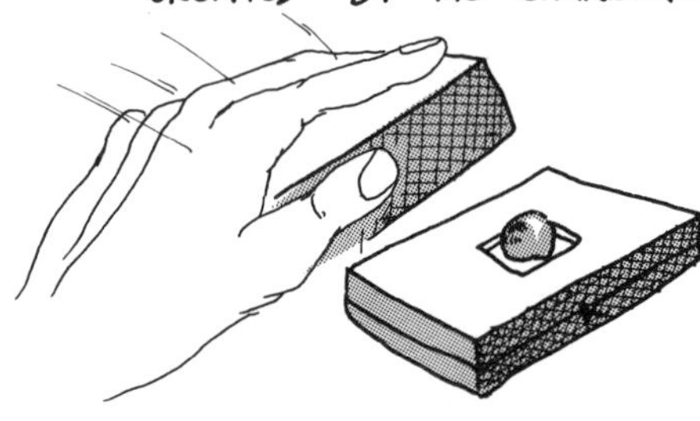

3. AS SOON AS THE METAL IS MOLTEN BRING A SECOND CHARCOAL BLOCK DOWN ON THE FIRST WITH EVEN PRESSURE. AVOID SLAMMING AND SPLASHING THE METAL – WORK WHILE STANDING TO AVOID A LAPFUL OF HOT METAL.

   MOLD DETERIORATES WITH EACH USE BUT CAN USUALLY PROVIDE 3 OR 4 CASTINGS.

method two

1. CARVE A DEPRESSION IN FLAT CHARCOAL BLOCK OR BLOCKS. IF BOTH SIDES ARE CARVED ALIGNMENT WILL BE APPROXIMATE: CAREFUL MEASUREMENT WILL HELP.
2. CARVE A SPRUE FUNNEL.

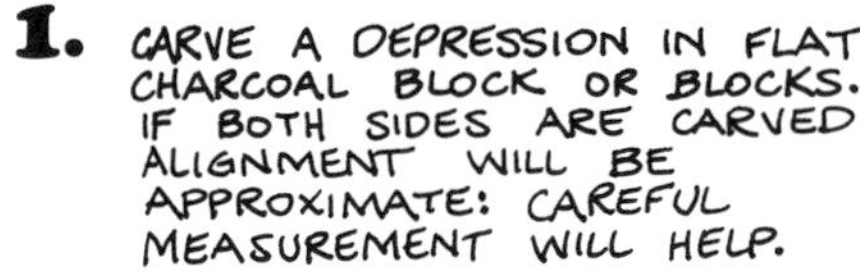

3. TIE THE BLOCKS TOGETHER WITH BINDING WIRE.

4. POUR MOLTEN METAL FROM A POURING CRUCIBLE,

   OR

   CARVE A MELTING RESERVOIR IN THE TOP OF ONE OF THE CHARCOAL BLOCKS AND CONNECT TO THE SPRUE WITH A CHANNEL. WHEN THE METAL IS MOLTEN, GRIP THE WHOLE ASSEMBLY IN TONGS AND TIP TO POUR.

OTHER MATERIALS CAN BE USED FOR EITHER OF THE METHODS SHOWN HERE. NATIVE AMERICAN INDIANS IN THE SOUTHWEST USE A ROCK CALLED TUFF OR TUFA (COMPACTED VOLCANIC ASH) TO MAKE CASTING MOLDS.

PLASTER OR INVESTMENT CAN BE FORMED INTO BLOCKS BY POURING INTO BOXES OR WOODEN FRAMES LAID OUT ON A SHEET OF GLASS OR PLASTIC. WHEN THOROUGHLY DRY THESE CAN BE CARVED AND USED AS ABOVE.

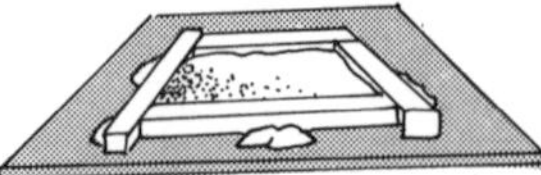

Water Casting:

METALS MAY BE POURED INTO WATER TO CREATE UNUSUAL SHAPES AND TO REDUCE LARGE PIECES TO SMALLER, EASIER MELTING PIECES. USE A DEEP BUCKET OR THE PIECES WON'T COOL BEFORE HITTING THE BOTTOM.

THE USE OF A CUTTLEFISH SKELETON AS A MOLD IS AN ANCIENT TECHNIQUE.

ADVANTAGES: PROVIDES RICH TEXTURE, IS QUICK, AND REQUIRES VERY LITTLE EQUIPMENT.

DISADVANTAGES: LACKS EXACTNESS AND IS APPROPRIATE FOR ONLY THIN, SMALL PIECES. AVERAGE SIZE OF BONE IS ABOUT 3" WIDE AND 7" LONG.

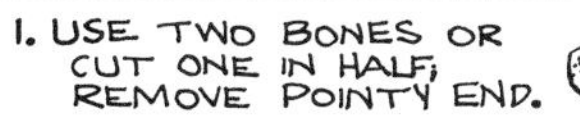

1. USE TWO BONES OR CUT ONE IN HALF; REMOVE POINTY END.

2. RUB PIECES TOGETHER, SOFT SIDE TO SOFT SIDE, IN A CIRCULAR MOTION TO MAKE FLAT SURFACES.

3. CARVE INDENTATION FOR DESIRED FORM. REMEMBER: DEPTH OF CUT EQUALS THICKNESS OF FINAL PIECE. LOCATE SHAPE ABOUT 3/4" FROM THE BIGGER END.
4. CARVE SPRUE FUNNEL IN BOTH SIDES.

5. IF GRAIN PATTERN IS TO BE EMPHASIZED, THE MOLD IS BRUSHED WITH A FINE BRUSH.
6. SCRATCH VENTS UPWARD TO ALLOW ESCAPE OF GASES FROM INSIDE THE MOLD.

7. TIE MOLD HALVES TOGETHER WITH BINDING WIRE OR MASKING TAPE.
8. SET MOLD INTO SAND OR PUMICE; POUR METAL.

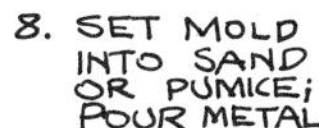

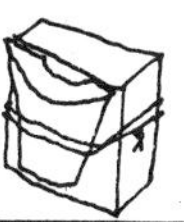

FOR ECONOMY, THE BACK – IF FLAT – CAN BE FIREBRICK, MASONITE, OR CHARCOAL.

WHEN USING A MODEL:

NOTE: MODEL MUST NOT HAVE UNDERCUTS.

1. PREPARE FLAT-SIDED MOLD HALVES.
2. PUSH STUBBY PIECES OF WOOD – DOWEL, PENCIL, MATCH – INTO ONE SIDE. AVOID MODEL AND SPRUE AREA.

3. LAY MODEL (WOOD, WAX, METAL) ONTO MOLD HALF, SET OTHER HALF IN PLACE AND CAREFULLY PRESS TWO SIDES TOGETHER UNTIL THEY MEET. TO AVOID BREAKING CUTTLEFISH, DISTRIBUTE FORCE WITH HANDS.

EVEN PRESSURE USING KNEES

4. CAREFULLY OPEN, REMOVE MODEL, BRUSH TO SHOW GRAIN IF DESIRED, CARVE SPRUE AND VENTS.
5. SET HALVES TOGETHER USING PINS TO SHOW PROPER ALIGNMENT.
6. TIE, AND POUR THE METAL.

three part molds

FOR COMPLICATED MODELS:

LIKE THIS

1. CUT OFF THREE BONE PIECES

2. RUB Ⓑ AND Ⓒ TOGETHER UNTIL FLUSH. TIE WITH TAPE.
3. RUB Ⓐ ALONG TOP EDGE OF (BC) UNTIL FLUSH.
4. OPEN (BC), SET PINS AND MODEL:
5. PRESS Ⓑ & Ⓒ TOGETHER; TIE WITH TAPE.

NOTE THAT THIS SECTION PROJECTS ABOVE THE MOLD.

6. PRESS Ⓐ DOWN ON PART OF MODEL THAT EXTENDS OUT OF MOLD. MARK LOCATION WITH TAPE OR INK LINES.

SPRUE GOES HERE

7. OPEN, REMOVE MODEL, CARVE SPRUE AND FUNNEL.
8. PUT MOLD BACK TOGETHER, TIE, AND CAST.

casting a heavy ingot

1. PUT SAND IN PAN THAT IS 2" DEEPER THAN THE INGOT YOU INTEND TO MAKE.
2. FILL PAN HALF FULL, PACKING FIRM, NOT HARD.
3. PREPARE MODEL OF WOOD (DOWEL), PLASTIC, OR METAL BY COATING WITH POUNCE.*
4. SLIDE MODEL INTO SAND LEAVING ABOUT AN INCH OF SAND BELOW THE MODEL.

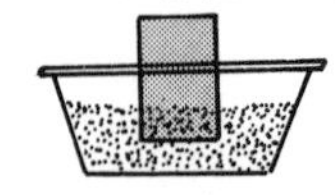

5. ADD SAND TO COVER MODEL AND PROVIDE FUNNEL. PACK HARD.

6. CAREFULLY SLIDE MODEL OUT. POUR METAL INTO MOLD.

DESCRIPTION: IN THIS ANCIENT TECHNIQUE MOIST SAND IS TIGHTLY PRESSED AROUND A MODEL WHICH IS THEN REMOVED, LEAVING A MOLD CAVITY TO BE FILLED WITH MOLTEN METAL.

CASTING A FLAT BACKED OBJECT

YOU WILL NEED A CASTING FRAME. THESE TWO PART UNITS CAN BE BOUGHT OR MADE. SEE A TECHNICAL TEXT FOR PLANS.

1. SET DRAG (FRAME PIECE WITHOUT PINS) ONTO A FLAT SURFACE LIKE GLASS. FILL WITH DAMPENED SAND, PACKING IT DOWN FIRMLY WITH A BLOCK OF WOOD. STRIKE OFF, I.E. MAKE FLAT, AS SHOWN.

2. FLIP THE DRAG OVER AND SET INTO PLACE ON IT THE OTHER FRAME PIECE, CALLED THE COPE. DUST THE PACKED SAND WITH POUNCE* AND LAY THE MODEL INTO PLACE.
3. SPRINKLE SAND OVER THE MODEL AND PACK IT LAYER BY LAYER UNTIL THE COPE IS FULL. STRIKE OFF AS BEFORE.
4. CAREFULLY LIFT THE COPE AND REMOVE THE MODEL WITH TWEEZERS. REMOVE SAND WITH A BRUSH TO CREATE A SPRUE.

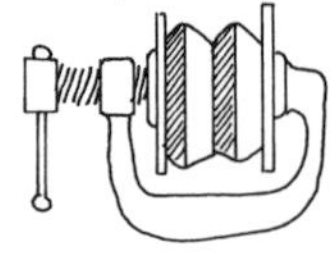

5. SET COPE AND DRAG BACK TOGETHER, COVERING EACH WITH A PIECE OF WOOD, GLASS, OR PLASTIC. TIE WITH WIRE, TAPE, OR USE A CLAMP AS SHOWN.
6. POUR MOLTEN METAL INTO MOLD USING A POURING CRUCIBLE.

THERE IS ONLY ONE SUCCESS— TO BE ABLE TO SPEND YOUR LIFE IN YOUR OWN WAY.

CHRISTOPHER MORLEY

*** POUNCE:** A POWDER USED TO KEEP MOLD SECTIONS FROM STICKING TOGETHER. TALC, CORNSTARCH, CHALK DUST, OR GRAPHITE CAN BE USED.

IT IS OFTEN KEPT IN A BAG OF LOOSE WOVEN MATERIAL LIKE GAUZE FOR 'DUSTING' ONTO MOLD.

Preparing Sand

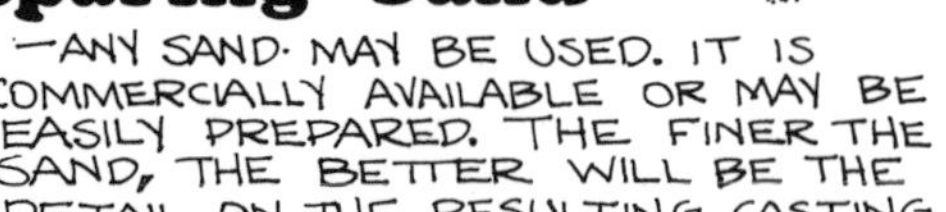

—ANY SAND MAY BE USED. IT IS COMMERCIALLY AVAILABLE OR MAY BE EASILY PREPARED. THE FINER THE SAND, THE BETTER WILL BE THE DETAIL ON THE RESULTING CASTING.

① GET A BUCKET OF SAND FROM THE HARDWARE STORE, HIGHWAY DEPT. OR PLAYGROUND.

② SIFT THROUGH SIEVE SEVERAL TIMES.

③ SIFT THROUGH SCREEN OR CHEESECLOTH.

④ MIX WITH WATER BY MASSAGING. AVOID MAKING SAND TOO WET. IF YOU GOOF SPREAD SAND ON A BOARD AND DRY IN SUN OR OVEN. IT IS RIGHT WHEN IT HOLDS TOGETHER AFTER SQUEEZING INTO A BALL.

⑤ PUMICE POWDER MAY ALSO BE USED.

A VARIATION—

FOR A THIN BUT NOT FLAT MODEL

AS ABOVE EXCEPT THAT AT STEP ② PRESS THE MODEL INTO THE SAND UP TO THE PARTING LINE OF THE MODEL.

FOR THICK MODELS LIFT A LITTLE SAND AWAY TO PREPARE A CAVITY. BE CAREFUL NOT TO TAKE TOO MUCH: MODEL MUST PRESS IN FIRMLY.

TIPS ON WORKING...

SOFT WAX CAN BE FOLDED, TWISTED, STAMPED, PINCHED, PIERCED, BUILT UP, OR PRESSED TO RECEIVE A TEXTURE.

ALL KINDS OF WAX CAN BE USED TOGETHER.

TO ADD WAX TO AN AREA, HEAT A NEEDLE, TOUCH A WAX WIRE TO THIS AND ALLOW THE WAX TO SLIDE DOWN AND DROP OFF THE END.

STORE WAX SHEETS WITH PAPER BETWEEN THE PIECES TO PREVENT THEM STICKING TOGETHER. STORE IN A COOL PLACE.

BEFORE WORKING ON SHEET WAX, SOFTEN IT BY DIPPING INTO WARM WATER OR BREATHING ON IT.

SHEET WAX CAN BE CUT WITH A RAZOR BLADE KNIFE. BECAUSE WAX IS TRANSPARENT, A DESIGN DRAWN ON PAPER CAN BE TRACED.

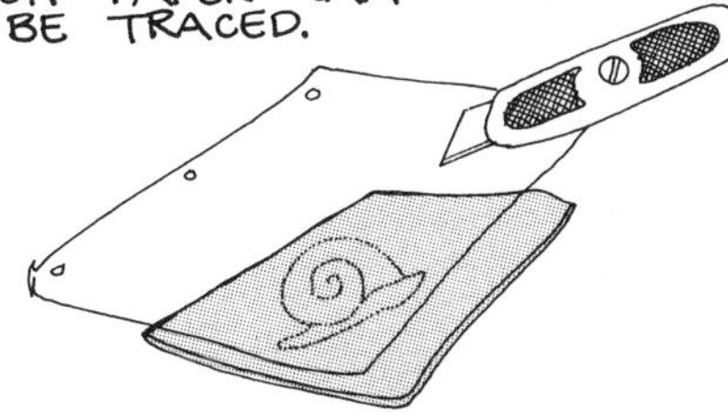

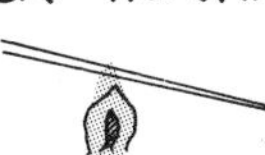

WHEN HEATING A NEEDLE, HOLD IT AS SHOWN, WITH THE FLAME HITTING IT AT MID-LENGTH. THIS WILL PRESERVE THE TIP AND KEEP THE NEEDLE WARM FOR A LONGER TIME.

To Make a Lamp

USE A GLASS JAR WITH A METAL CAP. A PIECE OF ROPE OR SHOE LACE CAN BE USED FOR A WICK.

TO MAKE THE HOLE FOR THE WICK, POUND A NAIL THROUGH FROM THE INSIDE. THIS MAKES A SHARP BUR THAT WILL GRIP THE WICK.

FUEL:

1. COMMERCIAL LAMP FUEL.
2. METHYL ALCOHOL —ALSO CALLED: WOOD ALCOHOL, METHANOL, CARBINOL
3. ALCOHOL-BASED SOLVENT
4. DUPLICATOR FLUID

Establishing a Ring Size

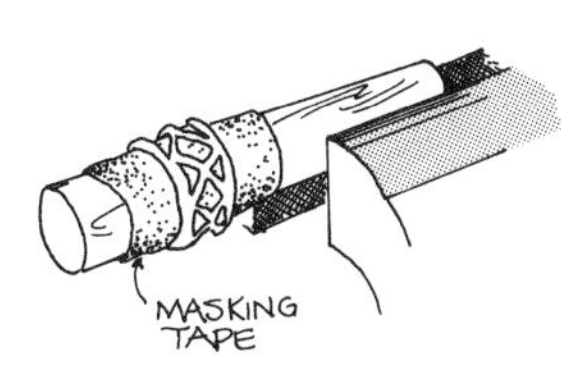

WRAP TAPE AROUND A DOWEL UNTIL THE CORRECT SIZE FORM IS MADE. TO ALLOW THE WAX PATTERN TO SLIP OFF EASIER, LUBRICATE THE TAPE WITH VASELINE OR OIL. THE MANDREL MAY BE HELD IN A VISE IF NEEDED.

BE YOURSELF, BECAUSE SOMEBODY HAS TO, AND YOU'RE THE CLOSEST.

— JACK KENT

WAX EXTRUSION

WIRE OF UNIQUE CROSS SECTION CAN BE MADE BY EXTRUDING SOFT WAX THROUGH DIES YOU MAKE YOURSELF. BUY 3 SIZES OF TELESCOPING BRASS TUBING AT A HOBBY SHOP. CAP THE SMALLEST TO BE THE RAM. DIE IS SOLDERED ONTO A 3" PIECE OF THE LARGEST TUBE AND HELD IN PLACE WHILE EXTRUDING. WEAR GLOVES.

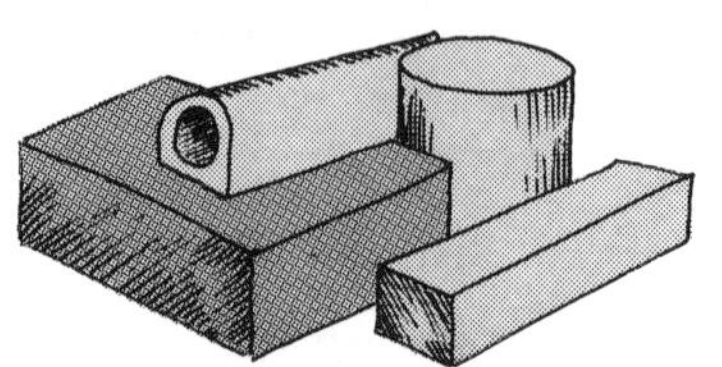

AVAILABILITY

CARVING WAXES ARE SOLD BY MOST JEWELRY SUPPLY COMPANIES AND BY SOME HOBBY SHOPS. THOUGH OTHER WAXES (LIKE CANDLE WAX) CAN BE USED, IT DOES NOT CARVE AS WELL AS WAXES BLENDED FOR THIS PURPOSE.

SCRAPS

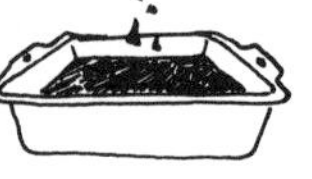

LEFT OVER BITS OF WAX MAY BE REMELTED AND FORMED INTO A BLOCK SUITABLE FOR CARVING.

PUT CLEAN SCRAPS INTO A PAN OR CARDBOARD BOX AND HEAT SLOWLY IN OVEN OR KILN TO ABOUT 300° F. LET COOL SLOWLY.

WAX MELTED DURING BURN-OUT MAY BE CAUGHT IN A TRAY. BE SURE TO REMOVE THE TRAY BEFORE TURNING KILN TO FULL BURN-OUT TEMPERATURE.

PIECES MAY BE WELDED BY HEATING BOTH SECTIONS UNTIL GOOEY AND PRESSING THEM TOGETHER.

CARVING THIS RING FROM A WELDED BLOCK WOULD REQUIRE LESS TIME THAN STARTING FROM A HUGE MASS.

Tools

CARVING TOOLS CAN BE PURCHASED OR MADE FROM DISCARDED DENTAL TOOLS, STEEL WIRE (PAPER CLIP, COAT HANGER, E.G) OR OLD SILVERWARE. HANDLES CAN BE DOWEL OR A PIN VISE.

A NON-CLOG WAX BUR IS MADE FOR A FLEXIBLE SHAFT. VERY HANDY.

COARSE FILES (SOFT METAL FILES) AND UTILITY KNIVES ARE USED.

CHAMOIS, COARSE PAPER TOWEL, OR FABRIC FOR REMOVING SCRATCHES.

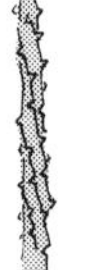

A SPIRAL BLADE FITTED IN A STANDARD SAWFRAME IS USED TO CUT OFF SECTIONS OF WAX.

IF AT FIRST YOU DON'T SUCCEED, TRY, TRY, AGAIN. THEN QUIT. THERE'S NO USE BEING A DAMN FOOL ABOUT IT.

W.C. FIELDS

MELTED WAX POURED ON WATER CREATES INTERESTING EFFECTS.

VARIATIONS INCLUDE POURING ONTO ICE, STEEL, WOOD, CONCRETE, ETC.

watch your weight

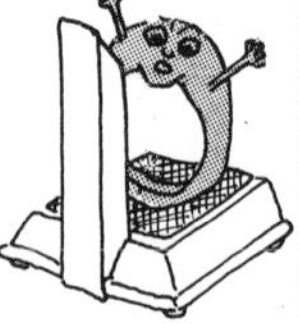

BECAUSE THE STARTING POINT FOR WAX CARVING IS USUALLY A LARGE BLOCK, A COMMON ERROR IS TO MAKE MODELS TOO LARGE.

FINAL WEIGHT MAY BE CALCULATED AS SHOWN ON PAGE 71

TO REDUCE, CARVE OUT INSIDE WITH CHISEL POINTS OR FLEXIBLE SHAFT BUR.

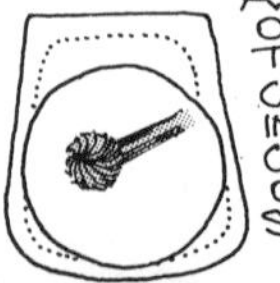

UNEVEN CARVED SURFACES MAY BE MADE MORE UNIFORM BY REPEATED WORKING WITH A SCRIBE,

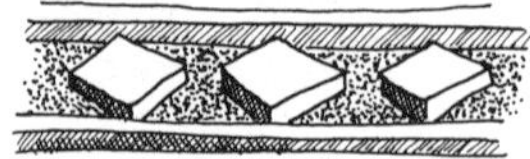

NEEDLE TOOL, OR BEADING PUNCH.

CASTING

ORGANIC

MANY ORGANIC OBJECTS, LIKE LEAVES, TWIGS, FLOWER PETALS, INSECTS, ETC. WILL BURN OUT COMPLETELY WHEN ENCASED IN AN INVESTMENT MOLD. THIS MEANS THEY CAN BE CAST DIRECTLY, OFTEN WITH VERY CLEAR DETAIL.

BURN OUT OFTEN TAKES LONGER FOR ORGANIC MATERIALS THAN FOR WAX. HIGHER TEMPERATURES MAY ALSO BE NEEDED. EXPERIMENTATION IS OFTEN HELPFUL.

POROUS MATERIALS—PAPER, CARDBOARD, POPCORN, ETC—SHOULD BE SEALED BY SPRAYING, PAINTING, OR DIPPING WITH LACQUER, WAX, OR THINNED WHITE GLUE (ELMER'S).

THIN AND DELICATE MODELS ALSO REQUIRE SOME TREATMENT. FLOWER PETALS OR INSECT WINGS CAN BE SPRAYED WITH SEVERAL COATS OF HAIR SPRAY, FIXITIVE, OR PAINT. OFTEN A MODEL MAY BE REINFORCED BY ADDING WAX WIRES ON THE BACK.

CLAY RELIEF

ATTRACTIVE WAX MODELS RICH IN TEXTURE CAN BE MADE BY POURING OR BRUSHING MOLTEN WAX OVER CLAY. THE RESULT WILL BE A REVERSE IMAGE OF THE SHAPE AND MARKINGS OF THE CLAY.

CLAY CAN BE SHAPED WITH FINGERS, PUNCHES, SCRAPS OF WOOD, KITCHEN UTENSILS, ETC.

INJECTING WAX OR AN EVEN MIX OF HARD AND SOFT WAX WILL GIVE BEST RESULTS.

USE EARTHEN CLAY: NOT PLASTICENE (KIDS' MODELING CLAY). IT WILL MELT.

USE A DOUBLE BOILER ARRANGEMENT TO MELT WAX. ON AN OPEN BURNER IT CAN IGNITE.

SAFETY ALERT

PLASTICS

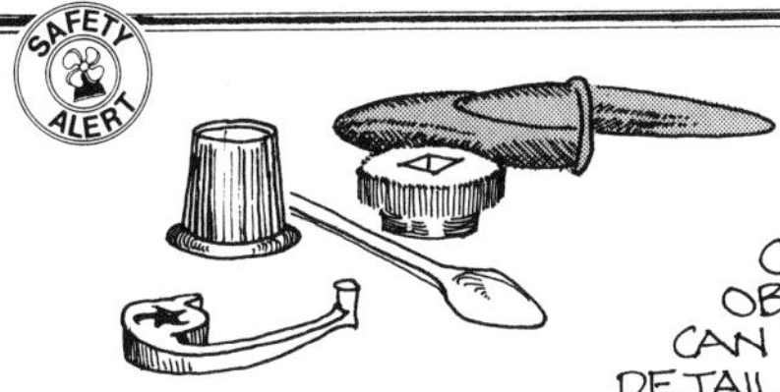

MOST PLASTICS WILL BURN OUT COMPLETELY, SO FOUND OBJECTS LIKE THESE CAN BE CAST IN ACCURATE DETAIL. PLASTICS MAY BE MODIFIED BY HEATING OR USED IN COMBINATION WITH WAX. PIECES MAY BE GLUED TOGETHER WITH WHITE GLUE OR STICKY WAX.

CAUTION

THE FUMES OF BURNING PLASTIC ARE TOXIC AND MUST BE VENTILATED.

STYROFOAM® CAN BE USED TO MAKE MODELS, TOO. PIECES ARE HELD DIRECTLY IN THE FLAME TO SHAPE, OR MAY BE CARVED WITH A HEATED NEEDLE.

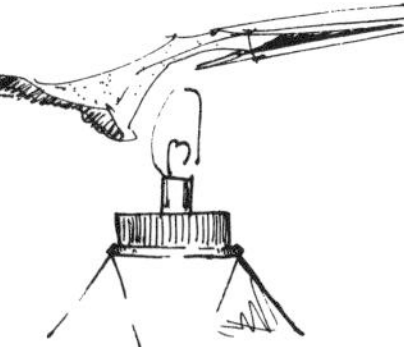

DEFINITION OF SPRUES:

- THE SUPPORTS THAT HOLD A MODEL IN CORRECT POSITION WHILE MAKING THE MOLD (INVESTING).
- PASSAGEWAY FOR ESCAPE OF MELTING WAX.
- ENTRY PASSAGE FOR MOLTEN METAL.

RULES

WRONG | RIGHT ①

1. ARRANGE SPRUES TO SUPPLY SUFFICIENT METAL TO EACH SECTION OF THE MODEL.
2. PLAN SPRUES TO AVOID FLOW-BACKS AND SHARP CURVES.
3. ATTACH SPRUES WHERE THEY WILL CAUSE THE LEAST DAMAGE TO THE MODEL'S SURFACE TEXTURE AND WHERE THEY CAN BE EASILY REMOVED.
4. AVOID SPRUING WORK DEAD LEVEL— SPRUES SHOULD NOT ENTER AT 90°
5. SPRUE TO THICKEST SECTION OF THE MODEL: SPRUE ITSELF SHOULD BE THICKEST MASS OF THE WHOLE ASSEMBLY.

CASTING

EXPLANATION

AS METAL COOLS IT CONTRACTS, SO MORE METAL IS NEEDED TO FILL A SPACE THAT WAS PREVIOUSLY FILLED BY MOLTEN METAL. IF NO EXTRA MATERIAL IS SUPPLIED AT THE INSTANT OF CONTRACTION THE METAL WILL CRYSTALIZE WITH VOIDS (PITS) AS IT TRIES TO FILL THE CAVITY. TO CONTROL THIS, THE SPRUE AND BUTTON SHOULD BE THE LAST AREA TO COOL (CONTRACT). IF PITS OCCUR HERE NO DAMAGE IS DONE. TO ACHIEVE THIS THE THINNEST—FIRST COOLING—AREA OF THE MODEL IS SET FURTHEST FROM THE SPRUE BASE, THE THICKEST IS ATTACHED TO THE SPRUE, AND THE SPRUE IS THICKER THAN ANY PART OF THE MODEL.

Vents

ALSO CALLED GATES

IN SMALL MODELS FUMES IN THE CAVITY ESCAPE INTO THE INVESTMENT AS METAL ENTERS. FOR LARGE MODELS IT IS WISE TO PROVIDE A VENT (BACK DOOR ESCAPE) FOR GASES. IN LARGE INDUSTRIAL CASTINGS SOMETIMES MANY VENTS ARE USED.

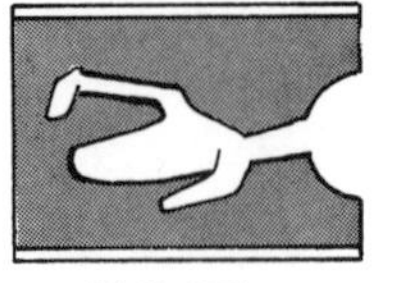

—RIGHT— —WRONG—

MODELS SHOULD BE CENTERED IN THE FLASK AND ORIENTED PARALLEL TO THE WALLS FOR SMOOTH EFFICIENT FLOW OF METAL.

Avoid

CONSTRICTED, PINCHED-NECK SPRUES: THESE WILL SPRAY THE METAL, CAUSING IT TO CHILL (HARDEN) PREMATURELY.

THE MAN WHO MAKES NO MISTAKES DOES NOT USUALLY MAKE ANYTHING.

W.C. MAGEE

DETERMINING HOW MUCH METAL TO USE

1. GUESS, PRAY, ASK A WISER PERSON.
2. ATTACH SPRUED MODEL TO WIRE, PUSH INTO CONTAINER OF WATER, AND NOTE RAISED LEVEL. REMOVE MODEL AND ADD METAL TO BRING WATER BACK UP TO MARKED LEVEL. A GRADUATED CYLINDER IS HANDY BUT NOT NECESSARY. TO THIS AMOUNT ADD 1/3 MORE FOR BUTTON.
3. MULTIPLY THE WEIGHT OF THE WAX* BY THE SPECIFIC GRAVITY OF THE METAL BEING USED. ADD **ABOUT** 10-20% MORE TO ALLOW FOR BUTTON.

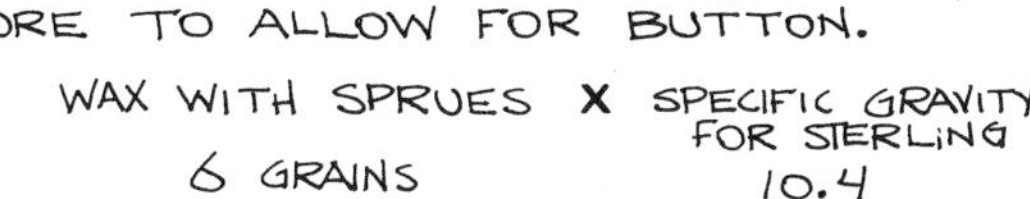

EXAMPLE: WAX WITH SPRUES 6 GRAINS X SPECIFIC GRAVITY FOR STERLING 10.4 = 62.4 GRAINS (OR 1 OZ/6 DWT.)

specific gravities

ALUMINUM	2.7
BRASS (70/30)	8.5
BRASS (88/12) NU-GOLD	8.7
18K YELLOW GOLD	15.5
14K YELLOW GOLD	13.4
10K YELLOW GOLD	11.6
IRON	7.9
LEAD	11.4
NICKEL SILVER	8.8
PLATINUM	21.4
FINE SILVER	10.6
STERLING	10.4

A MORE COMPREHENSIVE LIST MAY BE FOUND IN THE APPENDIX: PAGE 141.

TIMING:

INVESTMENTS HAVE 9-10 MINUTES OF WORKING TIME. IF YOUR PACE IS TOO SLOW THE INVESTMENT WILL HARDEN BEFORE IT CAN COAT THE MODEL. IF YOU WORK TOO QUICKLY AND THE INVESTMENT IS POURED INTO THE FLASK TOO SOON, WATER IN THE MIX IS FREE TO TRAVEL ALONG THE MODEL. THIS WILL RESULT IN WATER TRAILS (RAISED STREAKS) ON THE FINISHED CASTING.

TO AVOID THE PROBLEM TIME YOURSELF AS YOU INVEST AND ADJUST YOUR PACE ACCORDINGLY.

TO HELP INVESTMENT BETTER COAT A WAX MODEL

1. COAT MODEL WITH COMMERCIAL DEBUBBLIZER BY BRUSH OR BY DIPPING.
2. PAINT WITH ALCOHOL. LAMP WICK MAY BE USED AS SHOWN ABOVE.
3. TO MAKE A WETTING SOLUTION:

   50 % HYDROGEN PEROXIDE
   50 % LIQUID SOAP

IT IS NOT ENOUGH TO BE BUSY... THE QUESTION IS: WHAT ARE WE BUSY ABOUT?

HENRY DAVID THOREAU

* THE SPECIFIC GRAVITIES OF COMMON PLASTICS RANGE FROM 1.2 TO 1.7. TO CALCULATE THE METAL NEEDED FOR A PLASTIC MODEL ADD 10% TO THE SPECIFIC GRAVITIES LISTED, THEN MULTIPLY AS BEFORE.

HARD CORE METHOD

① MIX INVESTMENT; VIBRATE.

② PAINT INVESTMENT ONTO MODEL WITH FINE BRUSH. SPREAD MIXTURE SO AS TO AVOID BUBBLES, ESPECIALLY IN CREVICES.

③ SPRINKLE INVESTMENT POWDER ONTO COATED MODEL TO ABSORB MOISTURE & BEGIN SETTING OF THIS SHELL (CORE).

④ SET FLASK OVER MODEL, SECURE BASE, POUR IN INVESTMENT, KEEPING WEIGHT OFF MODEL BY POURING IT DOWN THE SIDE. BE SURE TO HOLD BASE WHILE POURING.

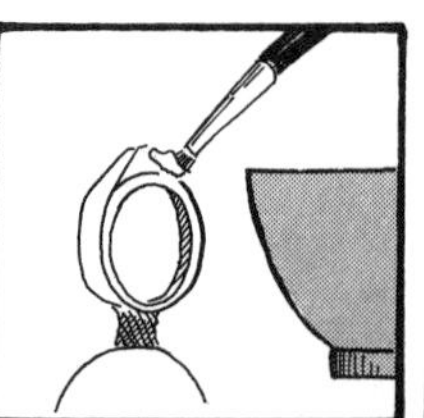

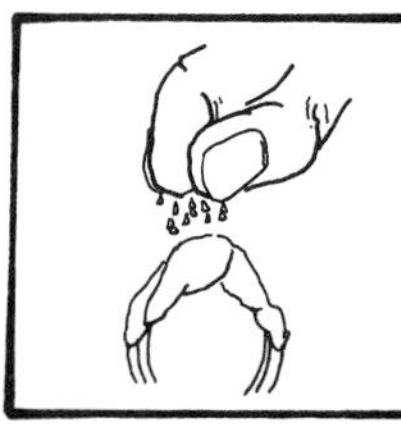

A BUFFING MACHINE CAN BECOME A VIBRATOR AS SHOWN. CUT THE CORNERS OFF A SQUARE ROD OF WOOD — LIKE A 2×2.

DRILL A HOLE TO FIT WOOD ONTO A TAPERED SPINDLE.

Substitute Bases

USE TIN CAN WITH BOTH ENDS CUT OUT. THE CAN MAY BE BENT, LIKE TO OVAL, AS LONG AS IT WILL FIT CASTING MACHINE. STEAM OR VACUUM METHODS CAN TOLERATE ODD FLASKS.

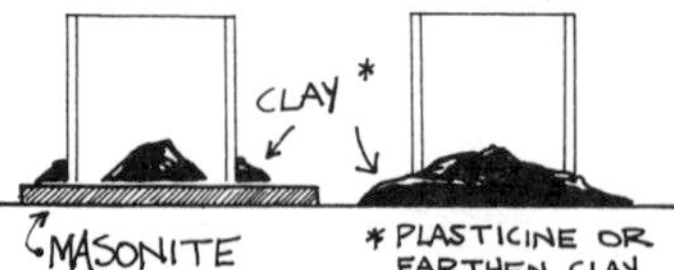

VACUUM METHOD

1. MIX INVESTMENT THOROUGHLY WITH HAND OR SPATULA. INVESTMENT SHOULD LOOK LIKE SOUR CREAM.
2. SET BOWL ON VACUUM TABLE. WET RIM OF BELL JAR AND SET OVER INVESTMENT.
3. TURN ON MOTOR AND DIRECT VACUUM TO TABLE. PRESS DOWN ON BELL JAR TO GUARANTY THAT SUCTION IS ACHIEVED.
4. LEAVE VACUUM AT MAXIMUM (25-28 FOR ABOUT ONE MINUTE. INVESTMENT WILL SWELL AND BUBBLE. WHEN IT "SPITS" AND JAR CONDENSES, TURN OFF VACUUM.
5. POUR CREAMY INVESTMENT INTO FLASK BY ALLOWING IT TO RUN DOWN THE SIDE. SET FLASK ON VACUUM TABLE AND REPEAT.

   * IF INVESTMENT IS STARTING TO THICKEN - LOOKS LIKE PUDDING - OMIT THE SECOND VACUUM OPERATION.

6. REMEMBER THAT THE INVESTMENT WILL SWELL IN THE FLASK. ALLOW FOR THIS BY NOT FILLING FLASK TO THE TOP —

OR BY USING A COLLAR OF RUBBER, PLASTIC, PAPER, OR MASKING TAPE.

THE PURPOSE OF BURNOUT IS TO
- CURE (HARDEN) THE MOLD
- ELIMINATE WAX OR OTHER MODEL MATERIAL
- HEAT THE MOLD FOR COMPATABILITY WITH THE MOLTEN CASTING METAL.

BURNOUT IS USUALLY DONE IN A SMALL ELECTRIC KILN THOUGH GAS KILNS MAY ALSO BE USED. BURNOUT IS BEST DONE WITHIN 48 HOURS OF INVESTING. IF CASTING MUST WAIT, REMOISTEN THE FLASK BY SOAKING IT IN WATER FOR A MINUTE OR TWO BEFORE BURNOUT.

RECENT DEVELOPMENTS IN INVESTMENT TECHNOLOGY HAVE CREATED A PRODUCT THAT WILL TOLERATE FASTER TEMPERATURE CHANGES THAN WHAT WAS AVAILABLE 10 YEARS AGO. THE PROGRESSION AND PACE OF BURNOUT WILL VARY DEPENDING ON THE SIZE AND NUMBER OF FLASKS IN THE KILN, THE TEMPERATURE OF THE KILN, AND THE PREFERENCES OF THE CASTER. AS A RULE OF THUMB, ALLOW 2½ HOURS FOR A TYPICAL JEWELRY-SCALE BURNOUT.

FLASKS ARE PLACED IN THE KILN WITH SPRUE HOLES FACING DOWN. PROP UP IN ANY OF THESE WAYS TO ALLOW WAX TO DRIP OUT.

HOT FLASKS ARE HANDLED WITH TONGS SOLD FOR THIS PURPOSE OR WITH HOUSEHOLD JAR LIFTERS SOLD FOR CANNING AND HANDLING BABY BOTTLES.

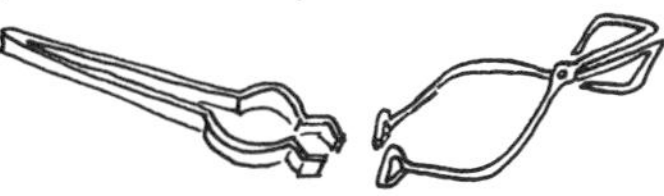

BURNOUT IS POSSIBLE WITHOUT A KILN BY USING A. FLOWERPOT AND HOT PLATE LIKE THIS → ALLOW AT LEAST 3 HOURS.

CLAY POT IS LINED WITH ALUMINUM FOIL AND VENTED AT THE TOP.

STEAM & GAS

SINCE ALUMINUM MELTS AT TEMPERATURES USED FOR BURNOUT (1220°F 660°C) IT CAN BE USED AS AN INDICATOR IN THE KILN. SET A SMALL PIECE NEAR BUT NOT TOUCHING THE FLASK. WHEN IT CURLS, FLASK IS READY.

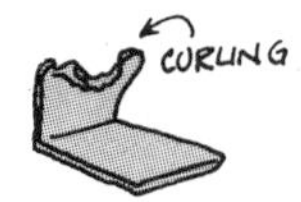

temperatures

APPROX. °F	°C	
300	150	WAX MELTS AND DRIPS OUT
450	235	WAX IGNITES
600	325	WOODY MATERIALS IGNITE
1000	550	PLASTICS VAPORIZE
1250	650	WAX RESIDUES VAPORIZE
1350	740	GYPSUM BINDER IN INVESTMENT BREAKS DOWN RELEASING SULFUR THAT WILL CAUSE OXIDATION. DO NOT GO TO THIS TEMPERATURE!

VENTILATE!

SAFETY ALERT

WAX FUMES ARE NOT GOOD FOR YOUR BODY. PLASTICS ARE WORSE. KEEP KILN IN LARGE ROOM, NEAR A WINDOW, AND IN A CROSS DRAFT. AN EXHAUST FAN IS NEEDED.

THE AMOUNT OF FUMES CAN BE REDUCED BY CATCHING THE WAX AT AROUND 300°F BEFORE IT IGNITES. THIS IS DONE IN A STAINLESS STEEL TRAY SOLD BY SUPPLIERS OR IN A SIMILAR AFFAIR MADE WITH A CAKE PAN.

THIS WAX USUALLY CONTAINS TOO MANY IMPURITIES AND HAS BEEN TOO CHANGED BY MELTING TO BE USED AGAIN.

TO KNOW THAT YOU DO NOT KNOW IS THE BEST. TO PRETEND TO KNOW WHEN YOU DO NOT KNOW IS A DISEASE.
-LAO TZU 500 BC

DESCRIPTION:

STEAM CASTING IS A LOST WAX, WASTE MOLD PROCESS IN WHICH THE HEAT OF THE MOLTEN CASTING METAL IS USED TO GENERATE STEAM PRESSURE. THIS PRESSURE THEN FORCES THE HOT METAL INTO THE MOLD.

IT IS FAST, ECONOMICAL AND VERSATILE.

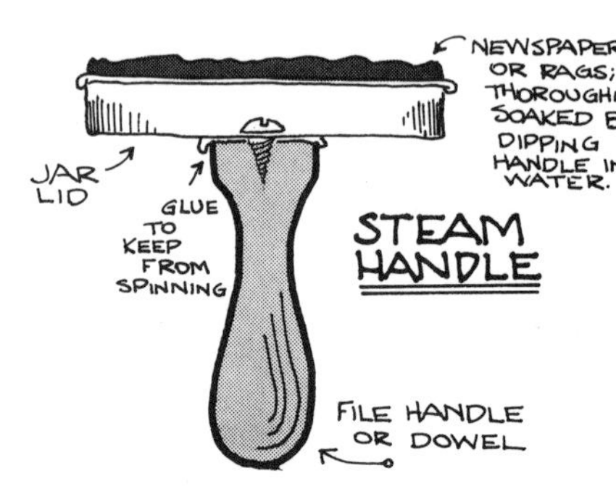

THE CARVING OF THE MODEL, LOCATION OF SPRUES AND MIXING OF INVESTMENT ARE THE SAME FOR STEAM CASTING AS FOR MORE CONVENTIONAL TECHNIQUES.

WHAT IS DIFFERENT IS THE SHAPE OF THE SPRUES. THEY MUST BE SMALL OR RECTANGULAR TO KEEP THE MOLTEN METAL FROM DRIPPING INTO THE MOLD. **SPRUE WITH 16 GA. WIRE OR STRIPS OF 16 GA. WAX SHEET.**

① A FUNNEL SHAPE IS CARVED IN THE TOP OF THE INVESTED FLASK IF THE SPRUE BASE DID NOT FORM A LARGE ENOUGH RESEVOIR.

② AFTER STANDARD BURN OUT, METAL IS MELTED IN MOUTH OF FLASK. FLUX AS USUAL.

③ WHEN METAL IS MOLTEN THE TORCH IS WITHDRAWN AS THE STEAM HANDLE IS CLAPPED FIRMLY ONTO THE FLASK. HOLD IT IN POSITION UNTIL METAL SOLIDIFIES. STEAM CASTING CREATES 10-14 POUNDS PER SQUARE INCH OF PRESSURE.

SLING

DESCRIPTION: AS ABOVE, SLING CASTING USES THE FLASK ITSELF FOR A MELTING CRUCIBLE. CENTRIFUGAL FORCE IS CREATED BY SWINGING THE FLASK BY HAND.

PROCEDURE:

1. MAKE A SLING IN ANY OF THE WAYS SHOWN. FOR SMALL FLASKS THE PAN CAN BE A JAR LID. FOR A LARGER SLING, SOLDER RING AND BASE FROM COPPER, BRASS, OR STEEL SHEET.
2. PREPARE FLASK WITH 16 GAUGE SPRUES.
3. AFTER BURNOUT SET FLASK INTO BASKET OF SLING. SET THIS ON A FIREPROOF SURFACE.
4. MELT METAL IN THE FUNNEL AT TOP OF FLASK. FLUX AS USUAL.
5. WHEN METAL IS MOLTEN SWING THE FLASK IN LARGE EVEN ARCS. A STEADY MOTION IS MORE IMPORTANT THAN A FAST OR MIGHTY SWING.

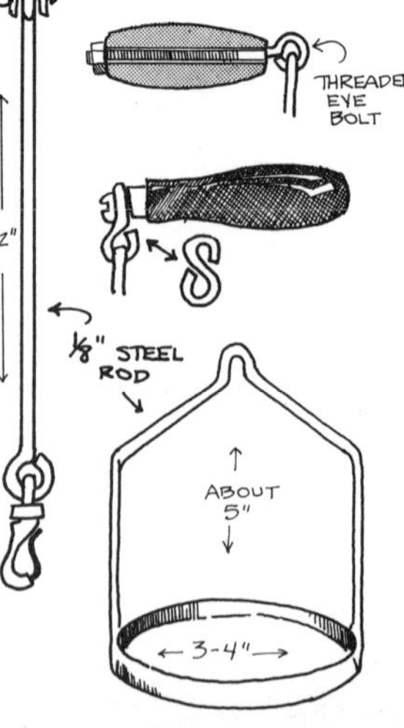

AN ARTIST NEVER REALLY FINISHES HIS WORK;

HE MERELY ABANDONS IT.

CASTING MACHINES MUST BE SOLIDLY MOUNTED AND SURROUNDED BY A SPLASH SCREEN.

A WASHTUB BOLTED TO BENCH.

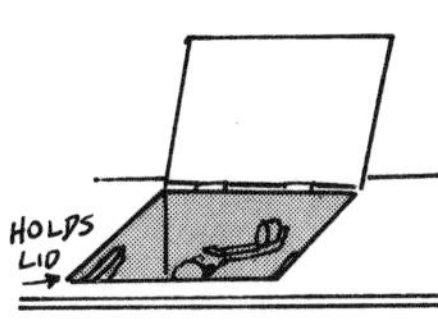

A GARBAGE CAN FILLED 2/3 FULL OF ROCKS AND CEMENT IS STABLE, SAFE AND CAN BE MOVED AROUND THE SHOP. SET BOLTS INTO WOOD TO INSURE PLACEMENT; SET WOOD INTO CEMENT.

THE CASTER CAN BE SET INO THE BENCH TO SAVE SPACE. HINGED LID IS PLYWOOD.

BECAUSE MOLTEN METAL HAS A STRONG SURFACE TENSION, SOME KIND OF PRESSURE IS NEEDED TO FORCE IT INTO THE FINE CAVITIES OF A DELICATE MOLD. THE MOST COMMON PRESSURE USED BY CRAFTSPEOPLE IS CENTRIFUGAL FORCE.

1. MACHINE IS BALANCED BY MOVING WEIGHT ALONG ARM; USUALLY THIS SCREWS ONTO A THREADED ROD. EITHER SET THE FLASK INTO PLACE BEFORE BURNOUT OR MAKE A GUESS BASED ON YOUR EXPERIENCE WITH SIMILAR SIZE FLASKS.
2. WIND CASTING MACHINE THREE OR FOUR TIMES. THE FORCE WITH WHICH THE METAL IS THROWN INTO THE MOLD IS NOT AS IMPORTANT AS HAVING THE PRESSURE EVEN AND CONSTANT. TOO MUCH FORCE WILL MAKE THE METAL RICOCHET OFF THE TOP OF THE MOLD.
3. HOLD ARM BY SLIDING PIN UP AGAINST IT.
4. MELT METAL: ADD FLUX AND HEAT CRUCIBLE MOUTH.
5. WHEN METAL IS MOLTEN, GET A GOOD GRIP ON THE ARM (DON'T HOLD THE WEIGHT), PULL TO RELEASE PIN. LIFT TORCH AND LET GO OF THE ARM SIMULTANEOUSLY.

Melting

CARE IN MELTING THE METAL IS IMPORTANT IN EVERY KIND OF CASTING:

- USE A FLAME THAT IS HOT ENOUGH TO BE EFFICIENT BUT NOT SO HOT IT WILL "BURN" THE METAL.
- USE A FUEL-RICH, REDUCING FLAME. THIS IS A BUSHY OR FEATHERY FLAME. IT SHOULD NOT MAKE A HISSING SOUND.
- WHEN THE METAL IS RED AND AGAIN WHEN IT IS MELTED SPRINKLE ON FLUX. (BORAX, BORIC ACID, POWDERED CHARCOAL, OR A COMMERCIAL FLUX).

SPICE CONTAINER

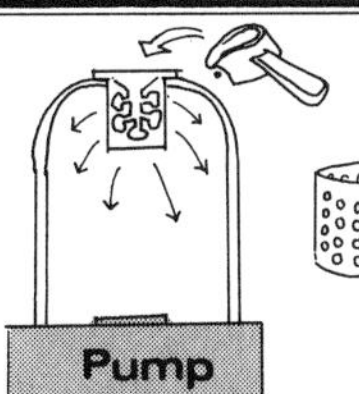

MANY COMMERCIAL CASTERS NOW USE VACUUM CASTING ASSISTED BY PERFORATED FLASKS. THESE ARE WRAPPED IN PAPER TAPE FOR INVESTING. THE TAPE BURNS OFF DURING BURNOUT. FOR THE CRAFTSPERSON A SIMILAR EFFECT MAY BE CREATED BY LEAVING A SLIGHT RECESS AROUND THE TOP OF THE FLASK. HANG BENT PIECES OF WIRE TO MAKE VENTS TO SPREAD THE VACUUM PRESSURE.

PULL OUT WIRES BEFORE BURN-OUT

VACUUM

A STRONG VACUUM PUMP — ONE THAT CAN PRODUCE A VACUUM EQUAL TO 25-29 INCHES OF MERCURY — IS NEEDED FOR INVESTING AND VACUUM ASSIST CASTING.

- AFTER BURNOUT THE HOT FLASK IS SET INTO PLACE ON A SILICONE RUBBER PAD.
- TURN ON VACUUM PUMP TO CHECK THE SEAL. IF PRESSURE GAUGE DOES NOT GO INTO THE 20'S, PRESS DOWN ON THE FLASK WITH TONGS.
- MELT METAL IN A POURING CRUCIBLE, FLUX, AND POUR INTO MOUTH OF FLASK WITH A SMOOTH EVEN FLOW. (IMAGINE THAT IT'S HONEY)
- FLASK IS COOLED AND QUENCHED AS USUAL.

Plaster Mold Method:

DRIP PLASTER OR INVESTMENT ONTO MODEL, TRYING NOT TO TRAP AIR BUBBLES. BUILD UP SEVERAL LAYERS, ENDING WITH CLOTH STRIPS.

WHEN PLASTER IS THOROUGHLY DRY, PRY MOLD PIECES APART AND REMOVE CLAY OR WAX. WASH THOROUGHLY. IF AIR BUBBLES ARE VISIBLE IN THE PLASTER, THEY CAN BE FILLED WITH FRESH PLASTER.

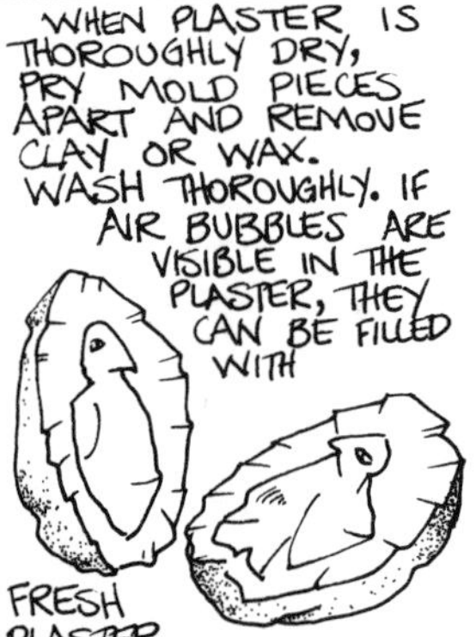

AFTER LUBRICATING MOLD WITH LIQUID SOAP OR VASELINE, PAINT MELTED WAX INTO IT. THE THICKNESS OF THE WAX WILL BE THE THICKNESS OF THE METAL WHEN YOU'RE DONE.

WHEN WAX IS HARD, PULL IT OUT, MAKE REPAIRS, SPRUE, JOIN HALVES WITH HOT NEEDLE & MODEL SEAM. INVEST AND CAST AS USUAL.

Rubber Mold Method:

THIS IS VERY SIMILAR TO THE TECHNIQUE SHOWN ABOVE. IT IS RECOMMENDED FOR A MODEL WITH INTRICATE TEXTURES OR MANY SMALL UNDERCUTS. IT WILL ALSO WORK FOR A HARD MODEL, LIKE SHELL OR METAL.

1. ATTACH MODEL TO A STAND OR HANDLE. DUST LIGHTLY WITH TALC POWDER.
2. COAT WITH AN RTV (ROOM TEMPERATURE VULCANIZING) RUBBER MOLD COMPOUND BY DIPPING THE MODEL REPEATEDLY OR BY DABBING IT ON. A SILICONE CAULKING COMPOUND MAY ALSO BE USED.
3. WHEN RUBBER HAS SET, CUT INTO IT WITH A SHARP BLADE, PULL MOLD APART AND REMOVE MODEL.
4. PAINT WAX INTO MOLD HALVES, THEN FOLLOW LAST STEP ABOVE.

USING CORES:

A CORE IS A LUMP OF MOLD MATERIAL (INVESTMENT) THAT IS ANCHORED WITHIN THE MOLD WHERE IT CREATES A CAVITY OR HOLLOW IN THE FINISHED CASTING.

① WITH COARSE FILES AND/OR KNIVES, CARVE A BLOCK OF HARDENED INVESTMENT TO THE SHAPE OF THE DESIRED INTERIOR. THIS WILL, OF COURSE, BE SMALLER THAN THE FINAL CASTING.

② COVER THIS CORE WITH WAX EITHER BY PAINTING HOT WAX OR BY DIPPING CORE INTO MELTED WAX TO BUILD UP SUCCESSIVE LAYERS.

③ FILE AND MODEL WAX AS USUAL.

④ TO ANCHOR THE CORE ONCE THE WAX HAS BURNED OUT, DRILL HOLES INTO CORE AND PUSH WIRES OF METAL TO BE CAST AS SHOWN.

IN LOST WAX CASTING IT IS OFTEN HELPFUL TO "IMPLANT" A STONE OR FINISHED PIECE OF METAL IN THE WAX MODEL. IF PROPER PRECAUTIONS ARE TAKEN THE OBJECT CAN REMAIN IN POSITION THROUGHOUT THE BURNOUT AND CASTING OPERATION.

BEZELS

IT IS DIFFICULT TO ACCURATELY SHAPE BEZELS IN WAX. A NEATER JOB IS LIKELY TO RESULT WHEN A BEZEL IS MADE OF THIN METAL STRIP. WHEN A SUBTLE TRANSITION BETWEEN THE BEZEL AND THE WORK IS DESIRED THE METAL BEZEL MAY BE SET INTO THE WAX. AS SHOWN, WAX CAN THEN BE MODELED AROUND THE BEZEL. SINCE BURNOUT SHOULD NOT GO ABOVE 1250 °F AND ALL THE METALS COMMONLY USED HAVE A MELTING POINT ABOVE THAT, THERE SHOULD BE NO PROBLEM WITH MELTING. SOMETIMES THE MOLTEN METAL WILL FUSE ONTO THE BEZEL BUT USUALLY THE OXIDES ACCUMULATED IN THE FLASK DURING BURNOUT WILL PREVENT A STRONG BOND. IT IS A GOOD PRACTICE TO SOLDER THE BEZEL IN PLACE IMMEDIATELY AFTER PICKLING THE CASTING, BEFORE THE PERFECT FIT HAS A CHANCE TO GET DISTORTED.

PROVIDING A GRIP

WHEN A MODEL IS MADE AND PIECES ARE SET INTO IT, THEY ARE OF COURSE HELD IN PLACE BY WAX. KEEP IN MIND THAT DURING BURNOUT THE WAX WILL BE REMOVED. IF PRECAUTIONS ARE NOT TAKEN THE SMALL PIECES MAY BECOME LOOSE AND DROP INTO THE MOLD CAVITY, RUINING THE CASTING. INVESTMENT HAS AN ADHESIVE QUALITY AND WILL PROBABLY GRIP SMALL PIECES AS LONG AS SUFFICIENT SURFACE AREA IS AVAILABLE. A SLIGHTLY ROUGH SURFACE WILL HOLD BETTER THAN A SMOOTH ONE. EACH SITUATION WILL REQUIRE ITS OWN SOLUTION: IN SOME CASES A DESIGN MAY BE MODIFIED TO PROVIDE A "FINGER" OF INVESTMENT TO GRIP THE IMPLANTED PIECE ON EACH SIDE AND HOLD IT IN PLACE. SOMETIMES A SMALL EXTENSION MAY BE SOLDERED TO A SMALL PIECE TO LOCK IT IN POSITION. THIS CAN LATER BE SAWN OFF THE FINISHED CASTING.

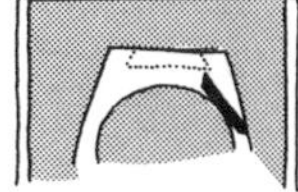

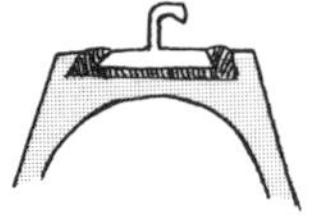

Double Metal

METALS OF CONTRASTING COLOR MAY ALSO BE CAST TOGETHER AS DESCRIBED ABOVE. IF A SMALL PIECE IS TO BE INLAID IT MAY BE SAWN OUT AND WORKED AS DESIRED. A BEVELED EDGE MUST BE FILED ON THE EDGE OF THESE PIECES TO LOCK THEM IN PLACE. SOFT WAX IS THEN DRIPPED OR PUSHED LIKE PUTTY OVER THE PIECE. GRIPPING AS DESCRIBED ABOVE, MUST BE PROVIDED.

GRIPS

WHEN A LARGE PIECE OF METAL IS BEING USED ITS WEIGHT CAN CAUSE THE SPRUE TO BEND OVER. TO PREVENT THIS A SECONDARY SUPPORT IS USED. SINCE THIS CONNECTS OUTSIDE THE FUNNEL AREA OF THE FLASK NO METAL WILL ENTER THIS CAVITY EVEN THOUGH ITS WAX HAS BURNED OUT.

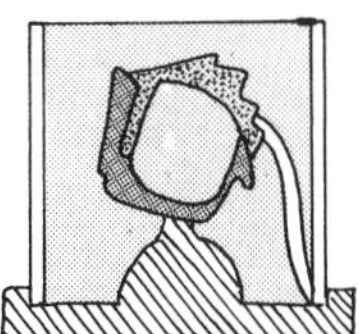

STONES

SOME STONES WILL WITHSTAND BURNOUT TEMPERATURES. AND THE IMPACT OF INRUSHING MOLTEN METAL. SUCH STONES MAY BE "SET" IN THE WAX AND CAST IN PLACE. THIS HAS ITS USES, ESPECIALLY FOR IRREGULAR STONES AND CRYSTALS BUT SINCE WAX CAN RARELY BE CONTROLLED AS ACCURATELY AS METAL, THE RESULT CAN BE LACKING IN GRACE AND SUBTLTY. THE METHODS USED ARE AS ABOVE AND IN THE PRECEDING PAGES EXCEPT THAT AFTER CASTING THE FLASK SHOULD BE AIR-COOLED RATHER THAN QUENCHED.

GEM MATERIALS THAT ARE LIKELY TO WITHSTAND THIS METHOD ARE:

- DIAMOND
- SAPPHIRE
- RUBY
- TOURMALINE
- TRANSPARENT SYNTHETICS

REMEMBER THAT IMPURITIES MAY CAUSE A TOUGH STONE TO CRACK.

SINCE CONVENTIONAL CASTING OF PRECIOUS METALS REQUIRES THAT THE INVESTMENT MOLD BE DESTROYED TO RETRIEVE THE FINISHED CASTING, A SUPPLEMENTARY STEP IS NEEDED TO PRODUCE WAX MODELS. THESE CAN BE MADE BY INJECTING MOLTEN WAX INTO A RUBBER MOLD THAT WILL FLEX SUFFICIENTLY TO ALLOW THE MODEL TO BE REMOVED WITHOUT DAMAGING THE MOLD. RUBBER MOLDS FALL INTO TWO CATEGORIES:

Vulcanized Molds

SLABS OF RAW UNCURED RUBBER ARE LAID INTO A STURDY RECTANGULAR MOLD FRAME MADE OF STEEL OR ALUMINUM. WHEN ROUGHLY HALF THE DEPTH IS FILLED THE OBJECT TO BE REPRODUCED IS LAID INTO PLACE. BECAUSE OF THE TEMPERATURES AND PRESSURES INVOLVED THIS MODEL MUST BE OF A HARD MATERIAL. METAL IS USUALLY USED BUT WOOD OR A HARD PLASTIC LIKE NYLON OR DELRIN WILL WORK. THE MOLD FRAME IS THEN PACKED WITH MORE RUBBER AND SET INTO A VULCANIZER. THIS MACHINE WILL MAINTAIN THE FIRM PRESSURE AND TEMPERATURE (310°F, 165°C) NEEDED TO CURE THE RUBBER. CURING TAKES ABOUT 15 MINUTES PER 1/4" OF MOLD, OR ABOUT AN HOUR AND A HALF FOR AN AVERAGE PIECE. WHEN VULCANIZED THE MOLD MATERIAL WILL BE SIMILAR TO RUBBER BAND RUBBER. THE MOLD IS COOLED AND CUT AS SHOWN BELOW. FOR FURTHER INFORMATION ON RUBBER MOLDS, CONSULT THE MANUFACTURER'S LITERATURE FOR YOUR EQUIPMENT.

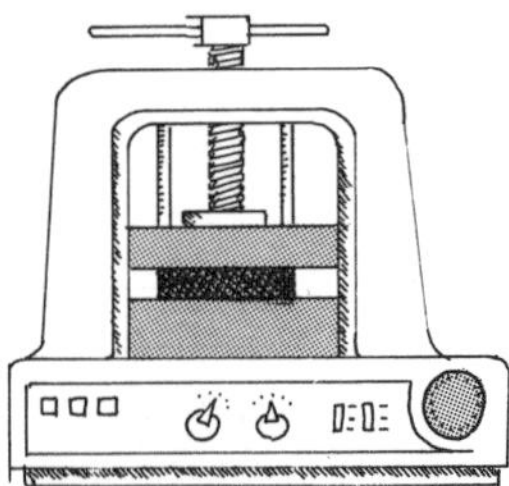

Room Temperature Vulcanizing (RTV)

THIS RELATIVELY NEW DEVELOPMENT IN MOLDMAKING USES A TWO-PART COMPOUND THAT CURES ("VULCANIZES") CHEMICALLY, WITHOUT SPECIAL EQUIPMENT. THESE MOLDS ARE NOT AS DURABLE AS THOSE DESCRIBED ABOVE AND CAN LOSE DETAIL BECAUSE OF BUBBLES, BUT BECAUSE OF THEIR SIMPLICITY AND RELATIVELY LOW COST THEY OFFER ADVANTAGES TO THE LOW-BUDGET METALWORKER.

MOLD FRAMES CAN BE BOUGHT OR MADE, USING A STRIP OF ALUMINUM AND GLASS OR PLASTIC SHEET. A SUITABLE ALUMINUM MOLDING IS AVAILABLE AT MOST HARDWARE STORES. TO AVOID WASTING THE MOLD COMPOUND IT IS A GOOD IDEA TO HAVE SEVERAL SIZE FRAMES ON HAND. THE SHEETS ARE HELD IN PLACE WITH RUBBER BANDS AFTER THE MODEL IS SET INTO POSITION ON A SPRUE. MODEL AND SPRUE MAY BE OF ANY NON POROUS (OR SEALED) MATERIAL. MIX THE COMPOUND CAREFULLY ACCORDING TO THE PROPORTIONS GIVEN ON THE CAN. ALLOW TO CURE (USUALLY 24 HOURS), AND CUT.

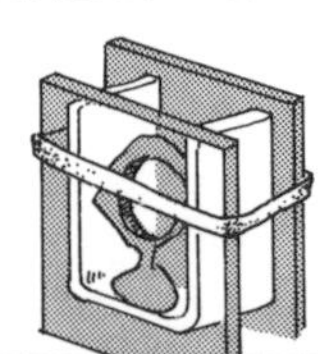

MOLD CUTTING

FOR EITHER OF THE ABOVE METHODS IT IS NECESSARY TO CUT THE MOLD SO THE ORIGINAL MODEL AND SUBSEQUENT WAX IMPRESSIONS CAN BE REMOVED. THIS IS A DELICATE AND DEMANDING SKILL, AND ONLY EXPERIENCE CAN REALLY TEACH IT. THE PARTING LINE, WHERE THE MOLD PIECES MEET, MIGHT BE VISIBLE ON THE FINISHED CASTING SO IT SHOULD BE LOCATED EITHER WHERE IT CAN BE EASILY REMOVED OR WHERE IT WON'T DAMAGE THE DESIGN. HAVING A CLEAR IDEA OF THE MODEL AND THE DESIRED LOCATION OF THE PARTING LINE WILL MAKE CUTTING EASIER. USE A SHARP BLADE AND WORK WITH SMALL SLICES WHILE PULLING THE MOLD APART: A BENT FORK OR CAN-OPENER IS A CONVENIENT THIRD HAND. TO REGISTER THE MOLD SECTIONS A ZIG-ZAG PATTERN OR A COUPLE OF RAISED BUTTONS ARE CUT NEAR THE OUTSIDE EDGE. SOME SHAPES REQUIRE COMPLEX MOLD DIVISIONS: FOR MORE INFORMATION ON THIS TOPIC I RECOMMEND:

<u>CENTRIFUGAL OR LOST WAX CASTING</u> BY MURRAY BOVIN.

STONES

LAPIDARY, THE ART OF WORKING WITH GEMSTONES, IS A COMPLEX FIELD OF STUDY ALL BY ITSELF AND FEW METALSMITHS CAN GIVE AS MUCH TIME AS THEY WOULD LIKE TO IT. THERE IS, THOUGH, A NEED FOR INFORMATION SINCE MOST WORKERS IN FINE METALS USE GEMS SOONER OR LATER. THE FOLLOWING PAGES MAKE AN ATTEMPT TO PROVIDE SOME 'WORKING KNOWLEDGE' FOR THOSE WHO DEAL WITH STONES AS A SECONDARY ASPECT OF THEIR CRAFT. IT DOES NOT PRETEND TO BE COMPLETE OR ACADEMIC: I HOPE ONLY THAT IT WILL LAY A FOUNDATION FOR FURTHER INVESTIGATION.

THERE ARE OVER 2000 MINERALS IN THE EARTH'S CRUST. TRYING TO ORGANIZE THIS MATERIAL HAS PROVEN DIFFICULT. COLOR AND HARDNESS, FOR INSTANCE, DON'T WORK SINCE A STONE MAY OCCUR IN SEVERAL SHADES AND KINDS OF CRYSTALS. CHEMICAL AND MINERALOGICAL DIVISIONS SIMILARLY CONFUSE RATHER THAN SIMPLIFY THE MATTER. I TURNED FINALLY TO THE ALPHABET: WHAT FOLLOWS IS AN ALPHABETICAL LIST OF FIFTY POPULAR STONES WITH SOME INFORMATION, HISTORY, OR TIPS FOR EACH ONE.

WHERE POSSIBLE, I HAVE INCLUDED FOLKLORE FOR EACH GEM. I DON'T MEAN TO IMPLY A PROVEN RELATIONSHIP BETWEEN THE STONE AND MAGICAL QUALITIES ASCRIBED TO IT. BUT THEN, I DON'T WANT TO DISCOUNT THE POSSIBILITY EITHER.

birthstones

TODAY THE COMMERCIAL JEWELRY INDUSTRY HAS EFFECTIVELY BLUNTED ANY CHARM OR SERIOUSNESS CONCERNING THE RELATIONSHIP BETWEEN EARTH MATERIALS AND THE SEASON OF ONE'S BIRTH. THERE WAS A TIME WHEN SUCH RELATIONSHIPS, OFTEN INVOLVING ASTROLOGICAL MOVEMENTS, PLAYED AN IMPORTANT PART IN DAILY LIFE.

THE LIST BELOW IS BORROWED FROM GEORGE FREDERICK KUNZ WHOSE BOOK THE CURIOUS LORE OF PRECIOUS STONES (DOVER 1971) IS RECOMMENDED FOR FURTHER INVESTIGATION.

JANUARY — GARNET, HYACINTH
FEBRUARY — AMETHYST, HYACINTH, PEARL
MARCH — BLOODSTONE, JASPER
APRIL — DIAMOND, SAPPHIRE
MAY — AGATE, EMERALD, CHALCEDONY, CARNELIAN
JUNE — EMERALD, AGATE, PEARL, CHALCEDONY, TURQUOISE
JULY — RUBY, CARNELIAN, ONYX, SARDONYX, TURQUOISE
AUGUST — CARNELIAN, MOONSTONE, TOPAZ, ALEXANDRITE, SARDONYX
SEPTEMBER — SAPPHIRE, LAPIZ LAZULI, CORAL
OCTOBER — OPAL, AQUAMARINE, BERYL
NOVEMBER — TOPAZ, PEARL
DECEMBER — TURQUOISE, RUBY, BLOODSTONE

Wedding Anniversary Tokens

1 ROSE BERYL - PAPER
2 CRYSTAL - COTTON
3 CHRYSOPRASE - LEATHER
4 MOONSTONE - SILK
5 CARNELIAN - WOOD
6 PERIDOT - SUGAR
7 CORAL - WOOL
8 OPAL - CLAY
9 CITRINE - WILLOW
10 TURQUOISE - TIN
11 GARNET
12 AMETHYST - LINEN
13 AGATE
14 IVORY - LACE
15 TOPAZ
25 SILVER
30 PEARL
35 JADE
40 RUBY
45 SAPPHIRE
50 GOLDEN
55 EMERALD
60 DIAMOND

THE ULTIMATE VALUE OF LIFE DEPENDS UPON AWARENESS, AND THE POWER OF CONTEMPLATION RATHER THAN UPON MERE SURVIVAL.

-ARISTOTLE (384-322 BC)

STONES

KNOWLEDGE OF STONES COMES ONLY AFTER SOPHISTICATED TRAINING AND YEARS OF EXPERIENCE. THE INFORMATION GIVEN HERE IS INTENDED ONLY TO OPEN THE DOOR TO A COMPLEX AND FASCINATING FIELD.

FROM THE METALSMITH'S POINT OF VIEW, GEM MATERIALS MAY BE CONSIDERED IN TERMS OF:

- COLOR: IN MANY CASES, LIKE AGATES, COLOR IS ENTIRELY A MATTER OF TASTE. IN OTHERS, LIKE EMERALD, A DEEP COLOR IS A MAJOR FACTOR IN VALUE.
- CUT: THE PLANES OR CURVES SHOULD BE SYMMETRICAL, WELL POLISHED, AND ARRANGED TO COMPLEMENT THE MATERIAL.
- HARDNESS: A GEM THAT WILL NOT RETAIN ITS POLISH IS OF LIMITED VALUE. IN SETTING, IT IS IMPORTANT TO KNOW THE HARDNESS OF THE MATERIAL BEING USED. SOFT STONES SHOULD BE SET IN A WAY THAT WILL PROTECT THEM.
- SPECIAL LIGHT PHENOMENA: CAT'S EYE, AND IRIDESCENCE ARE EXAMPLES OF THIS.
- LUSTER: BRIGHTNESS OF THE SHINE. SOME STONES HAVE A LESSER VALUE BECAUSE THEY WILL NOT POLISH.
- PURITY (INCLUSIONS): SOME STONES ARE VALUED FOR THEIR INCLUSIONS, LIKE RUTILATED QUARTZ OR MOSS AGATE. IN OTHER STONES, LIKE AMETHYST, INCLUSIONS LOWER THE VALUE.

Moh's Scale:

1 TALC
2 GYPSUM
3 CALCITE
4 FLUORITE
5 APATITE
6 ORTHOCLASE
7 QUARTZ
8 TOPAZ
9 CORUNDUM
10 DIAMOND

EACH MATERIAL WILL SCRATCH THOSE WITH A SMALLER NUMBER AND BE SCRATCHED BY THOSE WITH A HIGHER NUMBER.

THE STEPS ALONG THE SCALE ARE NOT REGULAR. #2 AND #3, FOR EXAMPLE, ARE CLOSE IN HARDNESS WHILE DIAMOND IS 80 TIMES HARDER THAN #9, CORUNDUM.

BRILLIANT CUT

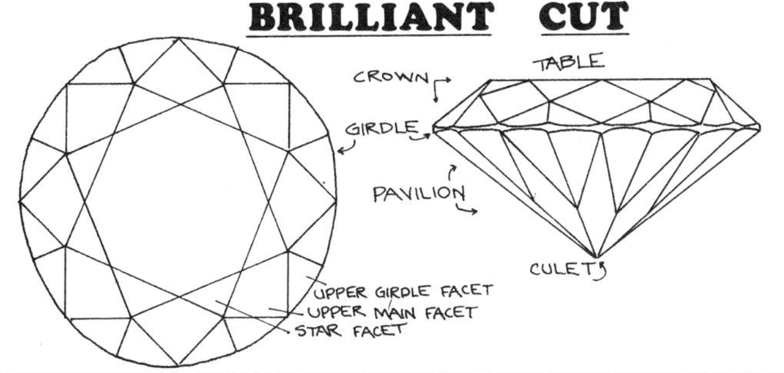

MISCELLANEOUS HARDNESSES:

2.5 FINGERNAIL
FINE GOLD
FINE SILVER
LEAD
3 COPPER
4 STERLING
5.5 WINDOW GLASS
6 KNIFE BLADE
6.5 FILE
9 SILICON CARBIDE

STONES

STONES ARE CUT BY FIRST SLICING A SLAB FROM THE ROUGH LUMP WITH A DIAMOND SAW. THE GENERAL SHAPE IS MADE BY CUTTING OFF CORNERS AND ABRASIVE WHEELS ARE USED TO CREATE THE DESIRED SHAPE. WHEELS OF PROGRESSIVELY FINER GRIT ARE USED, ENDING IN A BUFFING OPERATION.

TO EVALUATE THE QUALITY OF A CUTTING JOB, LOOK FOR REGULAR SYMMETRY ON A FACETED STONE AND A SMOOTH AND EVEN CURVATURE ON A CAB. HERE ARE EXAMPLES OF POOR CUTTING:

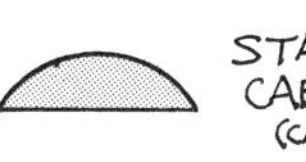
STANDARD CABACHON (CAB)

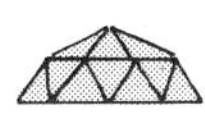
ROSE CUT

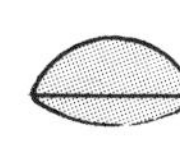
DOUBLE CAB (LENTIL)

MARQUISE

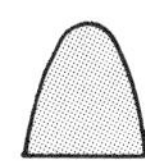
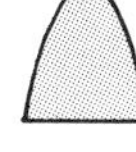
HIGH CAB (BULLET)

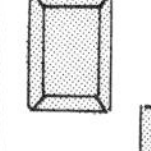
BAGUETTE

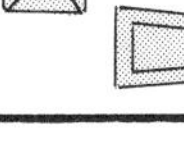
TAPERED BAGUETTE

BUFF TOP

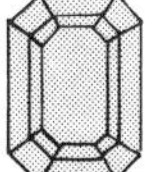
OCTAGON (EMERALD CUT)

Summary Chart **STONES**

FIFTY POPULAR GEM MATERIALS ARE BRIEFLY DESCRIBED IN THE FOLLOWING PAGES. THEY ARE SUMMARIZED BELOW WITH PARTICULAR THOUGHT TO THEIR USE IN JEWELRY. OF COURSE THESE MATERIALS EXHIBIT WIDE DIFFERENCES FROM ONE SPECIMEN TO ANOTHER SO IT IS IMPOSSIBLE TO BE PRECISE. IT IS HOPED THAT THIS INFORMATION WILL OFFER A STARTING POINT FOR PERSONAL EXPLORATION.

NAME	COLORS	COMMON CUTS	HARDNESS	OPAQUE/TRANSLUSCENT	HEAT SENSITIVITY	
AGATE	MANY	◠	7	O	YES	OFTEN BANDED
ALEXANDRITE	PURPLE-BLUE	▽	8	T		CHANGES COLOR
AMBER	YELLOW	◠	2	BOTH	VERY	ORGANIC
AQUAMARINE	LIGHT BLUE	◠ ▽	8	T	VERY	
AMETHYST	PURPLE	◠ ▽	7	T	YES	
AVENTURINE	GREEN-BROWN	◠	7	O	YES	SPARKLES
BERYL	MANY	▽	8	T	VERY	POSSIBLE CAT'S EYE
CARNELIAN	RED	◠	7	O	YES	
CHALCEDONY	BLUE	◠	7	O	YES	
CHIASTOLITE	PATTERNED			O		
CHRYSOBERYL	MANY	◠ ▽	8	BOTH		CAT'S EYE
CHRYSOCOLLA	BLUE-GREEN	◠	6	O	YES	ALSO CALLED ELAT
CHRYSOPRASE	LIGHT GREEN	◠	6	O	YES	
CITRINE	YELLOW	◠ ▽	7	T	YES	
CORAL	RED, PINK, BLACK	◠	3	O	VERY	ORGANIC
CORUNDUM	MANY	◠ ▽	9	T		
DIAMOND	CLEAR	▽	10	T		
EMERALD	GREEN	◠ ▽	8	BOTH	VERY	CLEAVES EASILY
GARNET	MANY	◠ ▽	7	T	YES	
HEMATITE	BLACK	◠	6	O		
IVORY	WHITE		2	O	VERY	ORGANIC, HAS GRAIN
JADE	MANY	◠	6	O	YES	GREASY LUSTER
JASPER	RED-GREEN	◠	7	O		
JET	BLACK	◠	4	O	VERY	ORGANIC
LABRADORITE	BLUE-BLACK	◠	6	O	VERY	IRIDESCENT
LAPIS LAZULI	DEEP BLUE	◠	6	O	YES	
MALACHITE	GREEN	◠	5	O	VERY	
MAGNETITE	DARK GRAY	◠	7	O		MAGNETIC
MOONSTONE	MANY	◠	6	T	YES	ADULARESCENT
ONYX	MANY	◠	7	O	YES	
OPAL	MIXED	◠	6	T	VERY	INTERIOR PLAY OF COLORS
PEARL	MANY		3	O	VERY	
PERIDOT	GREEN	◠ ▽	7	T	YES	
QUARTZ	MANY	◠ ▽	7	BOTH		
ROCK CRYSTAL	CLEAR	◠ ▽	7	T	YES	
RUBY	RED	◠ ▽	9	BOTH	YES	POSSIBLE STAR OR CAT'S EYE
RUTILE	RED	◠ ▽	7	BOTH		COMMON AS INCLUSION
SAPPHIRE	MANY	◠ ▽	9	BOTH		POSSIBLE STAR OR CAT'S EYE
SARDONYX	BROWN	◠	7	O	YES	
SERPENTINE	GREEN	◠	2-6	O		DUST CONTAINS ASBESTOS
SODALITE	BLUE	◠	6	O		
SPINEL	MANY	◠ ▽	8	T		
TIGER'S EYE	BLUE-BROWN	◠	7	O	YES	SILKY INTERIOR
TOPAZ	YELLOW	◠ ▽	8	T	YES	CLEAVES EASILY
TOURMALINE	GREEN, PINK	◠ ▽	7	T	YES	DICHROIC
TURQUOISE	BLUE	◠	6	O	YES	
ZIRCON	MANY	◠ ▽	7	T		CLEAVES EASILY
ZIRCONIA (CUBIC)	MANY	▽	9	T		RECENT SYNTHETIC

Agate

H7

- A TYPE OF CHALCEDONY; A CRYPTO-CRYSTALLINE QUARTZ.
- THE NAME COMES FROM AN ANCIENT, NOW UNTRACEABLE SICILIAN RIVER, ACHATES.

RED – PROTECTION FROM SPIDERS AND SCORPIONS.
GREEN – RELIEF FROM EYE TROUBLE.
GREEN W/ STRIPES – A WOMAN WHO DRINKS WATER IN WHICH SUCH A RING HAS BEEN WASHED WILL NEVER BE STERILE.
GREY – WORN ON THE NECK TO PREVENT A STIFF NECK.
MOSS AGATE – ALSO CALLED DENDRITIC (GREEK, "DENDRON", TREE)
- WORN BY A FARMER ON THE UPPER ARM TO INSURE A GOOD HARVEST.
- PLACED ON RIGHT HORN OF OXEN TO PROTECT THEM.

Alexandrite

H8½

- THIS NATURAL STONE IS A TYPE OF CHRYSOBERYL THAT SHOWS A RANGE OF TRANSPARENT COLORS FROM BLUE IN DAYLIGHT TO REDDISH-YELLOW IN ARTIFICIAL LIGHT.
- MORE WIDELY AVAILABLE IS A SYNTHETIC STONE, ACTUALLY A TREATED CORUNDUM, H9.
- THE STONE WAS NAMED FOR CZAR ALEXANDER II WHO, ACCORDING TO LEGEND, CAME OF LEGAL AGE ON THE DAY THE STONE WAS DISCOVERED.

Amber

H2-2½

- THIS IS NOT A STONE BUT THE NATURALLY HARDENED RESIN OF THE "PINUS SUCCINIFERA", THE AMBER PINE.
- TRANSPARENT AMBER IS 120-180 MILLION YEARS OLD. OPAQUE AMBER, CALLED COPAL, IS 60 MILLION YEARS OLD.
- THE NAME COMES FROM THE ARABIC "ANBAR". THE GREEKS CALLED IT "ELEKTRUM" FROM THE PHOENICIAN WORD FOR SUN/GOLDEN. BECAUSE AMBER WILL HOLD A CHARGE THIS GAVE US OUR WORD ELECTRIC.
- TO TEST A SAMPLE, BRUSH IT (OR SOAK IN) METHYL ALCOHOL OR ETHYLACETATE. NONFOSSIL RESINS – LIKE PLASTIC – WILL DISSOLVE.
- ANOTHER TEST IS TO SET INTO BRINE: REAL AMBER WILL FLOAT BUT ARTIFICIAL WILL SINK.
- IT CAN ALSO BE TESTED BY TOUCHING WITH A HOT NEEDLE. THE SMOKE THUS CAUSED WILL EITHER SMELL LIKE A PINE WOODS OR A PLASTICS FACTORY.
- SOME AMBER CONTAINS THOUSANDS OF TINY AIR BUBBLES. THIS IS CALLED BONE AMBER AND CAN BE CLEARED BY HEATING IN OIL.

MAGICAL USES:
- AMBER DUST MIXED WITH HONEY OR WATER WAS USED TO TREAT EARS, EYES, STOMACH, LIVER, AND KIDNEYS.
- THE SMELL OF BURNING AMBER HELPED A WOMAN IN LABOR.
- HOLDING AN AMBER BALL WILL KEEP ONE COOL ON A HOT DAY; IT WAS USED TO TREAT FEVER VICTIMS.
- AMBER BEADS PRESERVE THE WEARER AGAINST RHEUMATISM, TOOTHACHE, HEADACHE, RICKETS, JAUNDICE, AND GOITERS.

CAUTION: AMBER WILL DISSOLVE IN SOLVENTS LIKE ACETONE OR NAIL POLISH REMOVER. IT WILL BE WORN DOWN BY MECHANICAL BUFFING WITH COMPOUNDS LIKE TRIPOLI.

Aquamarine

H7½-8

- NAME COMES FROM THE LATIN, "BERYLLUS AQUAMARINUS", BERYL RESEMBLING SEAWATER.
- IT IS TRADITIONALLY A SAILORS' TALISMAN.
- THIS GEM INCREASED IN POPULARITY AROUND 1920 WHEN HEAT TREATMENT WAS DEVELOPED TO TURN PALE BLUE-GREEN STONES INTO DEEPER BLUE SHADES.

Amethyst

H7

- A FORM OF QUARTZ; TO BE TOP QUALITY THE COLOR SHOULD BE A DEEP PURPLE AND THERE SHOULD BE NO FLAWS LIKE INCLUSIONS, FEATHERS, ETC.
- FROM GREEK FOR "NOT DRUNKEN, WITHOUT DRUNKENNESS" THE GEM WAS BELIEVED TO PROTECT FROM THE EFFECTS OF WINE, ESPECIALLY IF HELD UNDER THE TONGUE WHILE DRINKING.
- WHEN HEATED TO 550-560 °C AMETHYSTS TURN DARK YELLOW OR REDDISH-BROWN AND ARE CALLED CITRINES. SINCE THEY ARE MORE RICHLY COLORED THAN NATURAL CITRINES THEY ARE MORE EXPENSIVE.
- THE COLOR CAN FADE IF THE STONE IS LEFT IN STRONG SUNLIGHT FOR A LONG TIME.
- PLACED UNDER THE PILLOW, AMETHYSTS INSURE PLEASANT DREAMS, IMPROVE MEMORY, AND MAKE ONE IMMUNE FROM POISON.
- SOME PEOPLE BELIEVE THAT A WEARER OF THIS STONE WILL BECOME GENTLE AND AMIABLE.

Aventurine

H7

- A FINE-GRAINED QUARTZ WITH MANY FLAKE INCLUSIONS, OCCURRING IN MANY COLORS, MAINLY GREEN, BROWN, AND GRAY.
- THE UBIQUITOUS SPARKLE OF THIS STONE IS CALLED AVENTURESCENCE.

Beryl

H7½-8

- THIS STONE OCCURS IN A WIDE COLOR RANGE: PINK, ORANGE, YELLOW, GREEN, BLUE-GREEN, AND BLUE.
- IT IS ALSO FOUND COLORLESS (CALLED GOSHENITE). THIS RESEMBLES ROCK CRYSTAL AND WAS USED FOR LENSES AND CRYSTAL BALLS.
- PINK BERYL IS KNOWN AS MORGANITE, AFTER THE BANKER AND GEM COLLECTOR J.P. MORGAN.
- BERYLS CAN SHOW SOME CHATOYANCY WHEN CUT AS CABACHONS.
- THIS STONE IS SAID TO PROTECT THE WEARER FROM HELPLESSNESS CAUSED BY FASCINATION. IT WAS ALSO USED TO TREAT DISEASES OF THE EYE (GREEN) AND JAUNDICE AND LIVER DISEASE (YELLOW STONES).

Carnelian

H6½-7

- A RED CHALCEDONY, ITS COLOR BEING DUE TO THE PRESENCE OF IRON.
- THE OPAQUE VARIETY IS CALLED SARD. WHEN IT OCCURS IN BROWN AND WHITE LAYERS IT IS CALLED SARDONYX.
- CARNELIAN WAS SAID TO STOP NOSEBLEEDS AND TO PREVENT BLOOD FROM RISING TO THE HEAD.
- IT IS A STRONG PROTECTION FROM THE EVIL EYE.

Chalcedony

H6½-7

- MINERALOGICALLY THE TERM REFERS TO A LARGE FAMILY OF CRYPTO-CRYSTALLINE QUARTZ; THAT IS, QUARTZ WITH VERY TINY CRYSTALS. CARNELIAN, ONYX, AGATES, AND CHRYSOPRASE ARE ALL KINDS OF CHALCEDONY.
- PRONOUNCED KAL·SED'·NE
- IN THE WORLD OF JEWELRY THE WORD REFERS TO A SOLID COLOR, TRANSLUCENT, LIGHT BLUE STONE. THESE MAY BE MADE BY DYING AGATES BUT THE NATURALLY OCCURRING VARIETY IS MORE DESIRABLE.

Chiastolite

- THIS IS AN OPAQUE FORM OF ANDALUSITE THAT GROWS IN CIGAR-SHAPED CRYSTALS. WHEN SLICED THESE SHOW A LIGHT COLORED CROSS OR SHAMROCK AGAINST A DARK GREEN BACKGROUND
- FOUND NEAR THE SHRINE OF ST. JAMES IN SANTIAGO DE COMPOSTELLA IN SPAIN WHERE RELIGIOUS POWER IS OFTEN ATTRIBUTED.

Chrysoberyl

H8½

THIS STONE OCCURS IN BOTH A TRANSPARENT AND A CLOUDY VARIETY AND CAN BE YELLOW, GREEN, OR BROWN. CLEAR STONES ARE USUALLY FACETED WHILE THE CLOUDY ARE CUT AS CABACHONS.

- CHRYSOBERYL HAS ONE OF THE MOST ATTRACTIVE CAT'S EYES OF ALL STONES. THIS OCCURS AS A BRIGHT SILVERY LINE THAT MOVES ACROSS THE CURVED SURFACE OF A POLISHED GEM AS IT IS MOVED. THE EFFECT IS CALLED "CHATOYANCY" FROM THE FRENCH "CHAT", CAT.

- PRONOUNCED: KRIS'·ə·BER·əL

Chrysocolla

H5-6

- A HYDROUS SILICATE FORMED BY THE DECOMPOSITION OF COPPER ORE NEAR THE SURFACE.
- FROM GREEK, "CHRYSOS", GOLD AND "KOLLA", GLUE. IN ANCIENT USAGE THE TERM INCLUDED MALACHITE. BOTH WERE USED AS A FLUX FOR SOLDERING (FUSING) GOLD.
- OCCURS IN VARIABLE SHADES OF BLUE AND GREEN AND CAN RESEMBLE TURQUOISE.
- CHRYSOCOLLA FROM THE SITE OF KING SOLOMON'S MINES IN EILAT, ISRAEL IS CALLED EILAT (ELAT) STONE.
- SINCE THIS IS A COPPER-BEARING ORE IT WILL BE DAMAGED BY PICKLES, LIKE SPAREX, DESIGNED TO ATTACK COPPER OXIDES.

Chrysoprase

H6½-7

- A LIGHT GREEN TRANSLUCENT CHALCEDONY, THE MOST VALUABLE OF THE CHALCEDONY FAMILY.
- FROM GREEK WORDS FOR "GOLD" AND "LEEK", REFERRING TO ITS GOLDEN-GREEN COLOR, WHICH IS CAUSED BY NICKEL SALTS.

Citrine

H7

- THIS YELLOW QUARTZ CAN BE FOUND NATURALLY OR MAY BE MADE BY HEATING AMETHYST (PURPLE QUARTZ) TO AROUND 550°C (1050°F). TREATED CITRINES HAVE A DEEPER COLOR AND ARE MORE EXPENSIVE THAN THE NATURAL.
- YELLOW-BROWN VARIETY IS CALLED CAIRNGORM AFTER THEIR PLACE OF ORIGIN IN SCOTLAND.
- DARK REDDISH-BROWN QUARTZ IS CALLED "SANG DE BOEUF," FRENCH FOR OX BLOOD.

Coral

H3½

- THIS IS NOT A STONE IN THE USUAL SENSE, BUT A ROCK-LIKE MATERIAL FORMED FROM THE UNDERWATER DEPOSIT OF MANY TINY SKELETONS OF INVERTEBRATE ANIMALS.
- FROM GREEK 'KORALLION' ORIGINALLY DERIVED FROM THE WORD FOR PEBBLE.
- CORAL CAN OCCUR IN MANY SHADES OF REDDISH-PINK, WHITE, AND BLACK, WHICH IS CALLED AKABAR.
- CORAL WAS THOUGHT TO STOP BLEEDING, GUARD AGAINST POISON, AND PROTECT DOGS FROM RABIES.
- THIS IS A SOFT STONE AND SHOULD BE TREATED GENTLY. IT WILL NOT TOLERATE HARSH CLEANSERS, ABRASION, OR HEAT.

Corundum

H9

- UNTIL THE MIDDLE AGES CORUNDUM WAS CALLED HYACINTH AND THOUGHT TO EXIST ONLY AS A BLUE STONE. WHEN IT WAS DISCOVERED THAT OTHER COLORS OF CORUNDUM EXISTED, THE NAME SAPPHIRE WAS USED FOR THE BLUE VARIETY.
- CORUNDUMS OF OTHER COLORS ARE USUALLY IDENTIFIED BY A COLOR NAME, AS; YELLOW SAPPHIRE, GREEN SAPPHIRE, ETC. EXCEPT FOR RED CORUNDUM WHICH IS CALLED RUBY.
- CORUNDUM OCCURS AS: YELLOW, GREEN, REDDISH-YELLOW, PINK, MAUVE, BROWN, AND BLACK.

Diamond

H10

- FROM GREEK "ADAMAS", UNBREAKABLE, INDOMITABLE.
- DIAMONDS WERE BELIEVED TO RENDER ALL POISONS HARMLESS.
- DIAMONDS WILL DRIVE AWAY MADNESS, NIGHT SPIRITS, AND EVIL DREAMS.

Emerald

H7½-8

- A BRIGHT GREEN BERYL, VERY VALUABLE IF FREE OF INCLUSIONS, AND OF STRONG COLOR.
- INCLUSIONS ARE CALLED THE "JARDIN" (FRENCH, GARDEN) OF THE STONE.
- EMERALDS ARE NOTORIOUSLY BRITTLE AND REQUIRE GREAT CARE IN SETTING. FOR THIS REASON FACETED STONES HAVING A THICK GIRDLE ARE PREFERRED.
- DO NOT CLEAN EMERALDS IN AN ULTRASONIC MACHINE. THE SOLUTION MAY PENETRATE THE STONE AND CAUSE IT TO SHATTER.

• MAGICAL ASSOCIATIONS:
- LINKED TO FERTILITY AND THE EARTH GODDESS; IT IS A BIRTHSTONE OF SPRING.
- SACRED TO THE GODDESS VENUS, WORN BY WOMEN TO EASE CHILDBIRTH.
- SAID TO STIFLE AN EPILEPTIC FIT.
- THE SIGHT OF AN EMERALD BROUGHT SUCH TERROR TO A VIPER OR COBRA THAT THEIR EYES LEAPED OUT OF THEIR HEAD.

Garnet

H6½-7½

- FROM LATIN "GRANUM", GRAIN OR PIP, WHICH IN TURN CAME FROM THE PHOENICIAN WORD FOR POMEGRANATE, "PUNICA GRANATUM."

TYPES OF GARNETS:

1. PYROPE - A DEEP RED COLOR. ITS NAME IN GREEK MEANS "FIERY EYE."
2. ALMANDINE - DARK RED WITH A TINGE OF MAUVE.
   - THE ESPECIALLY PURPLE VARIETY IS CALLED RHODOLITE.
3. SPESSARTITE - RED-ORANGE OR ORANGE-BROWN
   - SHOWS INTERNAL WAVY VEIL OF FLUID CONTAINED IN THE STONE.
   - RARE AND EXPENSIVE
4. GROSSULAR (GROSSULARITE) - A SPECKLED GREEN STONE RESEMBLING JADE.
   - HESSONITE IS A SUB-SPECIES.
5. UVAROVITE - RARE, INTENSELY GREEN STONE.
6. ANDRADITE - THIS CONTAINS IRON.
   - IT IS RARELY CUT.

MAGICAL PROPERTIES:

- WHEN ON THE BODY GARNETS ARE SAID TO PREVENT SKIN DISEASES.
- GARNET ASSURES THE WEARER OF LOVE, FAITHFULNESS, AND SAFETY FROM WOUNDS.
- WHEN DANGER APPROACHES THE STONE LOSES ITS BRILLIANCE.
- FOR OBVIOUS REASONS RED GARNETS HAVE BEEN ASSOCIATED WITH BLOOD. AS RECENTLY AS 1892 NATIVE SOLDIERS IN THE KASHMIR FOUGHT THE BRITISH WITH BULLETS MADE OF GARNET IN THE BELIEF THAT THESE WOULD FIND THEIR WAYS MAGICALLY TO THEIR TARGETS.
- WILL PROTECT THE WEARER FROM EVIL AND FROM TERRIFYING DREAMS.

Hematite

H5½-6½

- A LUSTROUS BLACK STONE OFTEN CUT WITH FACETS OR CARVED WITH A WARRIOR'S HEAD.
- THOUGH THE STONE IS BLACK IT WILL LEAVE A RED STREAK WHEN SCRATCHED ALONG A ROUGH SURFACE. THE STONE APPEARS TO "BLEED" AND SO TAKES ITS NAME FROM THE LATIN WORD FOR BLOOD, "HAIMA".
- HEMATITE (ALSO SPELLED "HAEMATITE") IS THE WORLD'S MOST IMPORTANT IRON ORE.
- POWDERED HEMATITE IS KNOWN AS RED OCHRE WHEN USED AS A PIGMENT AND AS CROCUS WHEN USED AS A POLISHING COMPOUND OR ABRASIVE.
- HEMATITE CAN FORM NATURALLY AS A CLUSTER OF THIN PLATES AND IN THIS CONFIGURATION IS KNOWN AS AN ALPINE ROSE OR IRON ROSE.

Ivory

H2½

- IVORY COMES FROM THE TUSKS OF ELEPHANTS AND IS BECOMING INCREASINGLY RARE AS ELEPHANT HUNTING IS RESTRICTED.
- OTHER SIMILAR MATERIALS SHOULD BE IDENTIFIED WITH AN ADJECTIVE, AS WHALE IVORY. TRUE IVORY IS MADE UP OF MANY TRANSLUCENT LAYERS AND HAS A SOFT SHEEN CAUSED BY THE PARTIAL PENETRATION OF LIGHT.
- IVORY CAN BE IDENTIFIED BY A CHARACTERISTIC GRAIN PATTERN. THIS WILL ONLY BE OBVIOUS FROM ONE AXIS, AND WILL DARKEN AND BECOME MORE OBVIOUS WITH AGE.
- IVORY CAN BE CARVED WITH GRAVERS AND FILES. POWER TOOLS ARE BEST AVOIDED SINCE THE VIBRATIONS THEY CAUSE CAN CAUSE CRACKING.
- POLISH IVORY WITH PUMICE FIRST AND THEN WITH CHALK, EACH BEING MIXED WITH MINERAL OIL TO MAKE A PASTE. TO PREVENT THE IVORY FROM DRYING OUT, RE-OIL YEARLY.

Jade

H6-7

- REFERS TO TWO DISTINCT MINERALS NOT DIFFERENTIATED UNTIL 1863. THESE ARE PROPERLY CALLED JADEITE AND NEPHRITE.
- SPANISH CONQUISTADORS FOUND MANY OBJECTS OF CARVED JADE AND, BELIEVING IT TO EASE KIDNEY PAINS, CALLED IT "PIEDRA DE IJADA" (LOIN STONE). EUROPEAN DOCTORS CALLED IT "PALIS NEPHRITICUS" FROM GREEK "NEPHROS," KIDNEY.
- BECAUSE OF ITS WAXY LUSTRE THE CHINESE CALLED IT <u>WET STONE</u> AND BELIEVED IT COULD SLAKE THIRST.
- OCCURS IN WHITE (MUTTON FAT JADE), YELLOW, LAVENDER, EARTHY BROWN, AND BLACK AS WELL AS THE FAMILIAR GREENS.
- JADE CAN BE CONFUSED WITH CALIFORNITE, GROSSULARITE, SAUSSERITE, PECTOLITE, CHRYSOPRASE, AND AVENTURINE.
- THIS STONE WAS BELIEVED TO PROTECT FROM LIGHTENING, TO AID IN BATTLE, BRING RAIN, DRIVE AWAY BEASTS AND EVIL SPIRITS, AND TO AID IN CHILDBIRTH.

Jasper

H6½-7

- FROM HEBREW "YASHPEH" AND ASSYRIAN "YASHPU"; REFERRED TO IN CUNEIFORM WRITINGS OF 1500 BC. ORIGINALLY THE WORD MEANT ANY GREEN STONE.
- JASPER OCCURS IN MANY COLORS AND PATTERNS, INCLUDING STRIPES AND PICTURES. THESE ARE REALLY FOSSILIZED ALGAE MADE WHEN DECOMPOSED ORGANIC MATTER WAS REPLACED BY SILICON OXIDE (I.E. JASPER).
- GREEN CHALCEDONY WITH FLECKS OF RED JASPER IS CALLED BLOODSTONE OR HELIOTROPE.
- IN ANCIENT EGYPT RED JASPER WAS ASSOCIATED WITH THE BLOOD OF ISIS.
- GREEN JASPER WAS ASSOCIATED WITH ST. PETER BY THE EARLY CHRISTIANS.
- SAID TO DRIVE AWAY NIGHT SPIRITS, STAUNCH BLEEDING, AND HELP DURING PREGNANCY. GREEN JASPER WAS USED IN RAINMAKING.

Jet

H3-4

- A DENSE BLACK COAL FOUND IN MANY PLACES AROUND THE WORLD; ESPECIALLY POPULAR IN BRITAIN UNDER THE REIGN OF QUEEN VICTORIA, WHEN MOST JET CAME FROM THE TOWN OF WHITBY.
- BURNT AND POWDERED JET IS SAID TO DRIVE AWAY SNAKES AND REPTILES AND TO HEAL TOOTHACHE AND HEADACHE.
- JET NULLIFIES SPELLS AND CHARMS.
- TRADITIONALLY IRISH HOUSEWIVES BURNT JET DURING THEIR HUSBAND'S ABSENCE TO INSURE HIS SAFETY.

Labradorite

H6

- THIS IS A BLUE IRIDESCENT FELDSPAR FOUND OFF THE COAST OF LABRADOR.
- A SIMILAR GEM MINED IN FINLAND SHOWS A WIDER RANGE OF COLORS AND IS CALLED SPECTROLITE.
- BLACK MOONSTONE IS USUALLY LABRADORITE FROM MADAGASCAR.

Lapis Lazuli

H5-6

- FROM LATIN "LAPIS", STONE AND ARABIC "LAZULI", BLUE.
- KNOWN FOR ITS DEEP BLUE COLOR; SOMETIMES FOUND WITH FLECKS OF GOLD-COLORED PYRITE OR WHITISH-GRAY MOTTLINGS OF CALCITE.
- THE PERSIAN NAME FOR THIS STONE, "LAJUWARD", IS THE ROOT OF OUR "AZURE".
- LAPIS IS STILL BEING MINED AT THE OLDEST MINES IN THE WORLD IN AFGHANISTAN. WHEN MINING BEGAN SIX OR SEVEN THOUSAND YEARS AGO THE COUNTRY WAS CALLED BABYLON.
- LAPIS WAS SENT TO EGYPT AS TRIBUTE. THERE IT WAS CARVED TO MAKE CYLINDER SEALS AND GROUND TO A POWDER FOR EYE MAKE UP.
- IN UR, KINGS SHARPENED THEIR SWORDS ON LAPIS IN THE BELIEF THAT IT WOULD MAKE WEAPONS INVINCIBLE.
- SUMERIANS BELIEVED THAT A WEARER CARRIED THE PRESENCE OF GOD WITH HIM.
- IN ANCIENT EGYPT THE STONE WAS SYMBOLIC OF TRUTH (MA) AND WAS WORN BY THE CHIEF JUSTICE.
- IN THE MIDDLE AGES PAINTERS MIXED OIL WITH POWDERED LAPIS TO MAKE THE COLOR AQUAMARINE.
- THE GEM IS BELIEVED TO EASE EYE TROUBLES, TREAT ASTHMA, INDUCE SLEEP, AND RELIEVE ANXIETY.

Malachite

H5-6

- A COPPER ORE MADE UP OF DEEP AND PALE GREEN STRIPES OR CONCENTRIC CIRCLES. SINCE IT IS FORMED IN THIN LAYERS, LARGE PIECES ARE RARE.
- MALACHITE POWDER WAS USED BY THE ANCIENTS AS EYE MAKE UP.
- IT WAS COMMONLY HELD TO EASE LABOR, PROTECT INFANTS AND CHILDREN, AND SOOTHE THEIR PAIN WHEN THEY WERE CUTTING TEETH.
- BECAUSE OF ITS HIGH COPPER CONTENT MALACHITE WILL BE DAMAGED BY JEWELERS' PICKLE.

Magnetite

H 6-7

- ALSO CALLED LODESTONE
- THIS IS A BLACK IRON ORE THAT IS VERY MAGNETIC. THOUGH WE USE LITTLE FOR JEWELRY TODAY, IN ANCIENT TIMES THIS WAS A VERY IMPORTANT STONE.
- IT WAS BELIEVED THAT THE STONE WAS ALIVE. TO CARE FOR IT THE OWNER SET THE STONE IN WATER ONCE A WEEK AND "FED" IT IRON FILINGS. IT WAS REGULARLY TAKEN TO MASS TO DRIVE THE DEVIL OUT OF IT.
- BELIEVED TO DISPEL MELANCHOLY, EASE LABOR, IMPROVE MEMORY WHEN WORN ON THE NECK, HEAL SORES, AND RELIEVE PAINS IN THE HANDS AND FEET.
- MAGNETITE ASSISTS SEXUAL ACTIVITIES WHEN ONE PARTNER IS ANOINTED WITH THIS STONE AND THE OTHER WITH IRON FILINGS.

Moonstone

H6-6½

- A FELDSPAR OF ORTHOCLASE WITH THIN LAYERS OF ALBITE. THIS YIELDS A PLAY OF LIGHT CALLED ADULARESCENCE AS LIGHT IS SPREAD BY THE FINE PARTICLES OR LAYERS. THE EFFECT IS A COOL FROSTY GLOW THAT ACCOUNTS FOR THE NAME OF THIS GEM.
- OCCURS IN WHITE, GRAY, PINK, GREEN, BLUE, CHOCOLATE AND AN ALMOST CLEAR VARIETY THAT LOOKS LIKE A WATER DROPLET.
- WHEN WORN AROUND THE NECK MOONSTONE PROTECTS AGAINST EPILEPSY AND SUNSTROKE. IT IS USED TO TREAT HEADACHES AND NOSEBLEEDS.
- WHEN HUNG ON FRUIT TREES IT PRODUCES ABUNDANT CROPS AND GENERALLY ASSISTS ALL VEGETATION.

Onyx

H6½-7

- A CHALCEDONY COMPOSED OF BLACK AND WHITE BANDS. IN COMMON USAGE THE TERM OFTEN REFERS TO AN AGATE DYED UNIFORMLY BLACK.
- ONYX WITH BROWN AND WHITE BANDS IS CALLED SARDONYX.
- WHEN CUT TO SHOW CONCENTRIC CIRCLES, ONYX FORMED AN EYE-LIKE AMULET THAT WAS WORN BY SUMERIANS, GREEKS, EGYPTIANS, AND ROMANS TO WARD OFF EVIL.
- THIS STONE WAS WIDELY DISFAVORED EXCEPT WHEN CUT AS A PROTECTIVE EYE. IT WAS SAID TO:
  - INCITE CONTENTION BETWEEN FRIENDS.
  - GIVE WEARER BROKEN SLEEP AND TERRIFYING DREAMS.
  - WHEN WORN ON THE NECK, COOL THE FIRES OF LOVE.

THE ARABIC NAME FOR THIS STONE – EL JAZA – MEANS SADNESS.

Opal

H5½-6½

- FROM SANSKRIT "UPALA", GEM
- A HIGHLY PRAISED STONE THAT SHOWS A RANGE OF COLOR FLASHES THAT INCLUDE RED, BLUE, GREEN, AND VIOLET.
- OPAL IS HYDRATED SILICON DIOXIDE. THE PLAY OF COLORS IS THE RESULT OF WATER (1-15% BY WEIGHT) TRAPPED IN THE STONE. CARE SHOULD BE TAKEN THAT OPALS DO NOT DRY OUT. A PERIODIC COATING OF OIL IS RECOMMENDED.
- OPALS FROM MEXICO AND BRAZIL USUALLY CONTAIN MORE WATER AND ARE LESS STABLE THAN AUSTRALIAN OPALS.

types of opal

- HARLEQUIN – A MOSAIC OF IRIDESCENT COLOR.
- PINPOINT – MULTITUDE OF MINUTE SPECKS OF MANY COLORS.
- FLASH – UNDIVIDED FLASHES OF A SINGLE COLOR AS THE STONE IS ROTATED.
- FLAME – AS ABOVE WHEN SHOWING RED.
- FIRE – BRIGHT ORANGE-RED; TRANSLUCENT TO TRANSPARENT.
- MATRIX – STONE CUT SO AS TO LEAVE THE OPAL ATTACHED TO THE ROCK IN WHICH IT WAS FORMED. THIS IS DONE TO ADD STRENGTH TO A SPECIMEN OTHERWISE DANGEROUSLY THIN.
- DOUBLET – OPAL GLUED TO A BACKING OF OBSIDIAN OR ONYX TO INCREASE COLOR PLAY.
- TRIPLET – DOUBLET WITH ROCK CRYSTAL GLUED ON TOP FOR LUSTRE AND STRENGTH.

- OPALS ARE THOUGHT TO POSSESS THE VIRTUES OF ALL THE STONES WHOSE COLORS APPEAR THERE. THE ROMAN SENATOR NONIUS, FOR INSTANCE, SO VALUED A LARGE OPAL THAT HE CHOSE EXILE RATHER THAN SURRENDERING THE GEM TO MARK ANTONY.

- OPALS ARE THOUGHT BY SOME TO BE UNLUCKY. THIS SUPERSTITION MIGHT HAVE ORIGINATED WITH LAPIDARIES DURING THE MIDDLE AGES. IT WAS THE CUSTOM THAT BREAKAGE DURING CUTTING HAD TO BE PAID FOR BY THE LAPIDARY AND SINCE OPALS FRACTURE EASILY IT IS UNDERSTANDABLE THAT BY SOME THEY WOULD BE CONSIDERED UNLUCKY.

Pearl

H2½-4

- A LUSTROUS DEPOSIT FORMED INSIDE A LIVING BIVALVE MOLLUSK, OFTEN IN RESPONSE TO AN IRRITATION FELT BY THE ANIMAL. THOUGH MANY MOLLUSKS FORM SUCH DEPOSITS MOST SPECIES DO NOT HAVE ATTRACTIVE SURFACES.
- PEARLS ARE FORMED IN SALT- AND FRESHWATER CLAMS. THEY ARE IDENTIFIED BY THEIR PLACE OF ORIGIN, AS, MISSISSIPPI RIVER PEARLS, ETC. THE LARGEST SOURCE OF PEARLS IS LAKE BIWA IN JAPAN WHERE EXTENSIVE PEARL FARMING IS DONE.
- PEARLS SOMETIMES GROW ATTACHED TO THE SHELL OF THE ANIMAL. THESE ARE CALLED BLISTER PEARLS.
- CULTURED OR CULTIVATED PEARLS ARE MADE INSIDE A MOLLUSK BUT HAVE A LITTLE HUMAN HELP TO GET STARTED. A BIT OF TISSUE OR A GLASS OR PLASTIC BEAD IS INSERTED IN THE ANIMAL AND ALLOWED TO COLLECT NACREOUS SECRETIONS FOR ABOUT FOUR YEARS.
- IMITATION PEARLS ARE MUCH LESS VALUABLE. THEY ARE MADE BY REPEATED DIPPING OF A PLASTIC BEAD INTO A COATING MADE OF GLUE AND GROUND SARDINE SCALES. WHEN LIGHTLY RUBBED ON THE FRONT OF A TOOTH THE IMITATION PEARL WILL FEEL SMOOTH; THE GENUINE (ORIENT) AND CULTURED PEARLS WILL FEEL SLIGHTLY ROUGH.
- PEARLS ARE ATTRIBUTED TO THE GODDESS VENUS AS THE SYMBOL OF INNOCENCE.
- CARE SHOULD BE TAKEN THAT PEARLS ARE NOT SUBJECTED TO SUDDEN TEMPERATURE CHANGES. WASH THEM IN LUKEWARM SOAPY WATER AND RE-STRING AS OLD CORD BECOMES WORN. KNOTS SHOULD BE TIED BETWEEN PEARLS TO KEEP THEM FROM RUBBING AGAINST ONE ANOTHER. APPLY PERFUME BEFORE PUTTING ON PEARLS.

Peridot

H6½-7

- A TRANSPARENT GEM, SOMETIMES CALLED CHRYSOLITE, OCCURRING AS PALE-TO-DEEP YELLOW-GREEN.
- PERIDOT IS ASSOCIATED WITH THE ASTROLOGICAL SIGN OF LIBRA (9/22-10/23) AND IS ASSIGNED TO THE SUN.
- IN ANCIENT HEBREW WRITINGS THIS STONE IS LINKED WITH THE TRIBE OF SIMEON.
- IT IS BELIEVED TO CURE LIVER DISEASE AND DROPSY, TO FREE THE MIND FROM ENVIOUS THOUGHTS, AND TO DISPEL TERRORS OF THE NIGHT. FOR FULL MAGICAL POWER IT SHOULD BE SET IN GOLD.

Quartz

H7

- QUARTZ IS THE MOST COMMON OF ALL MINERALS.
- INCLUDED IN THIS FAMILY ARE AMETHYST, CITRINE, FLINT, ONYX, AVENTURINE, JASPER, CARNELIAN, ROCK CRYSTAL, AGATE, AND CRYSOPRASE.

Rock Crystal

H7

- FROM GREEK "KRYSTALLOS," ICE.
- IT WAS ONCE BELIEVED TO BE HARDENED ICE AND TO HAVE THE ABILITY TO SLAKE THIRST.
- ROMANS DRANK FROM CRYSTAL GOBLETS AND EARLY PHYSICIANS USED THE STONE AS AN ICE PACK TO EASE INFLAMMATION OF THE GALL BLADDER.
- INDIANS THOUGHT IT WAS THE GLASS OF HEAVEN AND CARVED PRAYER BEADS OF IT.
- CRYSTAL HAS BEEN USED THROUGHOUT HISTORY AS A REFLECTIVE SURFACE TO CONCENTRATE A VIEWER'S CONSCIOUSNESS, ESPECIALLY IN THE FORM OF SPHERES; I.E. CRYSTAL BALLS. THE ACTIVITY OF STARING INTO CRYSTAL UNTIL THE MIND BECOMES RECEPTIVE TO SUPERNATURAL INFLUENCES IS CALLED SCRYING.

Ruby

H9

- A CORUNDUM THAT OCCURS AS A DEEP RED TRANSPARENT STONE AND AS AN OPAQUE REDDISH-GRAY MATERIAL. IN THIS FORM IT MAY EXHIBIT A STAR (ASTERISM) OR A SINGLE LINE CHATOYANCY (CAT'S EYE).
- WHEN FLAWLESS A RUBY IS MORE VALUABLE THAN DIAMOND.
- SYNTHETIC RUBIES ARE PRODUCED FOR JEWELRY, WATCH BEARINGS, AND LASER EQUIPMENT.
- HISTORICALLY THE RUBY IS ASSOCIATED WITH ROYALTY AND THE POWER OF LIFE AND DEATH.
- IT WAS ATTRIBUTED THE POWER TO PREVENT LOSS OF BLOOD, STRENGTHEN THE HEART, AND NEGATE POISONS.

Rutile

H6½-7

- FROM LATIN "RUTILUS," REDDISH.
- A BROWNISH-RED TO INTENSE RED STONE, OCCASIONALLY TRANSPARENT BUT MORE COMMONLY OPAQUE.
- IT IS BEST KNOWN AS AN INCLUSION IN OTHER MINERALS. IN QUARTZ RUTILE APPEARS AS NEEDLES OF A GOLDEN COLOR; THESE HAVE BEEN GIVEN THE ROMANTIC NAME OF "CUPID'S DARTS." IN RUBIES, NEEDLES OF RUTILE ARE CALLED SILK AND CREATE STAR EFFECTS.
- RUTILE WAS PRODUCED SYNTHETICALLY AS A PALE YELLOW MATERIAL THAT WAS USED TO SIMULATE DIAMONDS. THIS GAVE WAY TO SYNTHETIC STRONTIUM TITANATE, WHICH IS WHITER, WHEN IT WAS DISCOVERED IN 1953.

Sapphire

H9

- FROM HEBREW "SAPPIR"
- THIS FORM OF CORUNDUM CAN OCCUR AS BLUE, YELLOW, PINK, BROWN, BLACK, LILAC, AND GREEN, BOTH AS TRANSPARENT AND OPAQUE, THE LATTER SOMETIMES SHOWING A STAR (ASTERISM) OR CAT'S EYE (CHATOYANCY).
- UNTIL THE MIDDLE AGES SAPPHIRES WERE CALLED HYACINTHS BECAUSE OF THEIR PALE BLUE COLOR. WHEN IT WAS REALIZED THAT THE MINERAL OCCURRED IN OTHER COLORS THE TERM SAPPHIRE WAS ADOPTED FOR THE BLUE VARIETY WHILE OTHERS USE A COLOR DESCRIPTION; E.G. YELLOW SAPPHIRE.
- SAPPHIRES ARE TRADITIONALLY CONNECTED WITH THE EYE AND THE SKY, AND THEREFORE WITH VISION AND THE ABILITY TO READ THE FUTURE.
- SAPPHIRES RENDER BLACK MAGIC HARMLESS AND HELP THE WEARER DISCERN FALSEHOOD AND GUILE.
- BUDDHISTS BELIEVE THE SAPPHIRE BRINGS PURITY AND SPIRITUAL ENLIGHTENMENT.

Sardonyx

H6½-7

- A KIND OF CHALCEDONY MADE BROWN BY THE PRESENCE OF IRON. SPECIFICALLY THE NAME REFERS TO SPECIMENS THAT INCLUDE BANDS OF WHITE.
- SARDONYX WAS A POPULAR STONE IN ANCIENT TIMES AND WAS CREDITED WITH MANY POWERS. IT WOULD:
  - MAKE WARRIORS VICTORIOUS.
  - PROTECT AGAINST POISONOUS SNAKES.
  - MAKE A SUITOR MORE APPEALING.
  - NEUTRALIZE THE MALIGN INFLUENCE OF BLACK ONYX.
  - INCREASE INTELLIGENCE AND MAKE THE WEARER FEARLESS AND HAPPY.
  - PROTECT AGAINST WITCHCRAFT, SORCERY, & INCANTATION.

Serpentine

H2-6

- AN OPAQUE GREEN STONE WITH MOTTLED REDDISH-BROWN OR MILKY PATCHES. THIS WITH ITS WAXY APPEARANCE MAKES IT LOOK LIKE SNAKE SKIN, HENCE THE NAME.
- SERPENTINE IS COMMON AND OCCURS IN MANY COLOR AND HARDNESS VARIATIONS. IT IS USED ARCHITECTURALLY AND TO CARVE OBJECTS LIKE BOWLS AND SCULPTURES.
- THIS STONE WAS BELIEVED TO PROTECT AGAINST SNAKEBITE AND OTHER POISONS AND WAS THOUGHT TO BE MOST EFFECTIVE IF KEPT IN THE NATURAL, UNCUT STATE.
- MEDICINE DRUNK FROM A SERPENTINE VESSEL IS MADE MORE BENEFICIAL.

Sodalite

H6-7

- A POPULAR OPAQUE STONE MOST WIDELY KNOWN FOR ITS BLUE COLOR THAT SOMEWHAT RESEMBLES LAPIZ LAZULI. IT ALSO OCCURS IN LAVENDER, MAUVE, YELLOW-GREEN, GREEN, AND PINK. PURPLE SHADES TEND TO FADE IN SUNLIGHT.
- WHITE AND GRAYISH-WHITE MOTTLINGS ARE OFTEN FOUND IN SODALITE; IN POOR GRADE MATERIAL THESE WILL BE OBVIOUS.
- THE NAME OF THIS MINERAL COMES FROM ITS SODIUM CONTENT.

Spinel

H8

- A TRANSPARENT STONE OF RED (THE MOST VALUABLE), PALE PINK, GREY-GREEN, BLUE-GREEN, AND PURPLE.
- SYNTHETIC SPINEL IS PRODUCED IN LARGE QUANTITIES AND IS ASSOCIATED WITH INEXPENSIVE JEWELRY IN IMITATION OF DIAMONDS, AQUAMARINE, SAPPHIRES, AND OTHERS. AIR BUBBLES INSIDE THE STONE OFTEN BETRAY THESE SYNTHETICS.

Tiger's Eye

H6½-7

- BLUE, VIOLET, AND GOLDEN BROWN TRANSLUCENT STONES SHOWING A SILKY INTERIOR THAT OFTEN IRIDESCESES AS THE STONE IS ROTATED. THEY CAN SOMETIMES BE CUT TO SHOW A CAT'S EYE.
- THE FIBROUS NATURE IS THE RESULT OF ASBESTOS FIBERS THAT HAVE BEEN PARTIALLY REPLACED BY QUARTZ.
- WHEN THE FIBERS ARE COARSE THE STONE MAY BE CALLED A HAWK'S EYE.

Topaz

H8

- A TRANSPARENT STONE USUALLY OF GOLDEN YELLOW BUT ALSO OCCURRING AS PINK, RED, BLUE, GREEN, AND COLORLESS SPECIMENS.
- FROM SANSKRIT "TAPAS", TO GLOW.
- TOPAZ CLEAVES EASILY AND THEREFORE REQUIRES CARE IN CUTTING AND SETTING.
- SOME VARIETIES CAN FADE IN THE SUNLIGHT.
- IN ANCIENT TIMES THE WORD TOPAZ REFERRED TO SEVERAL OTHER STONES AND TODAY IT IS OFTEN MISTAKENLY USED FOR SMOKY QUARTZ AND CITRINE.
- RUBBING OR GENTLE HEATING OF TOPAZ ELECTRIFIES IT, CAUSING IT TO ATTRACT SMALL PARTICLES LIKE BITS OF PAPER OR HAIR.

Tourmaline

H7-7½

- A TRANSPARENT STONE OF MANY COLORS, MOST NOTABLY GREEN, BLUE-GREEN, AND PINK.
- OFTEN SEVERAL COLORS APPEAR SIDE BY SIDE; CRYSTALS CUT TO REVEAL A PINK SEMI-CIRCLE WITH A GREEN RIM ARE CALLED WATERMELON TOURMALINE.

GREEN
PINK

- NAME COMES FROM SANSKRIT "TURAMALI", A WORD ORIGINALLY USED TO DESCRIBE ANY RED OR REDDISH-BROWN STONE.
- TOURMALINE IS DICHROMATIC; IT SHOWS A BRIGHT COLOR FROM ONE DIRECTION BUT WILL LOOK ALMOST BLACK WHEN SEEN FROM THE SIDE. LIKE TOPAZ THIS STONE WILL HOLD STATIC ELECTRICITY IF IT IS RUBBED OR GENTLY HEATED. TOGETHER THESE TWO TESTS MAKE FOR IDENTIFICATION OF THIS STONE.

I AM SIMPLY IMPRESSED BY THE UNEXPECTED INSIGHTS THAT SHOWER DOWN ON ME WHEN MY JOB IS TO IMAGINE, AS CONTRASTED WITH THE WOODENLY FAMILIAR IDEAS WHICH CLUTTER MY DESK WHEN MY JOB IS TO TELL THE TRUTH.

KURT VONNEGUT, JR.

Turquoise

H5-6

- A BLUE OR BLUE-GREEN STONE, USUALLY OPAQUE BUT OCCASSIONALLY TRANSLUCENT AND GLASSY LOOKING.
- FROM FRENCH "PIERRE TURQUOISE", WHICH MEANS TURKISH STONE, A REFERENCE TO ITS POPULAR USE IN TURKEY. ARABS CALL IT "FAYRUŽ" OR "FIRŪZAJ", THE LUCKY STONE.
- BLUE MATERIAL WILL TURN GREEN WHEN IT ABSORBS OIL FROM THE SKIN. AFTER POLISHING, MOST TURQUOISE IS SEALED WITH A PLASTIC THAT SOAKS INTO THE STONE AND CLOSES THE PORES.
- RECONSTITUTED MATERIAL, BITS OF TURQUOISE COMPRESSED WITH ADHESIVE, IS OFTEN USED IN CHEAP JEWELRY. TO TEST A SAMPLE, LAY A HOT NEEDLE AGAINST THE STONE. IF IT CONTAINS ADHESIVE THE RESULTING SMELL OF PLASTIC WILL GIVE IT AWAY.
- SOME PIECES OF TURQUOISE ARE CUT SO AS TO CONTAIN SOME OF THE ROCK IN WHICH THEY WERE FORMED. THIS IS CALLED MATRIX TURQUOISE. SOME VARIETIES SHOW FINE DARK LINES RUNNING THROUGHOUT THE STONE; THIS IS CALLED SPIDERWEB TURQUOISE.
- TURQUOISE WILL PROTECT THE WEARER FROM POISON, BITES OF REPTILES, AND DISEASES OF THE EYE. SOME PEOPLE THINK THESE POWERS ARE IN FORCE ONLY IF THE STONE WAS RECEIVED AS A GIFT. GIVING A TURQUOISE IS ALSO SAID TO IMPROVE ITS COLOR.
- SINCE THE THIRTEENTH CENTURY THIS STONE WAS HELD TO GIVE SURE-FOOTEDNESS TO A HORSE. THE IDEA WAS LATER ENLARGED TO PROTECT AGAINST ALL FALLING.
- ACCORDING TO AMERICAN INDIAN TRADITION, A TURQUOISE ATTACHED TO A GUN OR A BOW WILL GUIDE THE SHOT TO ITS TARGET. IT WAS ALSO BELIEVED THAT THE STONE COULD BE FOUND IN THE DAMP EARTH AT THE END OF A RAINBOW.
- BUDDHISTS ASSOCIATE IT WITH BUDDHA BECAUSE, ACCORDING TO TRADITION, HE USED THE STONE TO DESTROY A MONSTER.

Zircon

H7-7½

- A TRANSPARENT BRITTLE STONE OCCURRING AS BROWNISH OR GREEN MATERIAL, USUALLY HEATED TO TURN PALE YELLOW AND BLUE. IT CAN BE FOUND NATURALLY COLORED AS ORANGE-RED (MOST VALUABLE), PURPLE, REDDISH-BROWN, AND BROWNISH-YELLOW.
- BECAUSE ITS BRITTLENESS MAKES IT DIFFICULT TO CUT, STONES OF MORE THAN A COUPLE CARATS ARE RARELY SEEN.
- ZIRCON IS SAID TO DRIVE AWAY EVIL SPIRITS AND BAD DREAMS, BANISH GRIEF AND MELANCHOLY, RESTORE APPETITE, INDUCE SLEEP, AND PROTECT AGAINST LIGHTNING.

Zirconium (Cubic)

H 9-9½

- A TRANSPARENT MAN-MADE GEM PRODUCED FROM THE ELEMENT ZIRCONIUM.
- IT IS AVAILABLE IN MANY COLORS, AS WELL AS A BRIGHT WHITE THAT RESEMBLES DIAMOND. BECAUSE OF ITS "FIRE" AND LOW COST IT HAS REPLACED YAG (SYNTHETIC GARNET), SPINEL, AND STRONTIUM TITANATE AS A DIAMOND SUBSTITUTE.

Some Tips...

HOLD WORK SECURELY—

IN A RING CLAMP

OR ON A SOFT WOOD JIG CLAMPED TO THE BENCH.

WHEN TRYING STONE IN PLACE IT MAY BE HELD WITH BEESWAX. TO LIFT STONE OUT LAY A PIECE OF THREAD BENEATH IT.

ALWAYS WORK FROM ALTERNATE SIDES.

sequence:

IT IS IMPOSSIBLE TO GIVE A FIXED FORMULA FOR STONE SETTING, BUT HERE IS A TYPICAL PROGRESSION:

1. COMPLETE ALL SOLDERING.
2. PICKLE, NEUTRALIZE, SAND, AND BUFF.
3. OXIDIZE (IF DESIRED).
4. PUT STONE IN PLACE; BE SURE IT IS LEVEL AND SEATED.
5. PUSH BEZEL OR PRONGS OVER STONE.
6. BURNISH METAL TO POLISH AND TOUGHEN IT.
7. POLISH WITH LEATHER-COATED STICK OR, FOR PRONGS, SMALL BRISTLE BRUSH.

tools

THE FACE OF A BEZEL PUSHER SHOULD NOT BE POLISHED SINCE THAT MIGHT MAKE IT SLIP. TO PROVIDE A LITTLE GRIP, SAND FACE WITH MEDIUM GRIT PAPER.

TOOL WILL BECOME WORN WITH USE. FILE PERIODICALLY TO RESTORE SHAPE.

ON SOFT OR DELICATE MATERIALS (SHELL, ENAMELS, ETC) A PLASTIC PUSHER IS RECOMMENDED. THIS MAY BE MADE FROM A TOOTHBRUSH AS SHOWN.

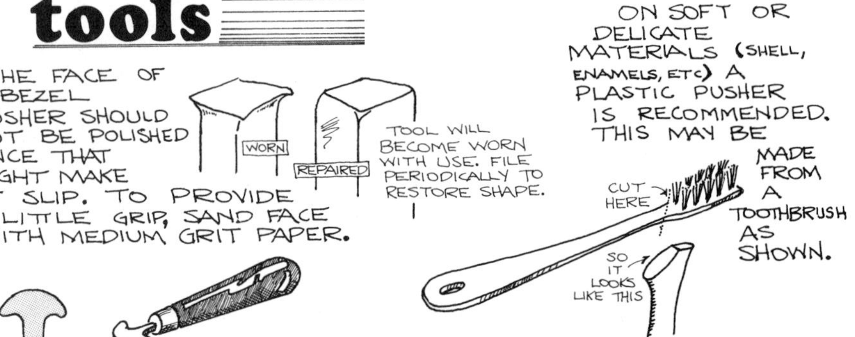

TO MAKE A BEZEL ROCKER, SAW THIS SHAPE FROM 12 OR 14 GAUGE BRASS AND MOUNT IN A FILE HANDLE.

TO USE, HOLD ROCKER LIKE PENCIL AND ROCK VERTICALLY AT SEVERAL POINTS AROUND STONE. THEN TURN TOOL FACE TO HORIZONTAL AND SWING BACK AND FORTH TO SMOOTH BEZEL.

TO USE A HAMMER..

GRIP WORKPIECE SECURELY AND TAP WITH SMOOTH PUNCH.

THE LARGER THE ISLAND OF KNOWLEDGE, THE LONGER THE SHORELINE OF WONDER.

R.W. SOCKMAN

what to avoid:

WITH A BEZEL BE SURE METAL LAYS DOWN ON STONE.

WITH BUFF TOP DON'T GO OVER THE TOP.

EACH PRONG SAME SIZE.

1

WRAP WIRE - BEZEL OR SIMILAR THIN STRIP - AROUND STONE; MARK; CUT.

FOR SMALL STONES BEND LOOP BY EYE; FIT TO STONE.

2

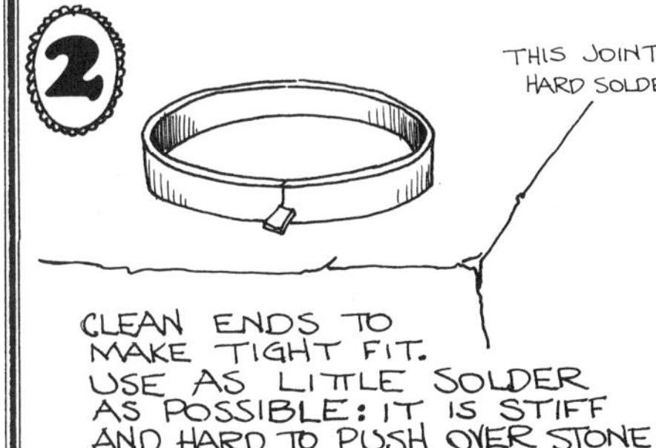

CLEAN ENDS TO MAKE TIGHT FIT. USE AS LITTLE SOLDER AS POSSIBLE: IT IS STIFF AND HARD TO PUSH OVER STONE.

3

- CHECK FIT -

IF BEZEL IS TOO SMALL STRETCH AS SHOWN OR BY PLANISHING WITH A STEEL HAMMER.

IF TOO LOOSE, REMOVE A PIECE OF BEZEL AND RESOLDER.

4

FILE OR SAND TO CORRECT HEIGHT

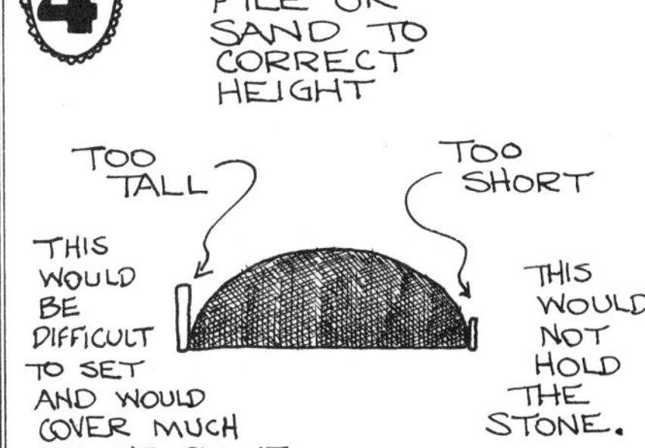

THIS WOULD BE DIFFICULT TO SET AND WOULD COVER MUCH OF THE STONE.

THIS WOULD NOT HOLD THE STONE.

5

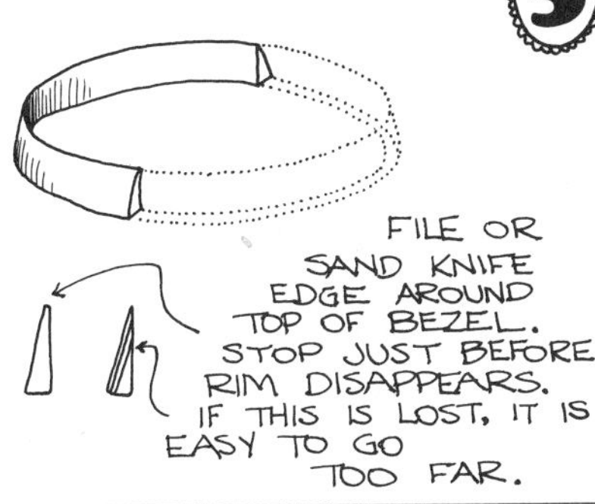

FILE OR SAND KNIFE EDGE AROUND TOP OF BEZEL. STOP JUST BEFORE RIM DISAPPEARS. IF THIS IS LOST, IT IS EASY TO GO TOO FAR.

6

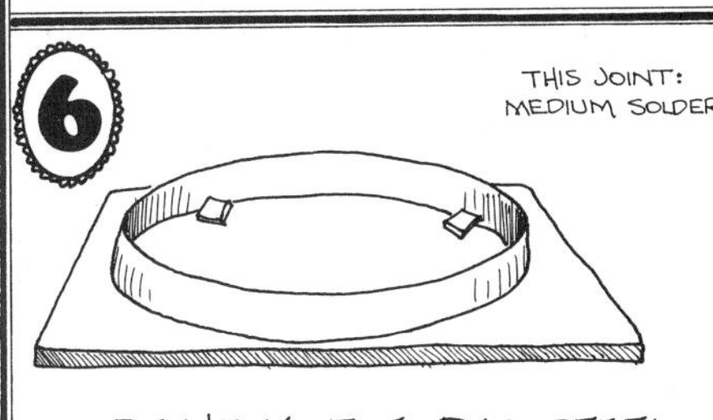

RECHECK FIT, RUB BEZEL ON SANDPAPER TO TRUE AND CLEAN BOTTOM EDGE.

AFTER SOLDERING TO FLAT SHEET, PICKLE AND CHECK JOINT.

7

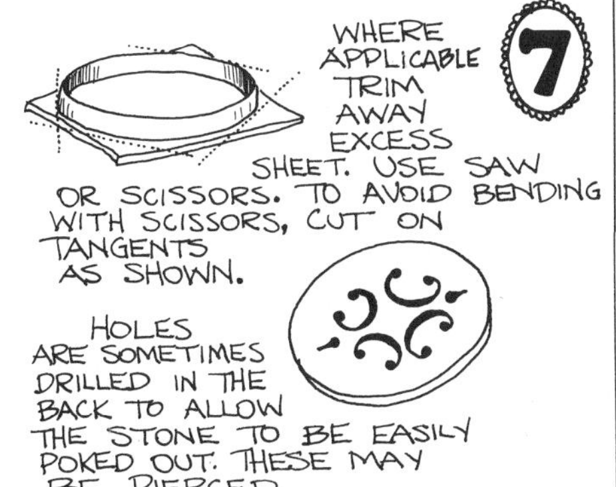

WHERE APPLICABLE TRIM AWAY EXCESS SHEET. USE SAW OR SCISSORS. TO AVOID BENDING WITH SCISSORS, CUT ON TANGENTS AS SHOWN.

HOLES ARE SOMETIMES DRILLED IN THE BACK TO ALLOW THE STONE TO BE EASILY POKED OUT. THESE MAY BE PIERCED.

8

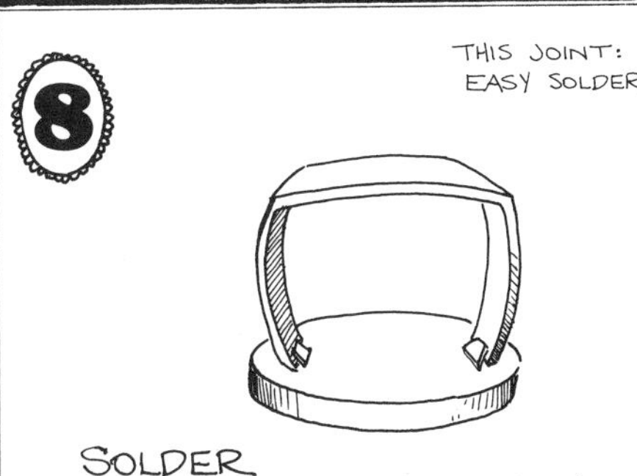

SOLDER BEZEL INTO POSITION ON WORKPIECE.

THE ADVANTAGES OF THIS KIND OF BEZEL ARE:
- IT USES LESS MATERIAL, SAVING COST AND REDUCING WEIGHT.
- IT CAN BE FASTER THAN A BOX BEZEL TO MAKE, DEPENDING ON THE TYPE USED.
- IT ALLOWS LIGHT TO SHOW THROUGH THE STONE.

styles:

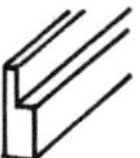

STEP BEZEL MAY BE BOUGHT. IT IS GENERALLY AVAILABLE ONLY IN FINE SILVER AND 14KY GOLD. SOME DISTRIBUTORS WILL ALSO HAVE GALLERY STEP BEZEL.

ANOTHER METHOD IS TO CREATE THE STEP BEFORE BENDING THE BEZEL AROUND THE STONE. TO AVOID AN EXCESS OF SOLDER, FILE THE ANGLE SHOWN.

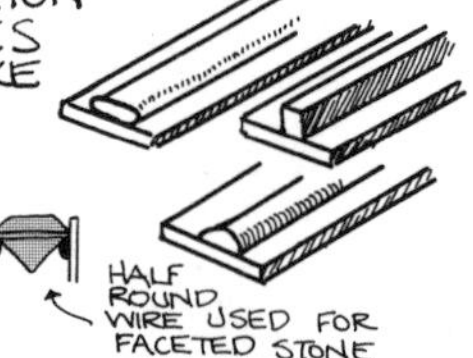

A VARIATION ON THIS USES WIRE TO MAKE A BEARING. PROVIDE A FLAT SURFACE FOR A NEAT SOLDER JOINT.

HALF ROUND WIRE USED FOR FACETED STONE

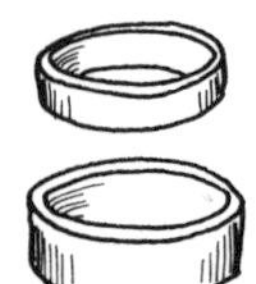

AFTER COMPLETING A BEZEL RING THAT FITS THE STONE, MAKE A SECOND RING (BEARING) THAT FITS INSIDE THE FIRST.

IF A BASE IS USED ON THE BEZEL THIS RING DOES NOT NEED TO BE SOLDERED IN.

FOR FACETED STONES, FILE A BEVEL ON THE INNER RING BEFORE SOLDERING THE TWO PIECES TOGETHER.

A BEARING MAY ALSO BE CUT: MAKE BEZEL OF HEAVY STOCK LIKE 16 OR 18 GAUGE. FLEX SHAFT TOOLS OR GRAVERS MAY BE USED.

ABOUT 1/3 OF THE THICKNESS SHOWS HERE

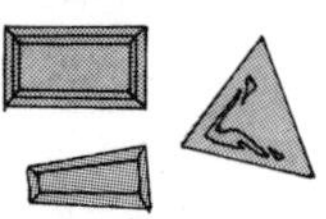

THE CORNERS OF STONES CUT AS RECTANGLES, SQUARES, ETC. ARE SUBJECTED TO A LOT OF PRESSURE IN SETTING. TO REDUCE THIS, CUT AWAY THE BEARING UNDER THE CORNER (SHADED) WITH GRAVER OR FLEX SHAFT.

PEARLS

PEARLS ARE USUALLY SET BY CEMENTING THEM ONTO A POST. TO MAKE A FIRM GRIP THE POST SHOULD BE PREPARED IN ONE OF THESE WAYS: NOTCHED WITH A FILE; BENT; DOUBLE STRAND TWISTED.

USE EPOXY OR PEARL CEMENT—NOT SUPER GLUE OR ANY OF THE OTHER CYANOACRYLATES.

PEARLS MAY BE DRILLED WITH CONVENTIONAL BITS. GO SLOW TO ALLOW PEARL TO COOL. WHEN DRILLING THROUGH, PUT MASKING TAPE ON BOTTOM TO PROTECT THE NACRE (OUTSIDE) FROM CHIPPING OFF.

DON'T PUT PEARLS OR SOFT STONES IN PRONGS WITHOUT CEMENTING THEM, TOO. ROTATION WILL CAUSE SCRATCHES LIKE THESE.

BEADS

SINCE GLUE DOES NOT ADHERE WELL TO STONES IT IS BETTER TO SET THEM MECHANICALLY. MANY METHODS MAY BE DEVISED, FROM TYING THEM ON TO THE USE OF TINY SCREWS AND NUTS.

SHOWN ABOVE IS A SAWN WIRE THAT HAS HAD A BEAD DRAWN BEFORE ASSEMBLY. THE OTHER DEVICE, A TUBE RIVET, IS NOT RECOMMENDED FOR PEARLS SINCE THE OUTWARD FORCE AROUND THE HOLE WOULD CAUSE CHIPPING.

WHEN NOTHING ELSE WILL WORK, STONES MAY BE NOTCHED WITH A SEPARATING DISK USED ON THE FLEX SHAFT. STONE CAN THEN BE WRAPPED OR TIED.

Gallery:

GALLERY WIRE IS A DECORATED STRIP OF WIRE USED AS MOLDING OR AS A BEZEL. IT CAN BE BOUGHT IN SEVERAL ORNATE PATTERNS.

THE DECORATIONS SHOWN HERE ARE MADE BY FILING, STAMPING, DRILLING, AND ENGRAVING. SOME PATTERNS CAN ONLY BE DONE WHEN THE STRIP OF METAL IS FLAT (LIKE STAMPING) BUT MOST ARE EASIER AFTER THE BEZEL HAS BEEN MADE.

ROUND NEEDLE FILE — TRIANGULAR FILE & DRILL BIT — HALF-ROUND FILE & SAW — STAMPED LINE, FILED EDGE, ENGRAVED ▲ — BARRETTE FILE, SCRIBE POINT.

AFTER A BEZEL HAS BEEN MADE AND SOLDERED ONTO A BASE IT CAN BE MADE MORE DELICATE BY CUTTING WITH A SAW.

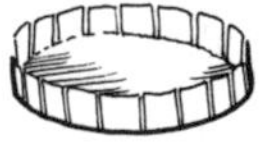

COLLAR

CONSTRUCT BEZEL AS FOR BOX BEZEL OR LEDGE BEZEL, BUT USE A THICKER WALL: 18 GAUGE FOR A SMALL STONE AND 16 GAUGE FOR A LARGE ONE.

SETTING IS DONE WITH A PUNCH SHAPED LIKE THIS, WITH ROUNDED CORNERS AND A POLISHED FACE. THE CHASING HAMMER IS USED WITH REPEATED LIGHT TAPS.

THE OBJECT MUST BE FIRMLY ANCHORED. USE A PITCH POT, SHELLAC OR SEALING WAX ON A BOARD, OR A VISE. SUPPORT RINGS ON A SOFT-WOOD WEDGE:

LEATHER

THE STONE IS 'LOCKED' IN PLACE WITH FOUR SHARP BLOWS EVENLY SPACED AROUND THE BEZEL. IN SUCCESSIVE COURSES AROUND THE STONE THE ANGLE OF THE TOOL IS RAISED UNTIL IT IS VERTICAL.

THE BEZEL IS PLANISHED WITH THE TOOL SO VERY FEW MARKS ARE MADE. THE SHAPE MAY BE DEFINED WITH A FILE —NO SANDPAPER!— AND THE COLLAR IS THEN BUFFED.

Tooled Edge:

THIS SETTING IS SIMILAR TO THE COLLAR BEZEL BUT HAS A ROUGHER LOOK BECAUSE OF THE TEXTURE LEFT BY THE TOOLS USED.

START WITH A BEZEL MADE OF THICK MATERIAL, KEEPING IT A LITTLE BIGGER THAN USUAL SO THE STONE MAKES A LOOSE FIT.

THE OBJECT IS SECURELY GRIPPED AS BEFORE AND THE BEZEL IS PUSHED IN WITH A TOOL SHAPED LIKE A SCREWDRIVER.

SEVERAL "STITCHES" CAN BE MADE AROUND THE BEZEL TO HOLD THE STONE FOR THE MORE CAREFUL EVENLY-SPACED TAPPING THAT WILL GIVE THE SETTING ITS FINAL APPEARANCE. THE TOOL IS HELD NEARLY VERTICAL, CREATING IN EFFECT MANY PRONGS THAT HOLD THE STONE.

millegriffe

1. MAKE COLLET (TUBE) WITH INSIDE DIAMETER SMALLER THAN STONE. THICK WALL IS IMPORTANT.

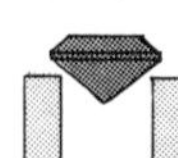

2. MARK LINE AROUND BASE AND FILE.

3. CUT BEARING WITH BUR OR GRAVER. FILE OFF INSIDE CORNER AT ARROW.

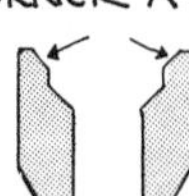

4. SOLDER TO PIECE, FINISH, SET STONE. NOTE THE SHAPE HERE.

5. MAKE A SERIES OF BEADS WITH PUNCH. USE A GRAVER &/OR FILE TO CUT NOTCHES BETWEEN THE BEADS.

For a Round or Oval Cabachon

1. MAKE A BEZEL OF THE USUAL SHAPE HAVING MOST OF THE BACK OPEN. THIS MAY BE DONE BY SOLDERING THE BEZEL TO A SHEET AND THEN CUTTING OUT THE INTERIOR SPACE, OR BY SOLDERING A RING OF SQUARE OR RECTANGULAR WIRE INSIDE THE BEZEL.

2. MAKE A SMALL SECTION OF CONE WHOSE LARGER DIAMETER IS THE SAME AS THAT OF THE BEZEL. THIS MAY BE DONE BY BENDING AN ARC → OR BY SOLDERING A LOOP CLOSED AND FORMING IT IN A DAPPING BLOCK.

3. SOLDER SPACERS (WIRE OR, AS SHOWN, SHORT LENGTHS OF TUBING) ONTO THE BACK OF THE BEZEL AT REGULAR INTERVALS.

FILE NOTCHES TO HOLD IN PLACE

4. RUB THE TUBING LIGHTLY ON SANDPAPER TO MAKE SURE EACH ONE HAS A FLAT FACE. SOLDER THE CONE INTO PLACE, THEN CUT AND FILE THE TUBING FLUSH WITH THE BEZEL. THE CONE MAY BE CUT TO ACCOMMODATE A RING SHAPE.

For a Square or Rectangular Stone

1. MEASURE AND MARK A LENGTH OF SQUARE WIRE. FILE NOTCHES AT THE CORNERS, A MITRE ON EACH END AND BEND INTO A NEAT FRAME. SOLDER.

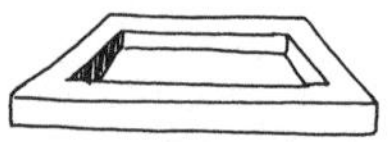

2. FILE A BEVEL ON THE INSIDE EDGE.

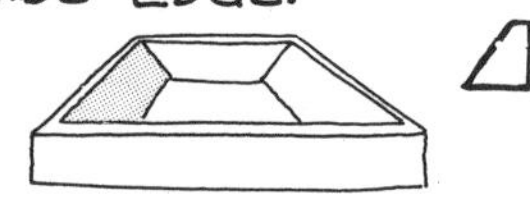

3. MAKE ANOTHER FRAME, A LITTLE SMALLER THAN THE FIRST. DO NOT FILE THE INSIDE EDGE.

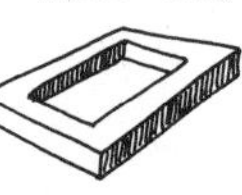

4. CUT SPACER BLOCK FROM SHEET OR WIRE. THE SIZE WILL DEPEND ON THE DEPTH OF THE STONE: DON'T LET IT PROTRUDE THROUGH THE BACK. SAW A GROOVE IN EACH SPACER TO HOLD IT IN POSITION AND SOLDER. 2 OR 4 SPACERS CAN BE USED.

5. KEEPING CAREFUL ALIGNMENT, SOLDER THE TOP IN PLACE.

6. CUT PRONGS FROM 22 GAUGE SHEET AND SCORE A GROOVE DOWN THE CENTER.

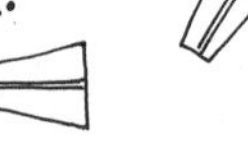

7. SOLDER PRONGS INTO POSITION, THEN SOLDER THE HEAD TO WORKPIECE. IN SETTING, THE PRONGS MIGHT NEED TO BE CUT SLIGHTLY AT THE TOP SO THEY WILL BEND OVER THE STONE NEATLY. BURNISH TO CLOSE THIS.

Box Setting

USE A SHEET THICKER THAN THE HEIGHT OF THE STONE. MARK CENTER AND DRILL A SMALL HOLE.

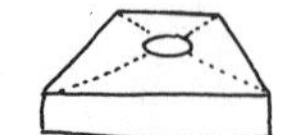

CUT A BEARING WITH GRAVER OR BUR. NOTE THE DEPTH; IT IS IMPORTANT.

CARVE A GROOVE WITH A ROUND GRAVER. THE LITTLE TRIANGLES WILL BECOME PRONGS.

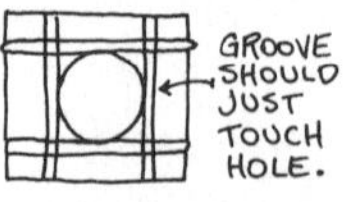

CUT AWAY THE SHADED AREA WITH FLAT GRAVER OR DECORATE WITH MILLGRAIN. SET STONE WITH BEADING TOOLS.

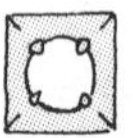

WHETHER SIMPLE OR COMPLEX, A SUCCESSFUL PRONG SETTING MUST FOLLOW THESE GUIDELINES:

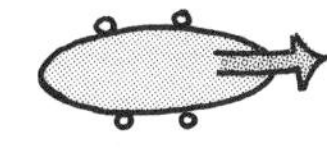

1. PRONGS MUST REACH OVER THE STONE'S GIRDLE TO HOLD IT SECURELY.
2. PRONGS MUST BE LOCATED SO THE STONE CANNOT SLIP OUT IN ANY DIRECTION.
3. THE STONE MUST BE SUPPORTED FROM BENEATH.
4. THE PRONGS MUST NOT COVER SO MUCH OF THE STONE THAT THE RESULT IS CUMBERSOME.

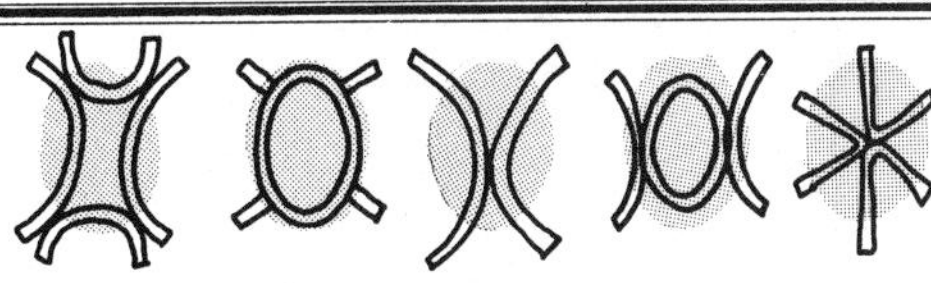

ALL THESE STYLES ARE MADE OF WIRE, SIMPLY LAID OUT FLAT AND SOLDERED. THE SHADED AREAS INDICATE THEIR SIZE IN RELATION TO THE STONE THEY WOULD HOLD. OFTEN THE TIPS OF THE PRONGS ARE PLANISHED TO MAKE THEM THINNER AND MORE GRACEFUL.

turtle

THIS IS CUT FROM SHEET METAL AND GETS ITS NAME FROM THE ANIMAL IT RESEMBLES. MANY VARIATIONS ARE POSSIBLE.

ARC BENDING JIG

THIS PIECE OF EQUIPMENT IS USED IN THE SETTING SHOWN BELOW AND IN ONE ON THE NEXT PAGE. IT IS NOT ESSENTIAL BUT IS HANDY IF MANY SETTINGS ARE TO BE MADE.

TO USE, CUT A STRIP OF SHEET AS WIDE AS THE CONE MUST BE. SLIDE THIS SNUGLY BETWEEN TWO POSTS AND BEND IN A SERIES OF SHORT CURVES.

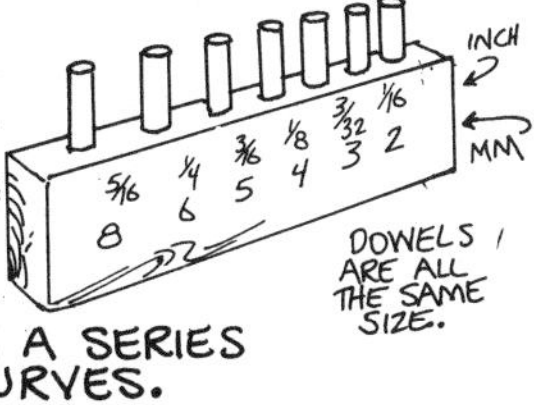

Basket Setting

① BEND TWO V-SHAPES OF WIRE AND PROP THEM UP ON THE SOLDERING BLOCK. SOLDER AT THE POINT OF CONTACT.

② MAKE A RING TO HOLD THE PRONGS TOGETHER. THIS CAN GO ON THE INSIDE (SUBTLE) OR OUTSIDE, (DECORATIVE). IN EITHER CASE MAKE THE RING AND SOLDER IT CLOSED SO IT CAN BE MADE ROUND AND STRETCHED TO THE RIGHT SIZE IF NECESSARY. THE INSIDE RING IS A LITTLE SMALLER THAN THE STONE'S DIAMETER; THE OUTSIDE RING WOULD BE MADE TO FIT AROUND THE STONE (AS IF IT WERE A BEZEL). WHEN THE SIZE IS RIGHT, SOLDER THIS IN PLACE. THE HEAD IS THEN ATTACHED TO THE WORKPIECE, PICKLED, AND POLISHED. A BEARING IS CUT, PRONGS ARE CUT TO THE RIGHT HEIGHT AND SHAPE, AND THE STONE IS SET AS USUAL.

Collet Prong Head

1. LAY OUT ARC; SEE PAGE 135. FOR SMALL STONES ACCURACY BY EYE IS SUFFICIENT.

2. BEND INTO A LOOP AND CLOSE WITH HARD SOLDER.

3. TRUE UP ON A SMALL MANDREL, OFTEN A SCRIBE OR CENTERPUNCH. SET ON VISE JAWS TO STRETCH.

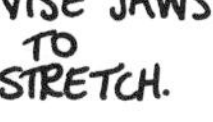

4. FILE A FLAT SURFACE FOR EACH PRONG. KEEP SPACING EVEN.

5. CUT TAPERED PRONGS FROM 18-22 GAUGE SHEET.

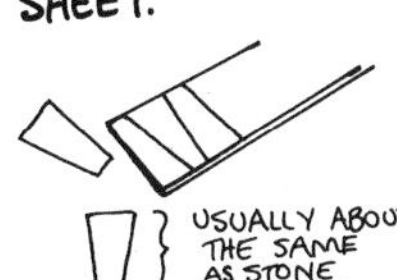

6. SOLDER PRONGS IN PLACE. THEY MAY BE POKED INTO A CHARCOAL BLOCK IF NEEDED.

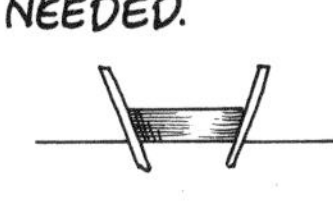

7. PICKLE AND CHECK FIT. STONE SHOULD NOT REST ON THE COLLET. IF IT DOES THE COLLET WAS TOO BIG OR THE ANGLE OF THE CONE WAS TOO STEEP.

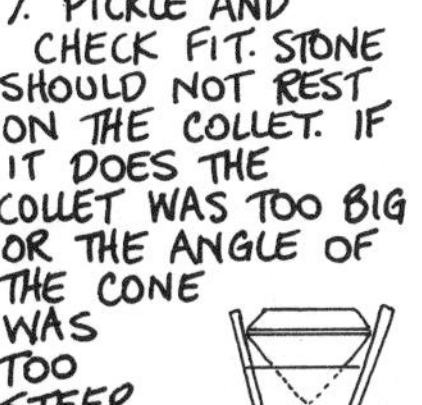

8. SOLDER TO PIECE, POLISH, AND CUT BEARINGS IN EACH PRONG.

Crown Setting

LAY OUT, AND BEND OR SAW FROM SHEET AN ARC TO MAKE A CONE THAT WILL ENCLOSE THE STONE.

BEND CONE WITH PLIERS AND BY TAPPING WITH A MALLET. TRUE THE SEAM BY RUNNING A SAWBLADE ALONG THE JOINT. SOLDER.

AFTER TRUING THE CONE ON A MANDREL. MARK OUT PRONGS, FIRST FROM A TOP VIEW AND THEN ON THE SIDES. USE DIVIDERS TO MARK A LINE PARALLEL TO THE BASE. WITH A SAW, CUT AWAY THE AREA BETWEEN THE PRONGS. THIS WILL BE EASIER IF THE CONE IS MOUNTED ON A DOWEL. USE SEALING WAX OR SHELLAC. FILE TO MAKE PRONGS NEAT AND EVEN.

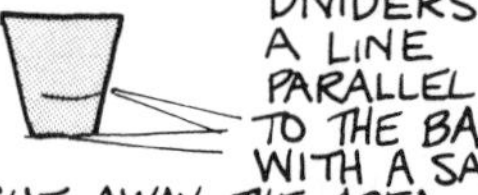

INVERT THE HEAD AND REPEAT, CUTTING AWAY THE AREA BETWEEN THE PRONGS. THIS IS NOT NECESSARY FOR SMALL STONES.

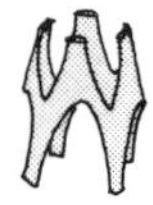

MANY VARIATIONS OF SHAPE ARE POSSIBLE:

BEND AND SOLDER A RING TO BECOME THE BASE OF THE SETTING. EITHER USE SQUARE WIRE OR FLATTEN THE RING SO IT WILL MAKE POSITIVE CONTACT WITH THE BASE OF THE CROWN.

SOLDER TO WORKPIECE. PICKLE AND FILE AS NECESSARY TO PERFECT SHAPE. CUT A BEARING WITH BURS, GRAVER OR A FILE. CUT PRONGS TO THE CORRECT HEIGHT, FILE TO SHAPE, AND SET STONE.

THOUGH MANY HAVE PRACTICED THE ART WITHOUT MAKING DRAWINGS, THOSE WHO MADE THEIR DRAWINGS FIRST DID THE BEST WORK.

BENVENUTO CELLINI

A Multi-Use Base

1. MEASURE AND MARK A STRIP ON MEDIUM-WEIGHT SHEET. DIMENSIONS WILL DEPEND ON USE; SEE #5. MAKE A DEEP NOTCH FOR CORNER BENDS AND FILE A 45° MITRE ON EACH END. WITH DIVIDERS MARK A LINE AT ABOUT ⅓ THE HEIGHT. SCRIBE THIS DEEPLY, FILE, OR CUT WITH A GRAVER.

2. BEND INTO SHAPE AND SOLDER JOINT. DON'T USE TOO MUCH SOLDER.

3. SAW OR FILE AWAY THE CORNERS, REMOVING THE ⅔ PORTION.

4. FOLD SIDES INWARD UNTIL THE CORNERS MAKE A NEAT JOINT. SOLDER.

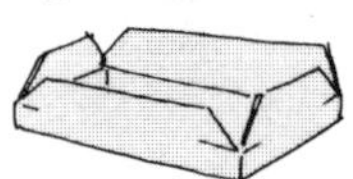

5. THERE ARE SEVERAL USES FOR THIS BOX:

a.

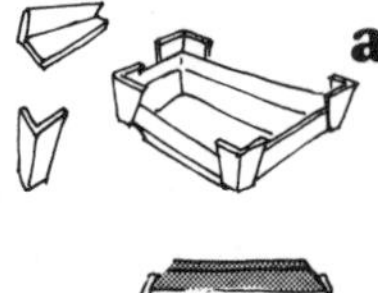

PRONGS ARE MADE OF SHEET METAL, CUT AND FOLDED BEFORE SOLDERING TO THE BASE.

b.

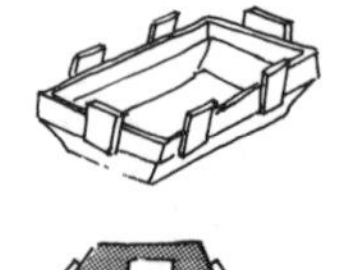

TO MAKE PRONG PLACEMENT EASIER START WITH THIS

c.

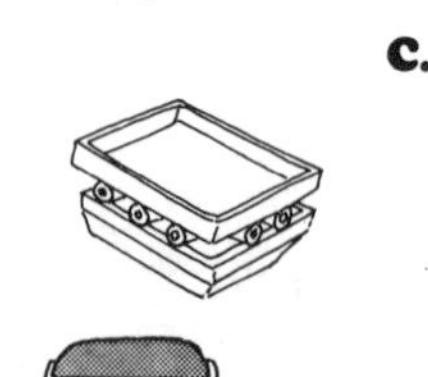

A RAISED BEZEL.

d.

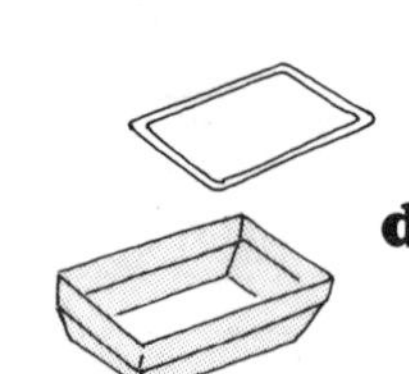

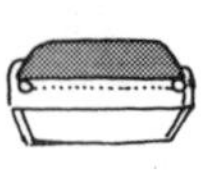

TO MAKE BOX INTO A BEZEL, SOLDER A WIRE IN PLACE AS A BEARING.

STONES

GYPSY

AS SHOWN FOR A RING —

1. MAKE RING SHANK, ALLOWING FOR STONE'S DIAMETER AND IF FACETED, FOR ITS HEIGHT.
2. DRILL A HOLE ABOUT HALF THE SIZE OF THE STONE'S DIAMETER.
3. ENLARGE THE SETTING WITH A BUR OR GRAVER TO HOLD THE STONE. WHEN CORRECT THE STONE'S GIRDLE SHOULD BE JUST BELOW THE SURFACE.
4. FOR A CABACHON THE HOLE WOULD LOOK LIKE THIS. USE A FLAT GRAVER OR CYLINDER BUR TO CARVE IT OUT.
5. FILE AWAY A SMALL BIT OF METAL JUST AROUND THE SETTING. THIS WILL CREATE A TINY RIM THAT WILL BE PUSHED OVER THE STONE.
6. WITH THE STONE IN PLACE SET FOUR "CORNERS" WITH A BEZEL PUSHER OR A CHASING HAMMER AND A SMALL PLANISHING TOOL. CHECK TO SEE THAT THE STONE IS LEVEL AND THAT THERE IS A CONSISTENT RIM TO COVER THE STONE. IF SO, CONTINUE WITH THE SETTING AND FOLLOW WITH A BURNISHER, PULLING THE METAL UP ONTO THE STONE.

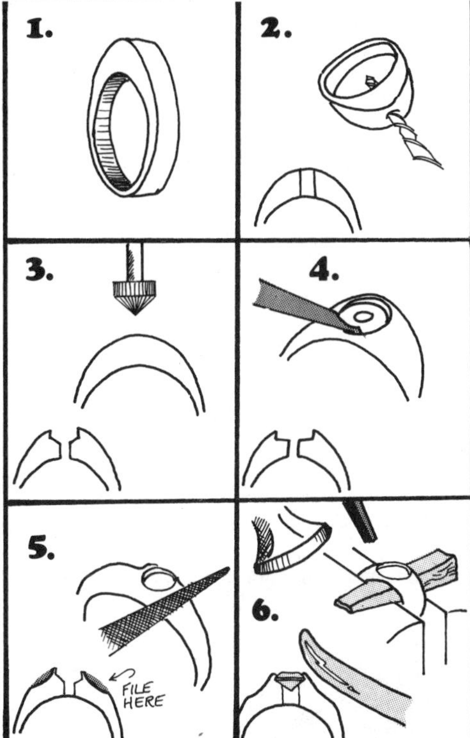

ROMAN
- A VARIATION -

WHEN A BEARING HAS BEEN MADE THE STONE IS SET INTO PLACE AND A BURNISHER IS USED CONCENTRICALLY AROUND THE SETTING TO WEAR DOWN A GROOVE. AS A TRENCH IS FORMED, PRESSURE IS INCREASED, PUSHING THE METAL OUTWARD AND ONTO THE STONE. IN EFFECT, THIS IS A COMBINATION OF STEPS 5 AND 6 ABOVE.

Pavé

DEFINITION: A SETTING IN WHICH THE STONES ARE SET SO CLOSE TOGETHER THAT LITTLE METAL CAN BE SEEN. THE OBJECT IS "PAVED" WITH STONES.

1. LOCATING STONES ACCURATELY IS CRITICAL. SET INTO PLASTI'CENE AND TAKE MEASUREMENTS FROM CENTER TO CENTER OF EACH STONE WITH DIVIDERS. TRANSFER THESE TO THE METAL; DRILL HOLES SLIGHTLY SMALLER THAN THE STONES.

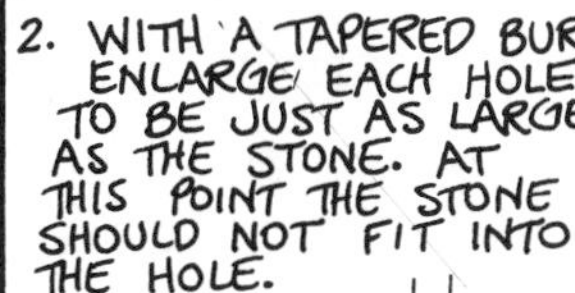

2. WITH A TAPERED BUR ENLARGE EACH HOLE TO BE JUST AS LARGE AS THE STONE. AT THIS POINT THE STONE SHOULD NOT FIT INTO THE HOLE.

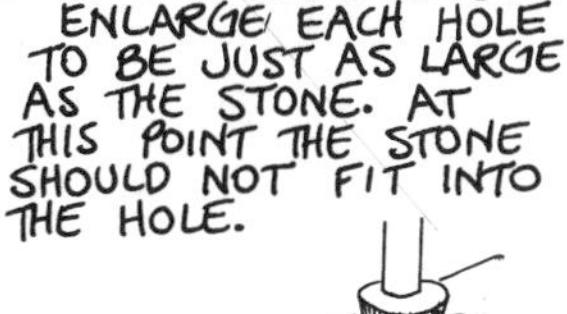

3. WITH A GRAVER OR BUR, CUT A BEARING FOR EACH STONE. NOW THE STONE SHOULD DROP INTO THE HOLE SO THAT ITS GIRDLE IS JUST BELOW THE SURFACE.

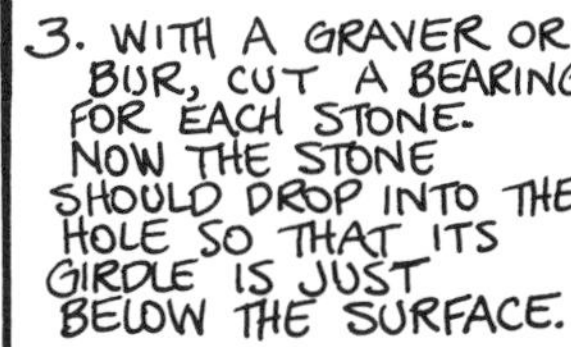

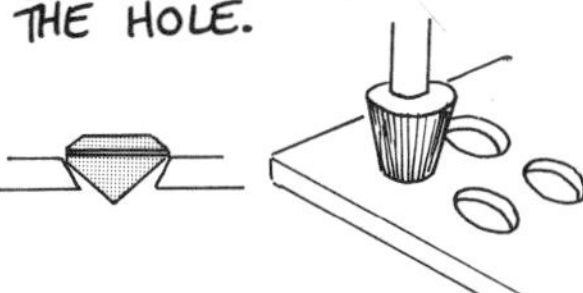

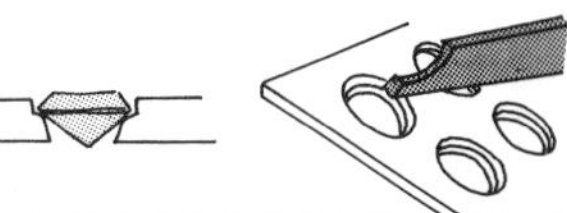

4. WITH A FLAT GRAVER CUT AWAY THE AREA AROUND THE HOLES, LEAVING 3 OR 4 TRIANGLES OF METAL EVENLY SPACED AROUND EACH.

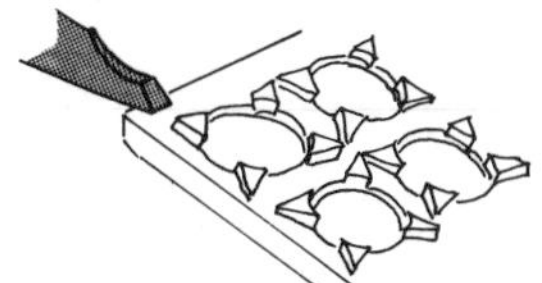

5. WITH A ROUND GRAVER LIFT EACH TRIANGLE UP BY CUTTING PART WAY UNDER IT. STONES SHOULD BE IN PLACE BECAUSE THE PRESSURE INWARD WILL START TO GRIP THEM.

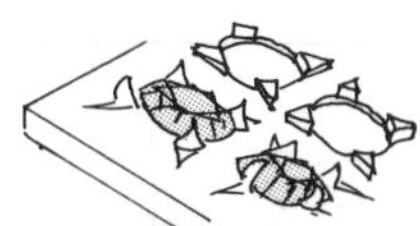

6. WITH A BEADING TOOL, GIVE A CLEAN HEMISPHERICAL SHAPE TO EACH PRONG. THIS IS USED IN A WIGGLING MOTION AND MAY BE LUBRICATED WITH SPIT.

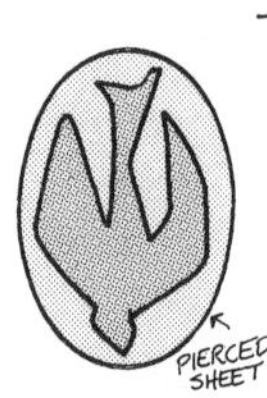

THERE ARE TIMES WHEN, FOR TECHNICAL OR AESTHETIC REASONS, A STONE IS BEST SET FROM THE BACK. A STANDARD BEZEL MAY BE USED, OF COURSE. ANOTHER POSSIBILITY IS THE FRAME-AND-STITCH METHOD SHOWN HERE.

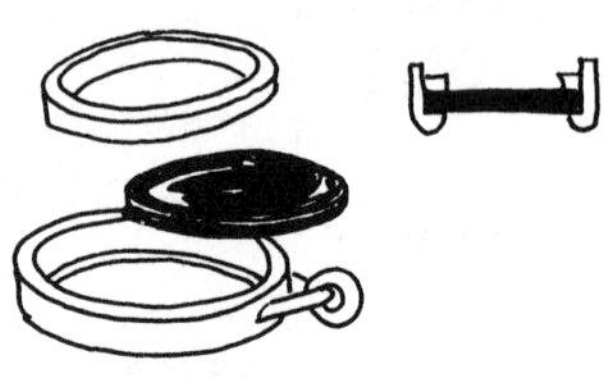

1. MAKE BEZEL TO SURROUND STONE, SLIGHTLY HIGHER THAN USUAL. TURN IN TOP EDGE WITH A BURNISHER OR SOLDER A RIM OR PIERCED SHEET IN PLACE.
2. MAKE A COLLAR TO FIT INSIDE THE BEZEL. SQUARE OR RECTANGULAR WIRE IS HANDY FOR THIS. A SNUG FIT WILL HELP IN THE NEXT STEP.
3. WHEN PIECE IS POLISHED AND CLEANED, THE STONE IS SET INTO PLACE AND THE COLLAR IS PUSHED IN AGAINST IT FROM THE BACK.
4. WITH A SQUARE OR ROUND GRAVER, SMALL CURLS OF METAL ARE CUT IN THE BEZEL WALL AND ROLLED DOWN ONTO THE COLLAR.

LAMINATION

CAREFULLY SAW AND FILE A SHEET TO MATCH THE SIZE AND SLOPE OF A STONE. THIS SHEET IS THEN SET OVER THE STONE AND HELD IN PLACE WITH RIVETS OR SCREWS. THE EDGE ADJACENT TO THE STONE CAN BE BURNISHED FOR A TIGHT FIT.

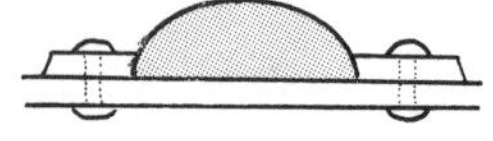

THE UNIVERSE IS FULL OF MAGICAL THINGS PATIENTLY WAITING FOR OUR WITS TO GROW SHARPER.
—EDEN PHILPOTS

FOIL

SILVER AND GOLD FOIL (OR LEAF) CAN BE USED LIKE A KIND OF METAL PUTTY TO PACK AROUND A STONE.
- CARVE HOLE FOR STONE, UNDER-CUT EDGES SLIGHTLY.
- SET STONE IN PLACE AND LAY A SMALL ROLL OF FOIL AROUND IT. PACK THIS DOWN INTO RECESS.
- CONTINUE ADDING AND PACKING UNTIL STONE IS SECURE.

A VARIATION ON THE ABOVE.... IS TO HOLD THE COLLAR IN PLACE WITH SCREWS OR RIVETS. NOTE HOW WELL THESE TECHNIQUES LEND THEMSELVES TO A 2-SIDED IMAGE.

PEDESTAL PRONG

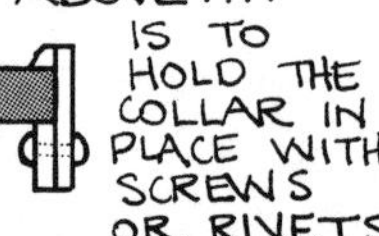

THIS SETTING CAN BE MADE WITH 3, 4, 5, OR 6 PRONGS AND CAN BE USED WITH CABS OR FACETED STONES.

1. MAKE A RING OF 16-20 GA. SHEET: OUTSIDE DIA. SHOULD EQUAL O.D. OF STONE.
2. FOR FACETED STONE, FILE BEVEL AROUND THE INSIDE EDGE.
3. CUT OVERSIZE LENGTHS OF SQUARE OR HALF-ROUND WIRE FOR PRONGS. FILE POINT ON EACH ONE.
4. PUSH WIRES INTO CHARCOAL AROUND THE RING (PEDESTAL). SOLDER.
5. AFTER COMPLETING PIECE, PRONGS ARE TRIMMED, SHAPED, AND PUSHED OVER STONE.

PLIERS

Cut Down Setting

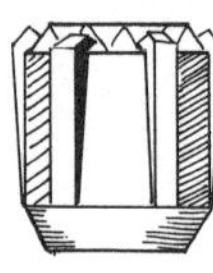

1. MAKE A COLLAR (TUBE) AS TALL AS THE DESIRED SETTING AND SLIGHTLY TOO SMALL FOR THE STONE TO FIT INSIDE. WALLS MUST BE THICK.
2. CUT A BEARING (SEAT) WITH BURS OR GRAVERS.
3. MARK LINE AROUND LOWER PORTION WITH DIVIDERS AND FILE TO VISUALLY LIGHTEN FORM.
4. LAY OUT PRONGS—4 OR 6—WITH A SCRIBE.
5. CUT AWAY METAL BETWEEN PRONGS WITH A FLAT GRAVER SO PRONGS ARE THE ORIGINAL THICKNESS AT THE TOP AND BLEND INTO TUBE AT THE LOWER LINE. COLLAR MAY BE HELD ON SHELLAC-COVERED PENCIL POINT WHILE CUTTING.
6. POLISH AND SET STONE.

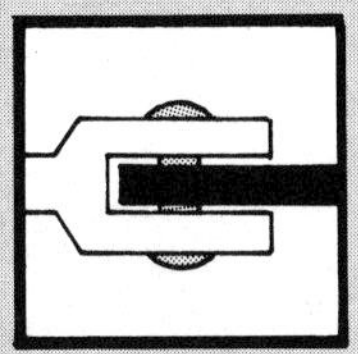

MECHANICS

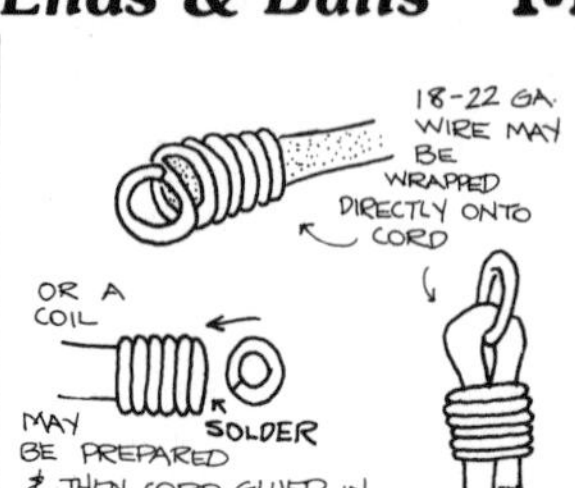

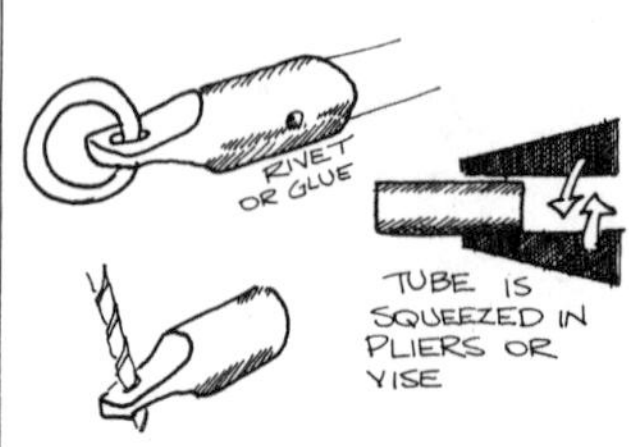

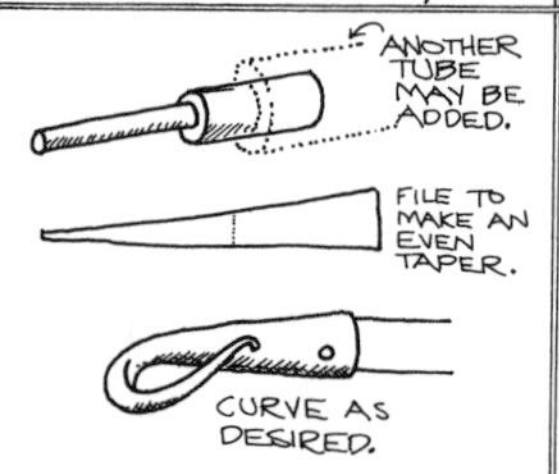

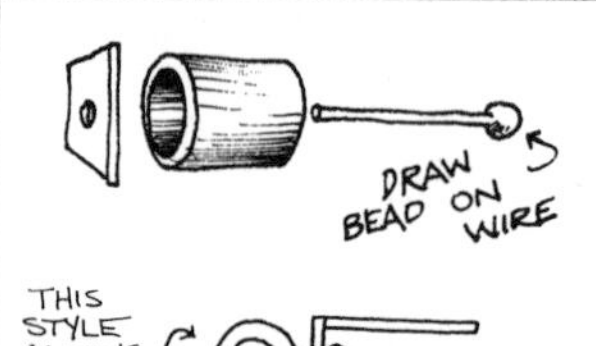

THIS STYLE ALLOWS ROTATION AS SHOWN.

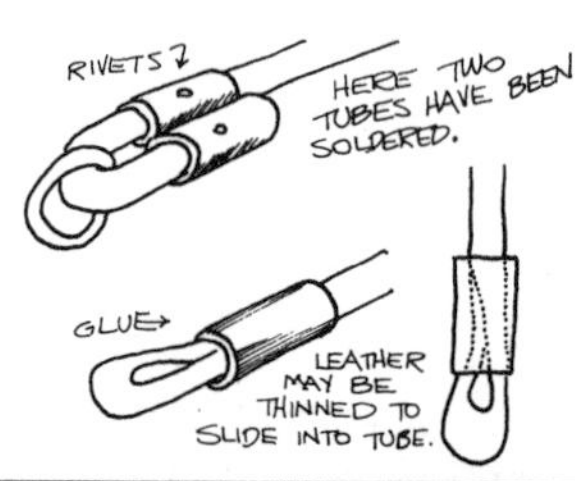

BAIL: (from Old Norse beygla; hook or ring) A SEMICIRCULAR HANDLE OF A PAIL, KETTLE, OR CANNON.

SOME SAMPLE BAILS:

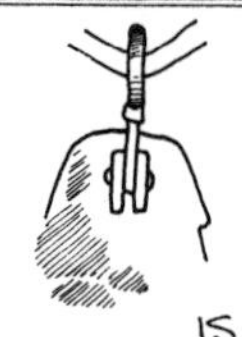

fold down bail

FOR A PIN THAT IS ALSO WORN AS A PENDANT.

1. WIRE IS BEADED, FLATTENED, & DRILLED
2. SHEET IS CUT & FOLDED AS SHOWN
3. LOOP IS FORMED AND RIVETED ONTO Ⓐ. BEND LOOP TO ONE SIDE TO DO THIS.
4. HOLDER Ⓑ IS SOLDERED INTO PLACE AND UNIT IS ASSEMBLED.

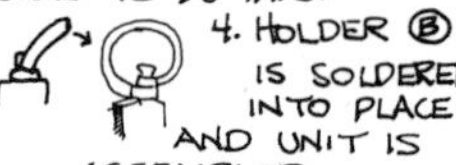

A SIMPLE TAB ARRANGEMENT MAKES A CLASP FOR THIS STYLE OF PIERCED EARRING.

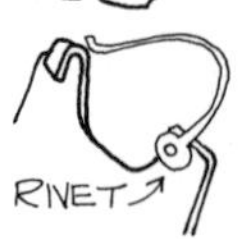

EARRINGS:

- THOUGH SOME PEOPLE CAN WEAR ONLY GOLD OR STAINLESS WIRES, MOST PEOPLE CAN WEAR STERLING.
- 22 GAUGE WIRE IS USUALLY A COMFORTABLE THICKNESS.
- WORK HARDEN WIRES BY TWISTING.
- STANDARD LENGTH OF POSTS IS 9MM - 3/8 INCH.

MAKING FRICTION BACKS

1. CUT STRIP OF 22 GA. STERLING (24 GA. GOLD) ABOUT 12 MM LONG.
2. POUND CENTER WITH SCRIBE TO FORM FUNNEL-LIKE DEPRESSION. DRILL HOLE.
3. MAKE GRIP WITH ▲ FILE.
4. POLISH.
5. BEND AS SHOWN.

THE HOOK IN A GRAVITY-HELD EARRING SHOULD BE ABOUT 25MM (1") LONG.

FRICTION NUTS SHOULD SIT IN A GROOVE AT THE END OF THE POST.

MECHANICS *Pins & Brooches*

THE PIN MECHANISM MUST BE LOCATED ABOVE THE CENTRAL AXIS TO PREVENT THE BROOCH FROM TIPPING FORWARD.

AT REST, THE PINSTEM SHOULD BE AS SHOWN – SLIGHTLY ABOVE THE CATCH. THIS WILL CREATE A TENSION THAT WILL HELP KEEP THE PIN CLOSED. A SIMILAR TENSION IS PUT ON THE ELEVATION OF THE STEM BY THE 'FOOT' OF THE PINSTEM.

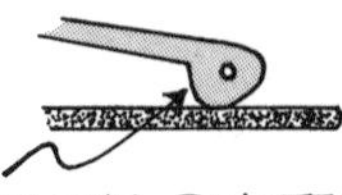

THE END OF THE PIN SHOULD NOT EXTEND BEYOND THE CATCH.

THE POINT OF THE PIN MUST BE SHARP AND SMOOTH TO PENETRATE FABRIC. SNIP TO SIZE, FILE, BURNISH.

CATCH IS POSITIONED WITH OPENING DOWNWARD.

PINHOLDER & CATCH ARE SWEAT SOLDERED CAREFULLY, USING A TINY PIECE OF EASY SOLDER. MELT SOLDER ONTO BROOCH AND SET PIN PIECES INTO PUDDLE.

INTEGRAL PIN:

BY CAREFUL SAWING, AN UNUSUAL PIN MECHANISM CAN BE "BUILT IN" TO THE PIECE.

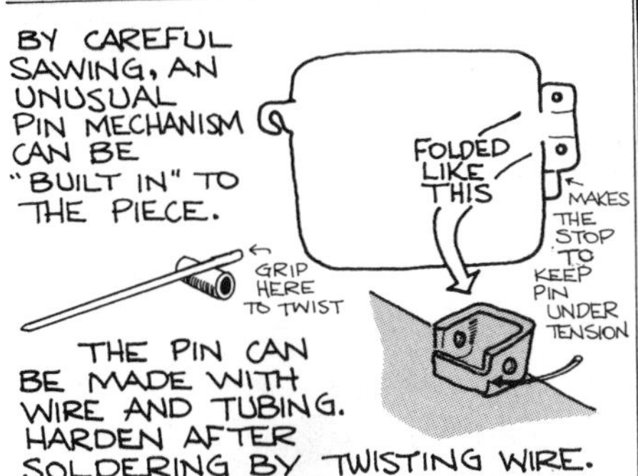

THE PIN CAN BE MADE WITH WIRE AND TUBING. HARDEN AFTER SOLDERING BY TWISTING WIRE.

tube-in-a-tube catch

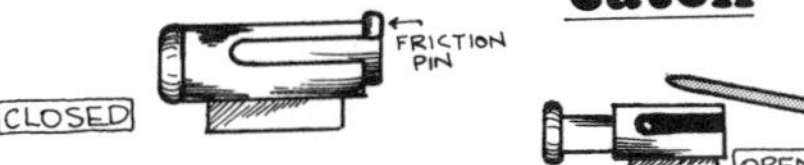

THIS CATCH IS HELD CLOSED BY THE FRICTION OF A SMALL PIN AGAINST THE END OF THE LARGER TUBE. TO OPEN, ROTATE SMALL TUBE UNTIL PIN ENGAGES SLOT. PIN KEEPS THE SMALLER TUBE FROM COMING ALL THE WAY OUT.

CONSTRUCTION DETAILS:

1. MAKE TELESCOPING TUBING. SMALLER I.D. MUST FIT PINSTEM.
2. SOLDER STRIP OF SHEET ONTO LARGER TUBE TO BE PEDESTAL.
3. CUT SLOT IN LARGER TUBE WITH SAW. PINSTEM MUST FIT HERE.
4. DRAW WIRE TO MATCH SLOT; SOLDER AT RIGHT ANGLE TO END OF SMALLER TUBE. CUT OFF ALL BUT ONE MM.
5. SLIDE TUBES TOGETHER AND WITH THE SMALL PIN TIGHT AGAINST THE END, SAW THE OTHER END OFF FLUSH.
6. WITH TUBES IN OPEN POSITION, SOLDER KNOB ON END OF SMALLER TUBE. KNOB IS SHEET, BEAD, OR BEZEL.

A VERY SIMPLE PIN:

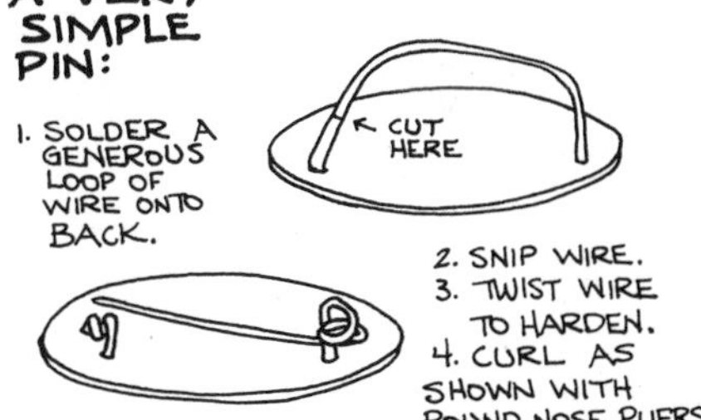

1. SOLDER A GENEROUS LOOP OF WIRE ONTO BACK.
2. SNIP WIRE.
3. TWIST WIRE TO HARDEN.
4. CURL AS SHOWN WITH ROUND NOSE PLIERS.

① PIN IS MADE BY SOLDERING WIRE TO PIECE OF SHEET.

② CATCH IS L-SHAPED PIECE OF SHEET OR FLATTENED WIRE.

③ PINHOLDER IS BENT OVER SHEET USED TO MAKE PIN END. AFTER SOLDERING, THE CURVE IS SAWN OFF AND HOLES ARE DRILLED.

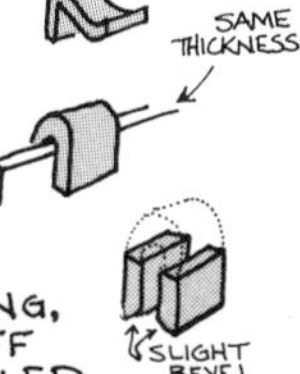

A SIMPLE WIRE PIN WITH ITS OWN FOOT OR BEARER

PRODUCTION SEQUENCE FOR A HINGE-TYPE PIN:

1. DIVIDE A SMALL TUBE INTO 3 MM SECTIONS. SAW PART WAY THROUGH.
2. REMOVE SECOND, FIFTH, EIGHTH, ELEVENTH PART.
3. SOLDER ONTO METAL STRIP WITH THE OPENINGS DOWN.
4. SAW AWAY THE REMAINING PARTS OF THE OPEN SECTIONS.
5. SOLDER A SECOND STRIP AT RIGHT ANGLES TO THE FIRST.
6. CUT INTO PINHOLDERS.

7. PIN IS MADE WITH PIECE OF SAME TUBE.

MECHANICS

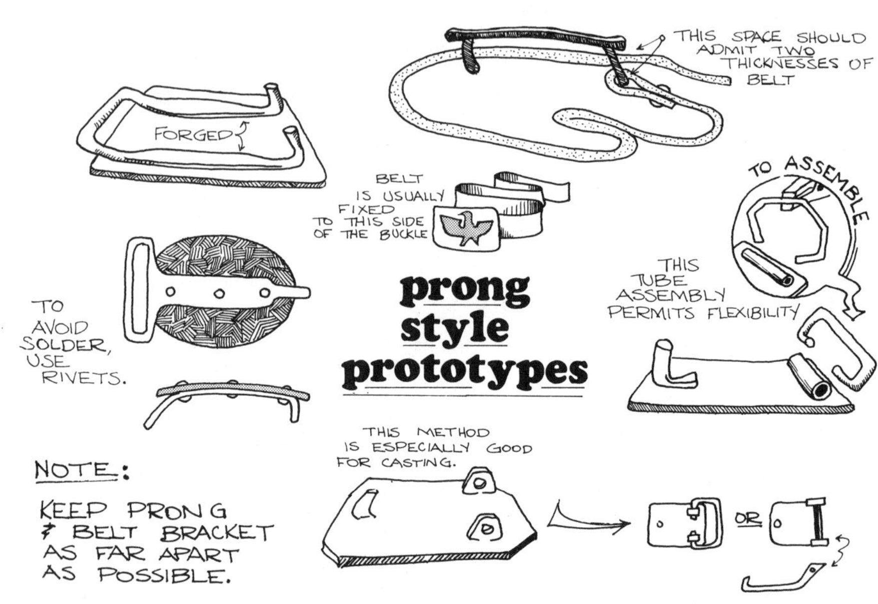

THE PIECES I MAKE ARE LIKE MILEPOSTS. IF I STOP AT ONE, IT MIGHT MEAN THAT I'LL PITCH A TENT AND THE TRIP WILL BE OVER.

HEIKKI SEPPÄ

cuff link prototypes

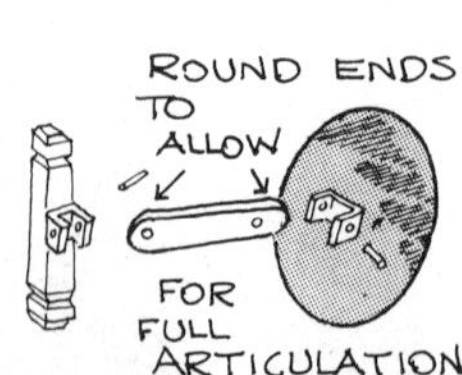

MECHANICS

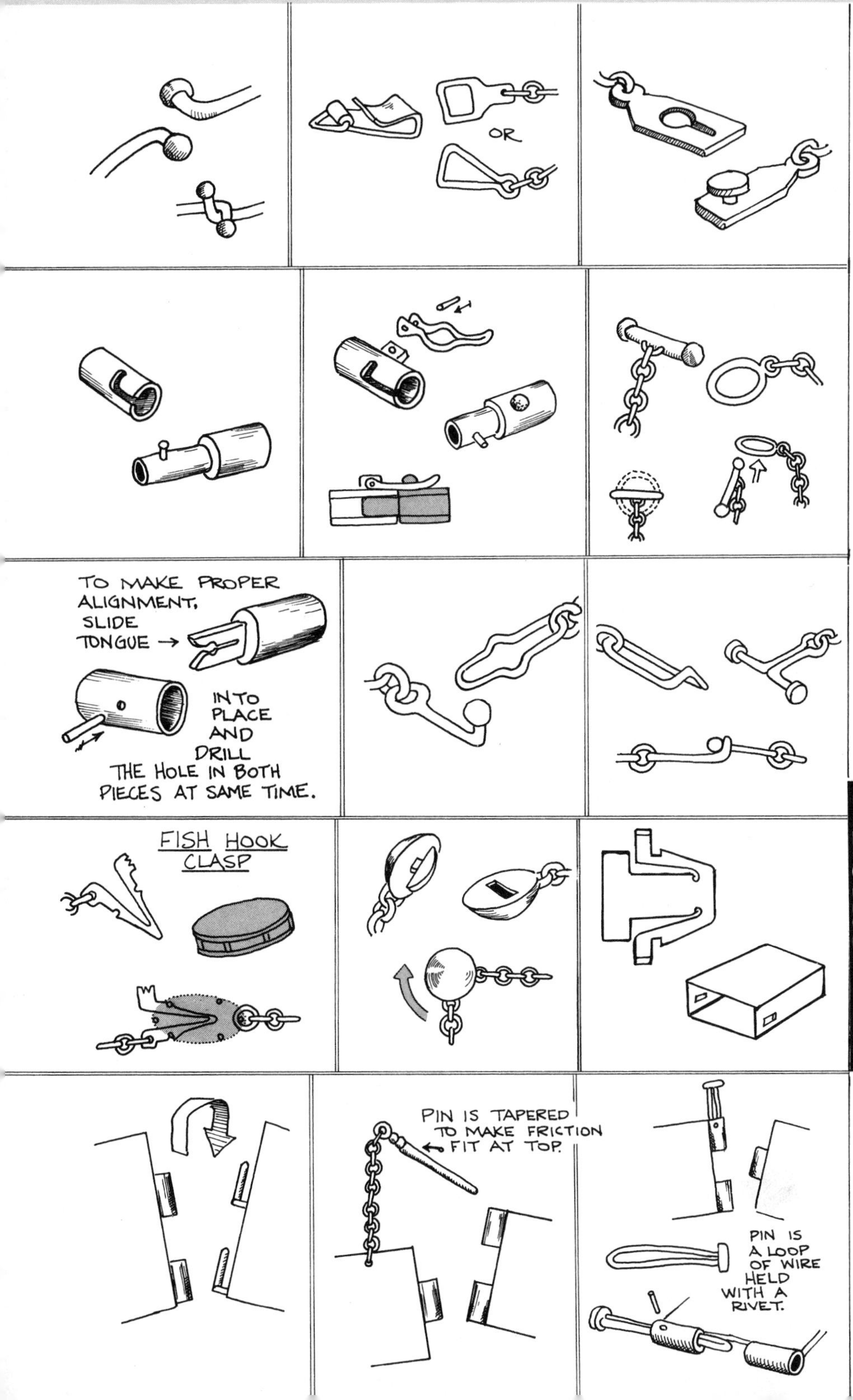

OR

TO MAKE PROPER ALIGNMENT, SLIDE TONGUE → INTO PLACE AND DRILL THE HOLE IN BOTH PIECES AT SAME TIME.

FISH HOOK CLASP

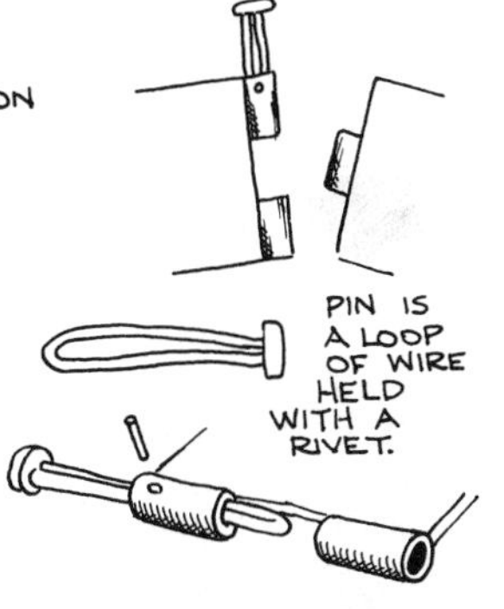

A **B**

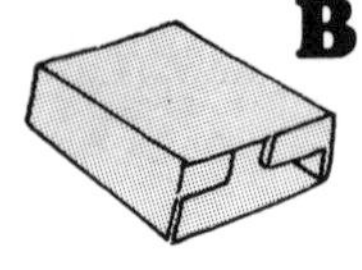
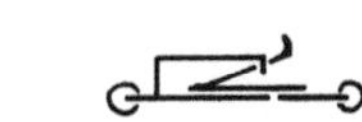

BY SOLDERING THE TONGUE ONTO ANOTHER SHEET, IT IS KEPT IN THE SAME PLANE AS THE BOX.

THESE TWO BASIC STYLES CAN BE MADE USING ANY OF THESE APPROACHES:

ANY OF THESE STYLES MAY BE GIVEN A DOUBLE LOCKING MECHANISM BY ADDING A BALL AND LOOP

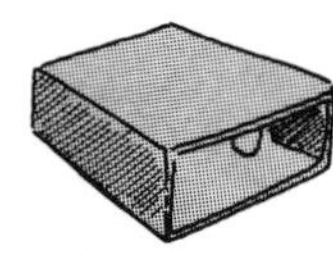
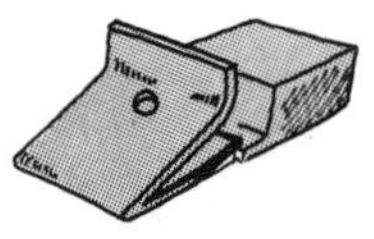

THIS FISH HOOK CLASP IS EASILY MADE FROM 14, 16, OR 18 GAUGE HALF-ROUND WIRE.

TO CLASP: HOOK AS SHOWN AND SLIDE UNTIL NOTCH CLICKS ON FIRST RING.

OPEN BY PUSHING DOWN ON TAB

THIS CLASP CAN BE CONSTRUCTED AS SHOWN ABOVE.

THE ADVANTAGE OF THIS STYLE IS THAT THE LOCATION OF THE HOLE CAN WAIT UNTIL THE CLASP IS COMPLETED. PAINT THE TONGUE AND SLIDE HALVES TOGETHER. DRILL AT THE END OF THE TRAIL LEFT IN THE PAINT.

- IN ALL THESE CLASPS IT IS VERY IMPORTANT TO <u>MEASURE CAREFULLY.</u> IT IS OFTEN HELPFUL TO MAKE THE RECEIVING SIDE FIRST AND THEN MAKE A TEST PIECE IN COPPER OR BRASS TO CHECK THE FIT.
- THE TONGUE SHOULD SLIDE SNUGLY INTO ITS BAY, WITH NO SLOPPINESS SIDE TO SIDE.
- THE TONGUE IS USUALLY 24 GAUGE METAL. AFTER FOLDING AND CHECKING THE FIT, PLANISH ON FOLD TO HARDEN.
- THE AMOUNT OF SQUEEZE NEEDED TO RELEASE SHOULD BE SLIGHT. THIS IS MADE BY HAVING A LONG TONGUE; 10 MM WOULD BE AVERAGE. TOO SHORT REQUIRES MORE PUSH TO RELEASE AND IS THEREFORE MORE LIKELY TO BREAK.

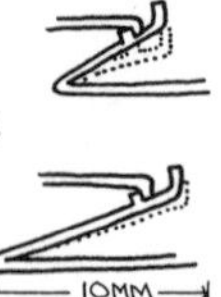

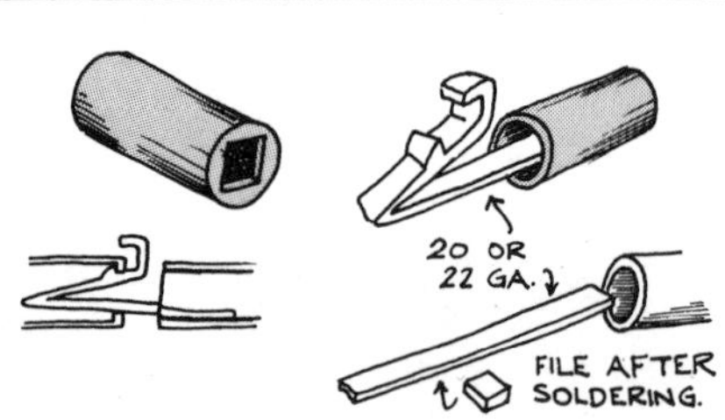

I HEAR AND I FORGET.
I SEE AND I REMEMBER.
I DO AND I UNDERSTAND.

CHINESE PROVERB

THIS SHORT CUT CLASP IS MADE BY DRILLING A HOLE, PIERCING A **U**-SHAPED LINE, AND SOLDERING A PEG INTO THE DRILLED HOLE. LIFT THE TONGUE WHILE SOLDERING.

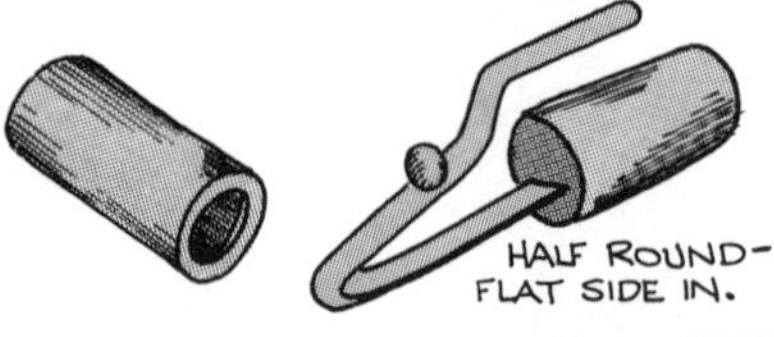

THIS CLASP HAS THE ADVANTAGE OF ROTATING, MAKING IT MORE COMFORTABLE AND EASIER TO FASTEN.

Spring Plunger Clasp

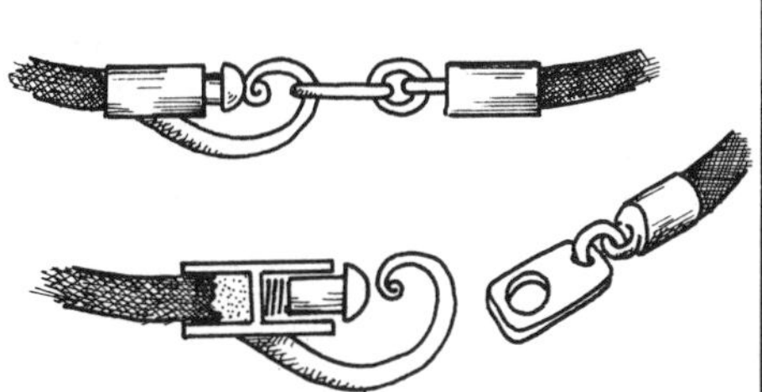

1. YOU WILL NEED A SHORT LENGTH OF TUBING AND A WIRE THAT FITS SNUGLY INSIDE IT. 14 GA. WIRE IS A CONVENIENT SIZE.

TO MAKE THE DIVIDING PARTITION IN THE TUBE, SAW OFF A THIN SLICE OF THE WIRE AND STRETCH IT SLIGHTLY BY HAMMERING. PUSH IT INTO THE TUBE AND SOLDER IT

2. FORGE OR FILE

SOLDER

BEND INTO A CURL.

BE SURE THAT THE END OF THE CURL IS "TUCKED UNDER", NOT LIKE THIS.

3.

MAKE A NAILHEAD ON THE WIRE; SEE PAGE 60.

FILE TO A SMOOTH AND SYMMETRICAL DOME.

4.

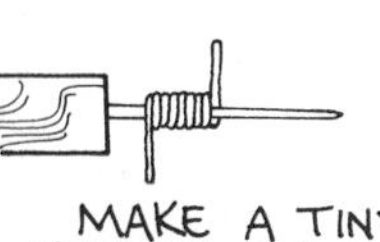

MAKE A TINY SPRING BY TIGHTLY WRAPPING HARD-DRAWN 28 GA. BRASS WIRE. FOUND SPRINGS FROM CLOCKS, TOYS, OR PENS CAN ALSO BE USED.

5. AFTER POLISHING, BEND THE CURL OFF TO THE SIDE TO ALLOW THE SPRING AND PLUNGER TO BE SLID INTO PLACE. BEND WIRE BACK TO HOLD THEM.

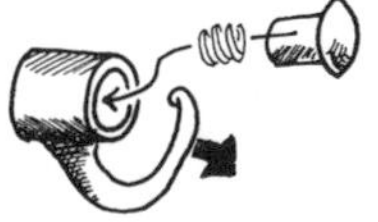

6. TO CLASP, PUSH THE RECTANGULAR TAB STRAIGHT IN; THE PLUNGER WILL SLIDE INTO THE TUBE.

TO UNCLASP, JUST PULL STRAIGHT OUT.

Basic Spring Clasp

FOR PINS, BARRETTES, EARRINGS, ETC.

1. SOLDER A BAR OF 14 OR 16 GAUGE SHEET ONTO THE BACK OF THE PIECE. FILE A GROOVE ON THE TOP EDGE.
2. SOLDER A TUBE IN THE CENTER OF THIS PIECE, RESTING IN THE GROOVE.
3. SOLDER HEAVY SQUARE WIRE ONTO TOP BACK EDGE OF TUBE. AFTER SOLDERING FILE TO THIS SHAPE:
4. CUT ANOTHER PIECE OF SHEET AS THICK AND WIDE AS THE BAR (#1) BUT SLIGHTLY LONGER. SOLDER PINS ON THE SIDES.
5. SAW TWO LINES IN FROM THE TOP EDGE, CENTERED AND THE SAME DISTANCE APART AS THE LENGTH OF THE TUBE IN #2.
6. BEND UP CENTER TAB AND CUT AWAY ABOUT 2 MM OF THE SIDE FLAPS.
7. FILE A GROOVE INTO THE SHORTENED ENDS.
8. SOLDER A PIECE OF THE SAME TUBE USED BEFORE ACROSS THE SHORT TABS. SAW OUT THE CENTER SECTION OF THE TUBE.

9. HARDEN PINS BY TWISTING; TAPER AND SHARPEN THEM. AFTER POLISHING, ASSEMBLE WITH A PIN AS YOU WOULD FOR A HINGE. THIS MIGHT REQUIRE PUSHING DOWN ON THE SPRING UNIT (PIECE WITH PINS). ADJUST CENTER TAB BY BENDING WITH PLIERS TO CREATE THE CORRECT AMOUNT OF TENSION.

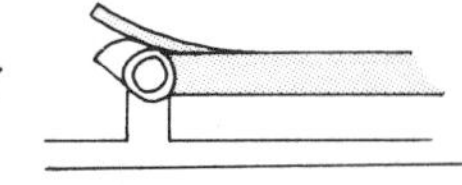

Making Jump Rings

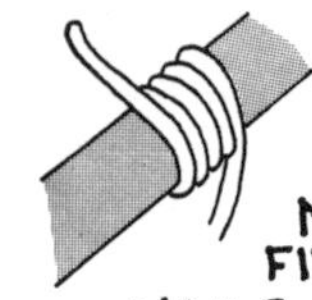

WRAP A WIRE AROUND A ROD OF THE CHOSEN SIZE. SOME HANDY MANDRELS ARE FILE HANDLES, NAILS, DOWELS, WIRE, KNITTING NEEDLES.

SLIDE THE COIL OFF AND SAW AS SHOWN OR CUT WITH A SEPARATING DISK.

WIRE WRAPPED AROUND A SQUARE OR RECTANGULAR MANDREL IS DIFFICULT TO SLIDE OFF. TO PROVIDE CLEARANCE, WRAP FORM WITH MASKING TAPE. AFTER WRAPPING WIRE, BURN TAPE AWAY—WIRE WILL THEN SLIDE OFF.

MANDREL MADE OF WOOD, PLASTIC OR HEAVY GAUGE METAL

ASSEMBLY SEQUENCE

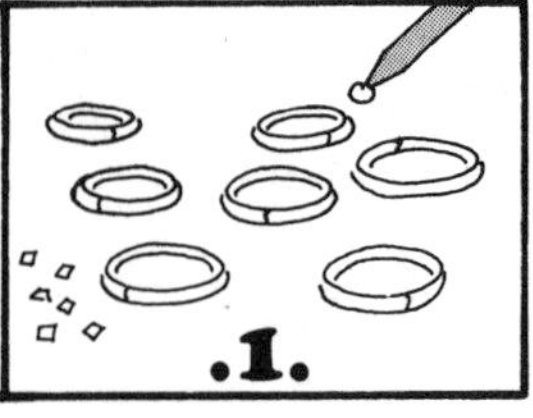

MAKE AS MANY RINGS AS YOU THINK WILL BE NEEDED. BEND HALF CLOSED AND JOIN WITH HARD SOLDER.

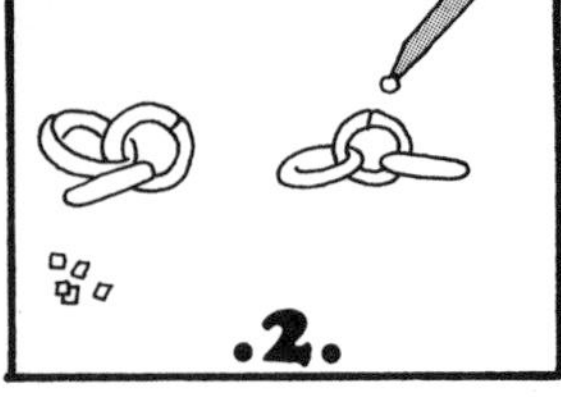

THREAD A PAIR OF CLOSED RINGS ONTO A FRESH RING. CLOSE IT AND SOLDER WITH MEDIUM.

CONNECT A PAIR OF 3-PIECE UNITS WITH A FRESH RING AND SOLDER IT WITH EASY. CONTINUE JOINING UNITS OF 7, 15, 31, ETC.

POLISHING CHAINS

NEVER POLISH CHAINS ON THE BUFFING WHEEL! UNLESS YOU HAVE AN OVER-SUPPLY OF FINGERS...

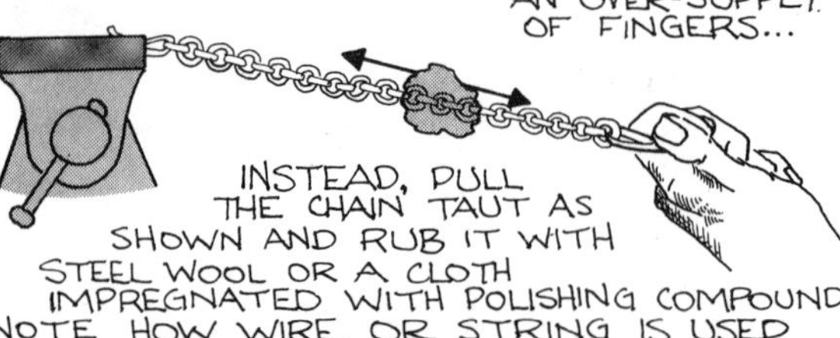

INSTEAD, PULL THE CHAIN TAUT AS SHOWN AND RUB IT WITH STEEL WOOL OR A CLOTH IMPREGNATED WITH POLISHING COMPOUND. NOTE HOW WIRE OR STRING IS USED AT EACH END TO HOLD THE CHAIN.

USING TWO PAIRS OF PLIERS AN OVAL LINK CHAIN MAY BE TWISTED, LINK BY LINK. THE RESULT LOOKS LIKE THIS:

A CHAIN MAY ALSO BE TWISTED BY HOLDING ONE END IN A VISE AND TWISTING: SOLDER JOINTS MUST BE GOOD.

Should this ring be soldered?

IT'S A GOOD QUESTION AND ONE THAT COMES UP OFTEN. WHERE IT IS POSSIBLE THE ANSWER IS PROBABLY YES. UNSOLDERED JUMP RINGS CAN LOOK MESSY AND WEAKEN A CHAIN.

WHERE YOU DO NOT WANT TO SOLDER THE PROPER QUESTION IS, HOW THICK A WIRE AND SMALL A RING DO I NEED TO PROVIDE ENOUGH STRENGTH? REMEMBER THAT JUMP RINGS CAN BE MADE OF WORK-HARDENED WIRE.

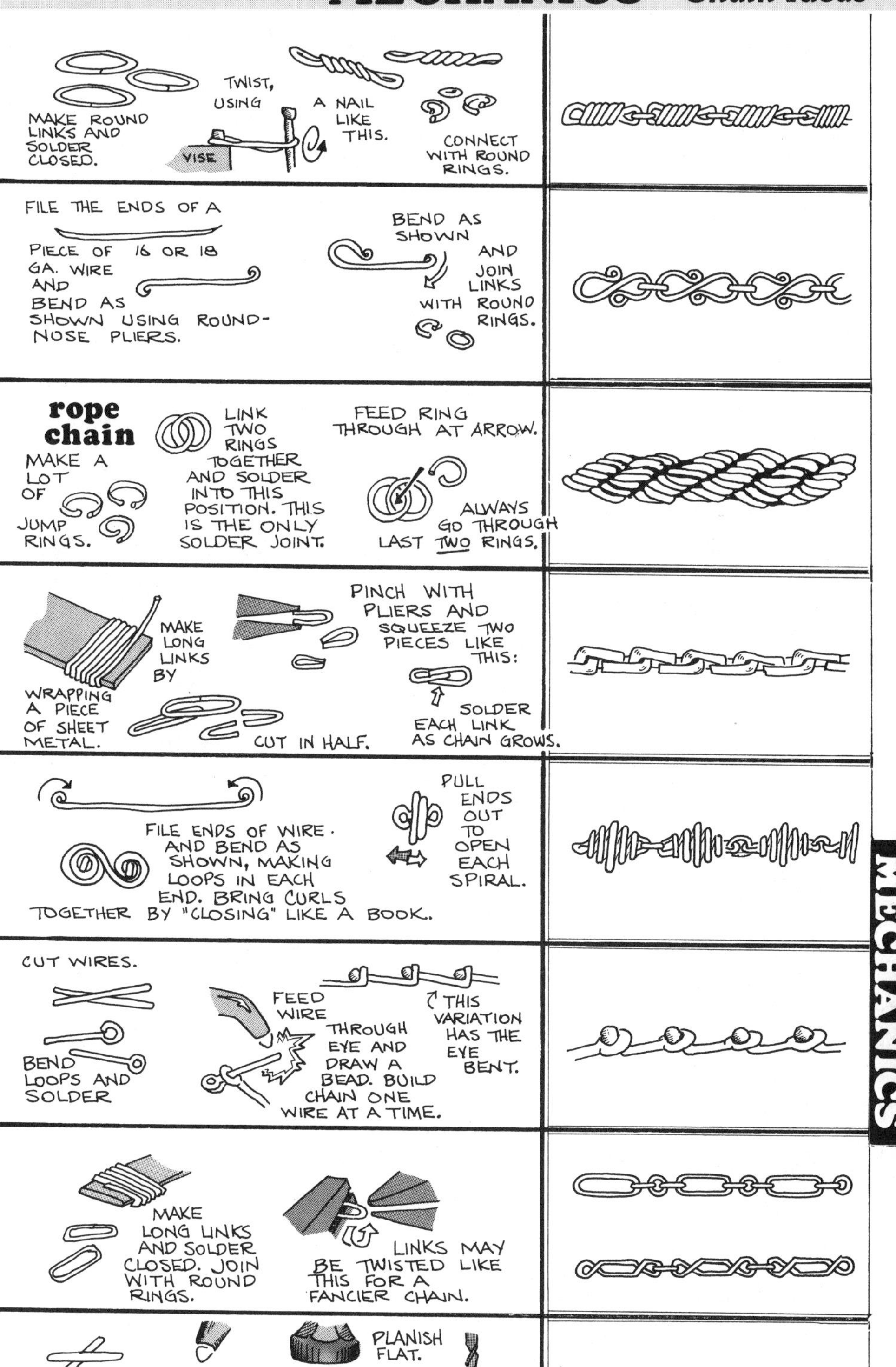
MAKE ROUND LINKS AND SOLDER CLOSED.
TWIST, USING
A NAIL LIKE THIS.
VISE
CONNECT WITH ROUND RINGS.
FILE THE ENDS OF A
PIECE OF 16 OR 18 GA. WIRE AND BEND AS SHOWN USING ROUND-NOSE PLIERS.
BEND AS SHOWN AND JOIN LINKS WITH ROUND RINGS.
rope chain
MAKE A LOT OF JUMP RINGS.
LINK TWO RINGS TOGETHER AND SOLDER INTO THIS POSITION. THIS IS THE ONLY SOLDER JOINT.
FEED RING THROUGH AT ARROW.
ALWAYS GO THROUGH LAST TWO RINGS.
MAKE LONG LINKS BY
WRAPPING A PIECE OF SHEET METAL.
CUT IN HALF.
PINCH WITH PLIERS AND SQUEEZE TWO PIECES LIKE THIS:
SOLDER EACH LINK AS CHAIN GROWS.
FILE ENDS OF WIRE. AND BEND AS SHOWN, MAKING LOOPS IN EACH END. BRING CURLS TOGETHER BY "CLOSING" LIKE A BOOK.
PULL ENDS OUT TO OPEN EACH SPIRAL.
CUT WIRES.
BEND LOOPS AND SOLDER
FEED WIRE THROUGH EYE AND DRAW A BEAD. BUILD CHAIN ONE WIRE AT A TIME.
THIS VARIATION HAS THE EYE BENT.
MAKE LONG LINKS AND SOLDER CLOSED. JOIN WITH ROUND RINGS.
LINKS MAY BE TWISTED LIKE THIS FOR A FANCIER CHAIN.
CUT WIRES AND
DRAW A BEAD ON BOTH ENDS.
PLANISH FLAT.
DRILL AND JOIN WITH ROUND LINKS.

GIMBALS

VARIATIONS ON THIS IDEA OF COMBINING PIVOTS IN TWO PLANES ARE ENDLESS. AS ABOVE, THE JOINT UNIT OF A MULTI-LINK CHAIN CAN COMBINE BOTH DIRECTIONS. ANOTHER APPROACH IS TO ALLOW EACH LINK TO ALTERNATE THE DIRECTION OF THE MOVEMENT.

GIMBALS CAN BE CAST OR FABRICATED.

Loop-in-Loop

THIS IS A VARIATION ON THE ROMAN CHAIN BUT THE EFFECT IS VERY DIFFERENT. IT LENDS ITSELF TO LARGE OR SMALL SCALE AND IS POSSIBLE IN BASE METALS BUT BETTER IN SILVER, STERLING, OR GOLD.

1. MAKE ROUND LINKS AND SOLDER OR FUSE CLOSED. FUSING IS PREFERRED SINCE IT LEAVES NO BUMP.
2. WITH ROUND-NOSE PLIERS PULL EACH LOOP INTO AN ELONGATED OVAL.
3. EACH LINK IS GIVEN ITS SHAPE INDIVIDUALLY AS THE CHAIN IS MADE. GRIP SCRIBE VERTICALLY IN A VISE, PUSH LOOP ONTO THIS AND A SECOND SCRIBE AND PINCH AT ARROWS WITH A ROUND-NOSE PLIERS.

TOP VIEW

PINCH AT ARROWS

A VARIATION IS TO SHAPE EACH END OF THE LINK SEPARATELY AND THEN FLATTEN THE LOOP BY PLANISHING. THE LINK IS THEN "FOLDED" OR BENT UPWARDS, SHAPED AS ABOVE, AND THE NEXT LINK IS INSERTED.

PERHAPS SOME DAY IT WILL BE PLEASANT TO REMEMBER EVEN THIS.

VIRGIL 50 BC

FIGURE 8

AFTER WINDING A COIL AROUND A ROD, PULL DOWN 4 RINGS AND CUT.

OPEN THE RINGS IN THE MIDDLE AS YOU WOULD OPEN A BOOK.

CURL OUT EACH END A FORM A SMALL LOOP AT EACH TIP.

HOOK THESE LINKS TOGETHER, EACH

ONE HOLDING THE LINK AHEAD & BEHIND.

Egyptian Spiral

AFTER EXPERIMENTING TO GET THE DESIRED SIZE CUT OFF PIECES OF WIRE, FILE ENDS AND BEND. BEGIN THE BEND LIKE THIS.

MAKE COIL ON EACH END & BEND IN THE MIDDLE LIKE THIS:

BEND OVER

CONNECT LINKS BY SLIDING EACH ONE HERE.

Coil Links

WRAP WIRE TO MAKE A COIL AND PULL DOWN ANY NUMBER OF RINGS (HERE FIVE). CUT.

WITH PLIERS THE END LINKS ARE PULLED OUT AND OPENED. AFTER CONNECTING THE LINKS THE RINGS ARE MADE SYMMETRICAL AND CLOSED.

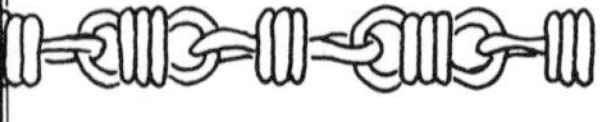

Woven Chain

1

BEGIN WITH THIS BEND IN SMALL (22-28 B&S) GAUGE WIRE.
USE A PIECE OF ANNEALED WIRE NO LONGER THAN TWO FEET.

2

FOLD SHORT END ACROSS LOOPS LIKE THIS.

3

GATHER IN THE LOOPS AND WRAP THEM WITH THE SHORT END. USING PLIERS, PULL THE LOOPS INTO A SYMMETRICAL ARRANGEMENT, LIKE THIS

4

FEED THE SUPPLY (LONG) END THROUGH ANY LOOP, GOING FROM INSIDE THE BUNCH OUTWARD.

INDICATES THAT WIRE IS LONGER.

5

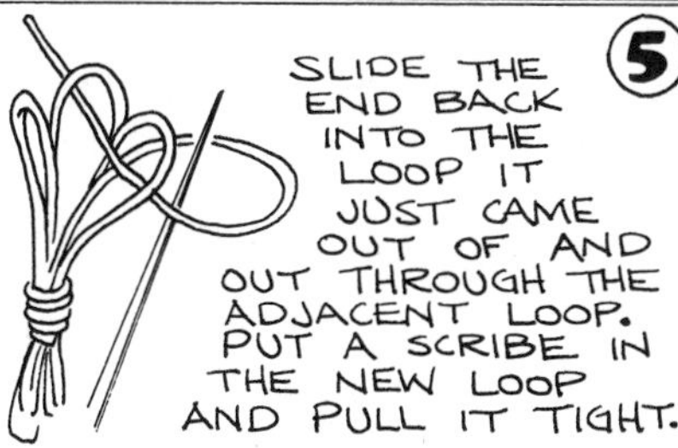

SLIDE THE END BACK INTO THE LOOP IT JUST CAME OUT OF AND OUT THROUGH THE ADJACENT LOOP. PUT A SCRIBE IN THE NEW LOOP AND PULL IT TIGHT.

6

REPEAT THIS PROCESS, FOLDING NEW LOOPS UPWARD (I.E. ALONG THE CHAIN'S AXIS) AS YOU GO. PULL EACH LOOP TIGHT ON THE SCRIBE.

7

TO ADD A NEW LENGTH OF WIRE, TWIST THE OLD AND NEW ENDS TOGETHER. KEEP THE TWIST INSIDE THE CHAIN

TOP VIEW

8

TO COMPRESS AND ELONGATE,

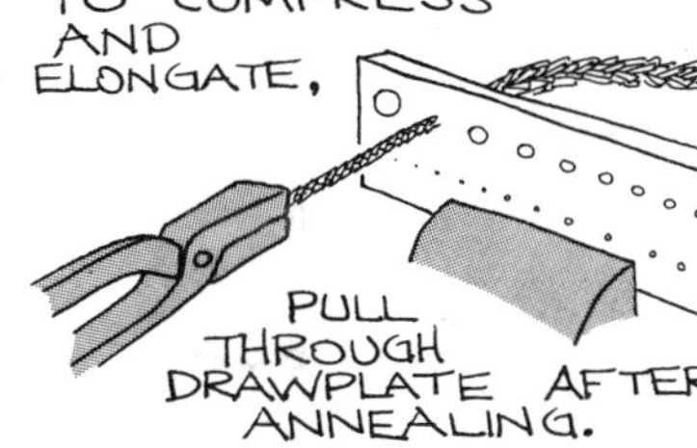

PULL THROUGH DRAWPLATE AFTER ANNEALING.

9

TO MAKE CHAIN FLEXIBLE, ANNEAL AND WRAP AROUND A DOWEL HELD IN A VISE. PULL BACK AND FORTH VIGOROUSLY. ANNEAL AND REPEAT.

note:

- THIS CHAIN MAY BE MADE WITH 3, 4, OR 5 LOOPS. ONLY STEP ONE IS DIFFERENT: MAKE THE NUMBER OF LOOPS YOU WANT TO USE.
- GRAB END OF WIRE WITH PLIERS TO PULL EACH NEW LOOP SNUG ON SCRIBE.
- TRY TO KEEP LOOPS UNIFORM IN SIZE.

THE INNER LIFE OF A HUMAN BEING IS A VAST AND VARIED REALM AND DOES NOT CONCERN ITSELF ALONE WITH STIMULATING ARRANGEMENTS OF COLOR, FORM, AND DESIGN.

—EDWARD HOPPER

Idiot's Delight

SOME PEOPLE THINK THIS CHAIN GOT ITS NAME BECAUSE IT IS SO EASY; EVEN AN IDIOT CAN MAKE IT. OTHERS MAINTAIN THAT THE NAME REFERS TO THE MENTAL DEGENERATION CAUSED BY TRYING TO FIGURE IT OUT.

LINKS ARE LEFT UNSOLDERED AND SO SHOULD BE WORK-HARDENED BY DRAWING THE WIRE DOWN OR TWISTING IT BEFORE COILING.

THE PROPORTION OF WIRE SIZE TO LOOP SIZE IS IMPORTANT FOR A COMPACT CHAIN IN EITHER STYLE.

WIRE SIZE	INSIDE DIA.		*
16 B&S	3/16"	4.8 mm	20
18	5/32"	4	24
20	1/8"	3.2	28
22	3/32"	2.5	33

* LINKS PER INCH

PARALLEL LINK

1. MAKE RINGS. OPEN ABOUT HALF, CLOSE THE OTHER HALF. ALWAYS OPEN BY TWISTING SIDEWAYS.

2. FEED OPEN RING THROUGH FOUR CLOSED RINGS AND CLOSE IT.

3. FEED A SECOND OPEN RING THROUGH THE FOUR AND CLOSE IT. SHADING INDICATES TWO RINGS, SIDE BY SIDE.

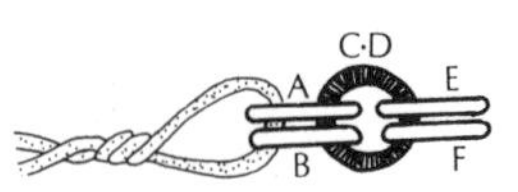

4. FLOP TWO RINGS BACK AND PUT A WIRE OR PAPER-CLIP THROUGH TO SERVE AS A HANDLE.

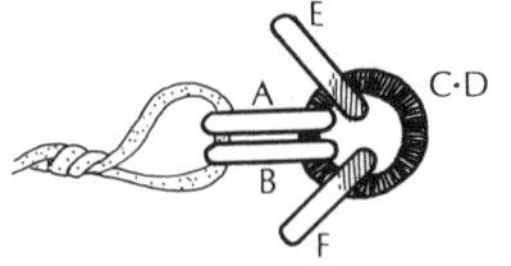

5. FLOP **E & F** TO THE LEFT AND RIGHT. FLOP **C & D** FORWARD AND BACKWARD TO EXPOSE SHADED AREA OF **E & F**. SLIDE A NEEDLE THROUGH HERE.

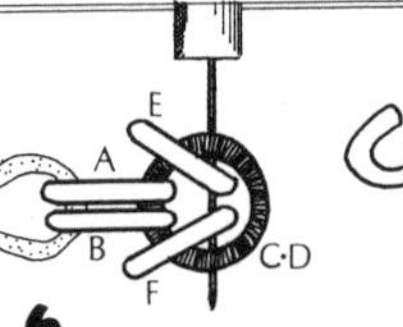

6. SLIP TWO CLOSED RINGS (G & H) ON AN OPEN RING (I) AND FEED IT THROUGH WHERE THE NEEDLE IS. LET NEEDLE DROP OUT, CLOSE RING. ADD A SECOND LINK BESIDE **I** AND CLOSE IT. **(J)**

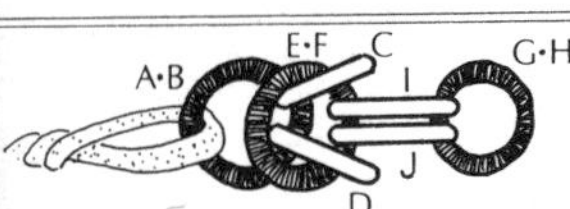

7. LET THE CHAIN DROOP TO ALLOW EACH LINK TO FALL INTO PLACE. WHEN YOU LAY IT OUT IT SHOULD LOOK LIKE THIS.

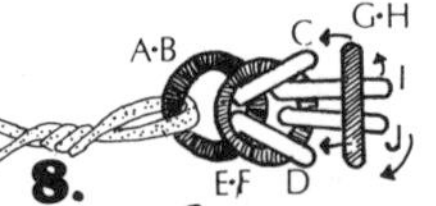

8. FOLD **G** FORWARD AND **H** BACKWARD TO EXPOSE LOWER PORTION OF **I** AND **J**. SLIDE THE NEEDLE THROUGH HERE. THIS IS A REPEAT OF 6.

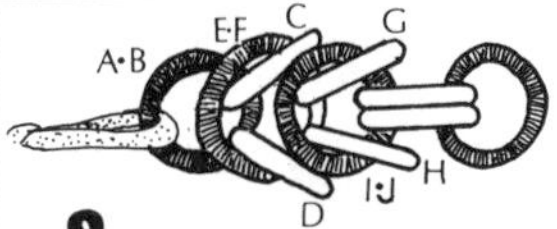

9. CONTINUE AS BEFORE, ADDING AN OPEN RING THAT ALREADY HAS TWO CLOSED RINGS ON IT. AND SO ON.....
DO NOT BUFF ON MACHINE.

SEQUENTIAL LINK

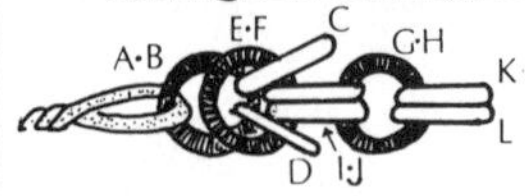

FOLLOW 1-7 ABOVE. AT THIS POINT ADD TWO MORE RINGS THROUGH **G & H** AND CLOSE THEM. THESE ARE MARKED **K & L**.

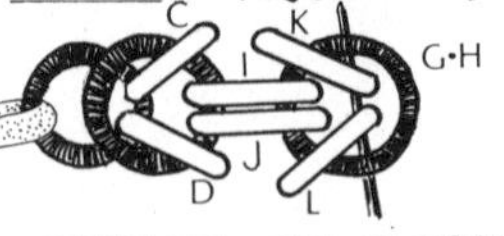

REPEAT THE FLOPPING OPERATION: **K & L** TO EITHER SIDE, **G & H** LAID APART TO EXPOSE THE BOTTOM SECTION OF **K & L**. INSERT NEEDLE HERE TO MARK THE SPOT.

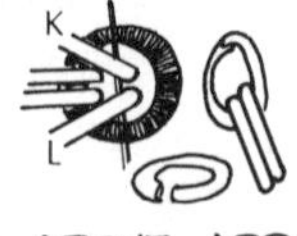

AS ABOVE ADD AN OPEN RING THAT HAS TWO CLOSED RINGS ON IT. DOUBLE THIS JOINT BY ADDING A SECOND RING IN THE SAME PLACE. NOW ADD TWO MORE (LIKE **K & L**) AND CONTINUE.

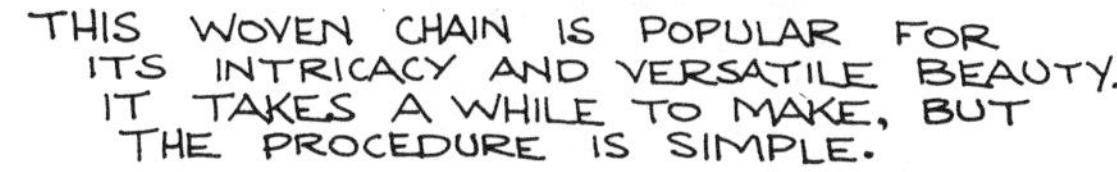

Roman

THIS WOVEN CHAIN IS POPULAR FOR ITS INTRICACY AND VERSATILE BEAUTY. IT TAKES A WHILE TO MAKE, BUT THE PROCEDURE IS SIMPLE.

1. CUT RINGS - SEE BELOW FOR SUGGESTED SIZES. FILE ENDS SQUARE AND BEND TO MAKE A GOOD JOINT.

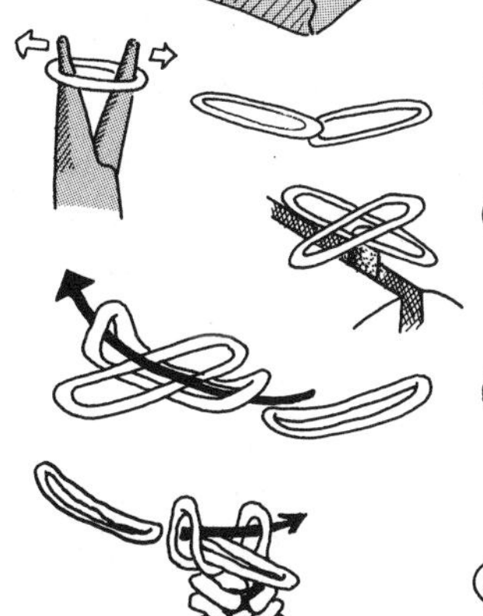

2. FUSE RINGS CLOSED. ONE METHOD IS TO SET RINGS INTO SLOTS CUT INTO A CHARCOAL BLOCK. SOLDER IS TO BE AVOIDED SINCE IT OFTEN FORMS A LUMP THAT WILL LEAVE THE CHAIN LOOKING SLOPPY.
3. WITH ROUND NOSE PLIERS, PULL EACH RING INTO A LONG OVAL. TRY TO AVOID STRETCHING THEM.
4. TO START A DOUBLE-LINK CHAIN, SOLDER TWO OVALS TOGETHER CROSSWISE. YOU MAY ALSO WANT TO SOLDER ON A LENGTH OF WIRE TO ACT AS A HANDLE.
5. TO WEAVE, THE TIPS OF THE LOWER OVAL ARE BENT UP AND A NEW LOOP IS FED THROUGH. IT MAY BE BENT UP A LITTLE TO HOLD IT IN PLACE.
6. NEW LOOPS ARE ADDED THIS WAY, ALWAYS GOING THROUGH THE LOWEST LOOP POSSIBLE. IT IS OFTEN NECESSARY TO STRAIGHTEN OR ENLARGE LOOPS WITH A SCRIBE OR SIMILAR SHARP TOOL.

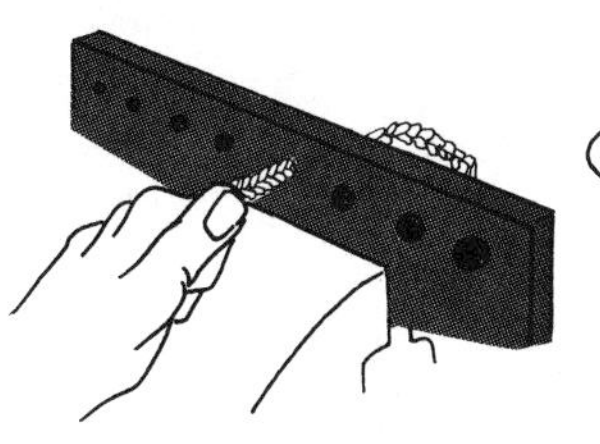

7. TO MAKE THE CHAIN DIAMETER UNIFORM THE FINISHED CHAIN IS PULLED LIGHTLY THROUGH A ROUND DRAWPLATE. DRAW TONGS ARE NOT USUALLY NEEDED FOR THIS. CHAIN SHOULD BE ANNEALED BEFORE DRAWING.

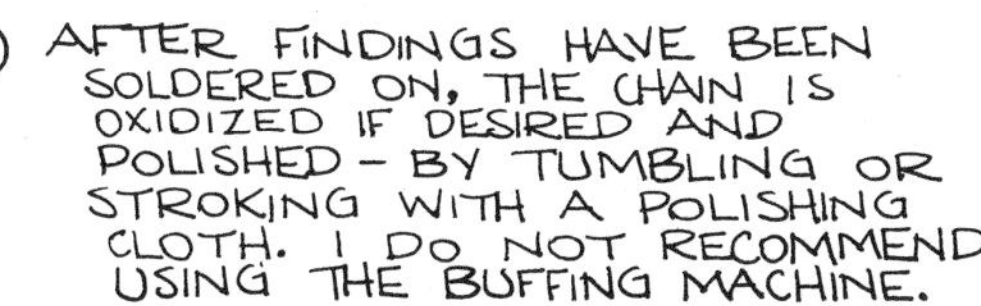

8. AFTER FINDINGS HAVE BEEN SOLDERED ON, THE CHAIN IS OXIDIZED IF DESIRED AND POLISHED - BY TUMBLING OR STROKING WITH A POLISHING CLOTH. I DO NOT RECOMMEND USING THE BUFFING MACHINE.

Size Specifications

FOR A DOUBLE-LINK CHAIN:
(ESTIMATE OTHERS FROM THIS)

WIRE (B&S)	LOOP INSIDE DIAMETER	FINISHED CHAIN DIAMETER	LINKS PER INCH
22	10	5 MM	13
24	7	4 MM	18
26	6	3.5 MM	20
28	5	3 MM	24

VARIATIONS

THE DESCRIPTION ABOVE IS FOR A DOUBLE-LINK CHAIN. TO VARY THE RESULT, ALTER THE NUMBER OF LINKS USED. THIS IS DONE BY STARTING WITH A SINGLE LINK

OR BY SOLDERING THREE OR FOUR LINKS TOGETHER:

Your Basic Hinge

IF THE MATERIAL IS TOO THIN TO TAKE THE STRESS OF A HINGE, BEARERS SHOULD BE SOLDERED ON, INSIDE OR OUT. THIS IS ESPECIALLY IMPORTANT FOR ROUND OR OVAL CONTAINERS.

PREPARE A SEAT (TROUGH) IN WHICH THE KNUCKLES, OR HINGE SECTIONS, WILL LIE. CARE IN THIS STEP IS IMPORTANT. A STRAIGHT ROUND FILE IS BETTER THAN A TAPERED ONE. A JOINT FILE IS MADE FOR THIS. A SCRAPER MAY BE MADE OF GROUND-OFF DRILL BIT.

JOINT FILE

STEEL ROD HELD IN PIN VISE

MEASURE AND CUT KNUCKLES, KEEPING ENDS SQUARE. IF ONLY THREE KNUCKLES ARE USED, THE SINGLE PIECE GOES ON THE LID AND IS SLIGHTLY LONGER THAN THE OTHER TWO.

FILE A BEVEL AT EACH END JUST AT THE SEAM TO MAKE A SOLDER-STOP. CUT NICKS IN THE SEAM TO HELP LOCATE IT DURING SOLDERING. IF TUBE HAS NO OPEN SEAM, SKIP THE NOTCHES.

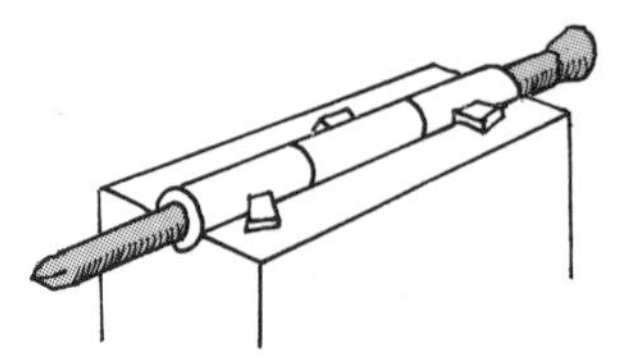

SLIDE THE KNUCKLES ONTO A SNUG-FITTING, OILED STEEL PIN (NAIL, BINDING WIRE, ETC.). FLUX AND SET INTO SEAT. TIE WITH BINDING WIRE IF NECESSARY. POSITION SEAMS TO CONTACT BOX AND PLACE SOLDER AS SHOWN.

HEAT ONLY UNTIL SOLDER FLOWS; QUICKLY REMOVE TORCH. QUENCH IN WATER. REMOVE BINDING WIRE AND STEEL PIN AND PICKLE. AFTER POLISHING AND WASHING, A TIGHT FITTING PIN (SEE BELOW) IS SLID INTO PLACE AND SLIGHTLY FLARED ON EACH END WITH A RIVETING OR CHASING HAMMER TO HOLD IT IN PLACE.

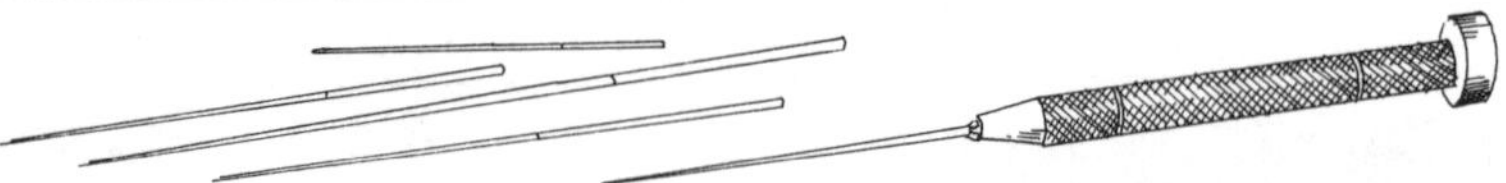

the use of the CUTTING BROACH

EVEN WELL MADE HINGES WILL PROBABLY HAVE KNUCKLES THAT ARE SLIGHTLY OUT OF ALIGNMENT. THIS WILL RESULT IN A SMALL AMOUNT OF PLAY OR SLOPPINESS IN THE HINGE. ONE SOLUTION USES A GRADUALLY TAPERED FIVE-SIDED STEEL ROD SHOWN HERE. THESE ARE USUALLY SOLD IN SETS OF A DOZEN.

WITH THE HINGE TOGETHER, THE BROACH, (HELD IN A PIN VISE) IS ROTATED GENTLY TO SCRAPE AWAY BITS OF METAL INSIDE THE HINGE. IT IS PULLED OUT AND WIPED OFF OFTEN AS WORK PROGRESSES.

WHEN CONTACT IS MADE WITH EACH KNUCKLE (LID WILL HOLD ITSELF OPEN) FILE WIRE TO A SIMILAR TAPER AND TAP LIGHTLY INTO PLACE.

1

PREPARE A TROUGH BY FILING OR SCRAPING.

2

BUY OR MAKE TWO TUBES THAT TELESCOPE TOGETHER. WHEN SOLDERED, TUBING MAY BE DRAWN LIKE WIRE.

3

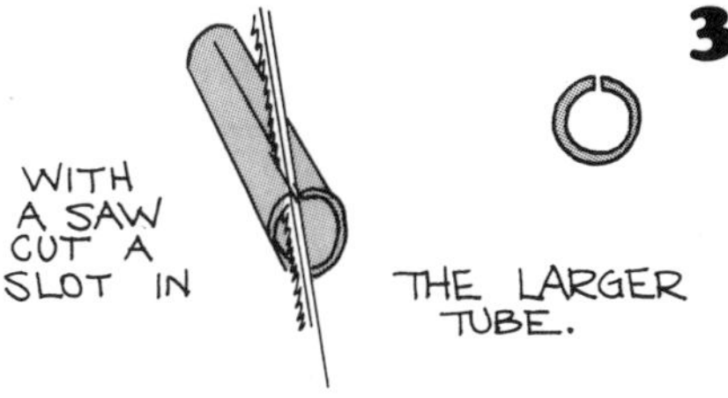

WITH A SAW CUT A SLOT IN THE LARGER TUBE.

4

SET THIS TUBE IN POSITION WITH THE SAWN SLOT LOCATED AS SHOWN, WHERE THE BOX AND LID COME TOGETHER. SOLDER BOTH SIDES.

5

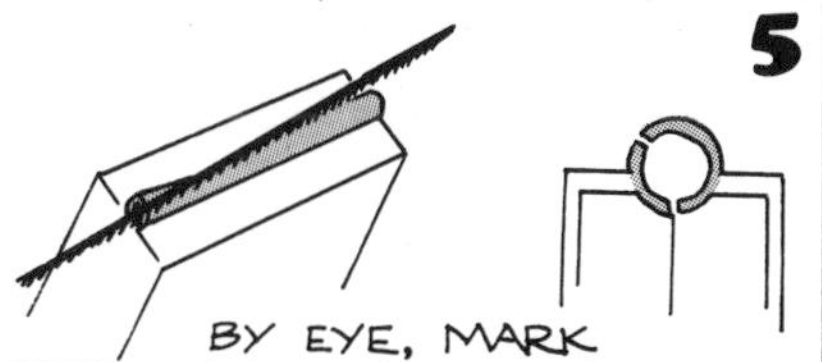

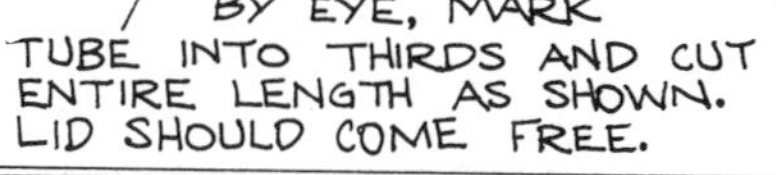

BY EYE, MARK TUBE INTO THIRDS AND CUT ENTIRE LENGTH AS SHOWN. LID SHOULD COME FREE.

6

REMOVE

REPEAT THE PROCESS, REMOVING THE TOP THIRD.

7

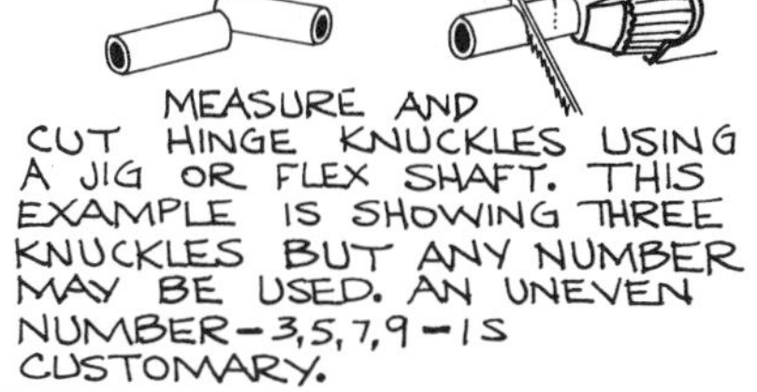

MEASURE AND CUT HINGE KNUCKLES USING A JIG OR FLEX SHAFT. THIS EXAMPLE IS SHOWING THREE KNUCKLES BUT ANY NUMBER MAY BE USED. AN UNEVEN NUMBER—3,5,7,9—IS CUSTOMARY.

8

SET A KNUCKLE INTO THE CRADLE AND SOLDER WITH EASY SOLDER.

9

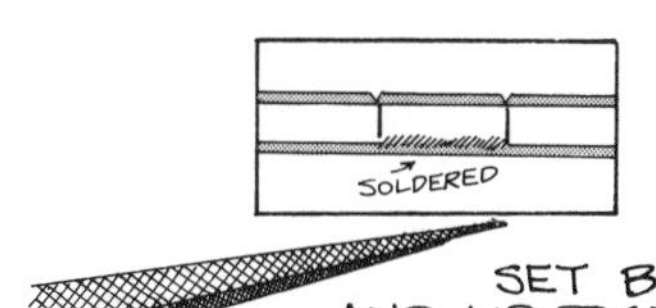

SET BOX AND LID TOGETHER AND MARK LOCATION OF FIRST KNUCKLE WITH FILE NOTCHES ON CRADLE.

10

USING NOTCHES AS GUIDES, SOLDER THE OTHER KNUCKLES INTO THEIR CRADLE.

11

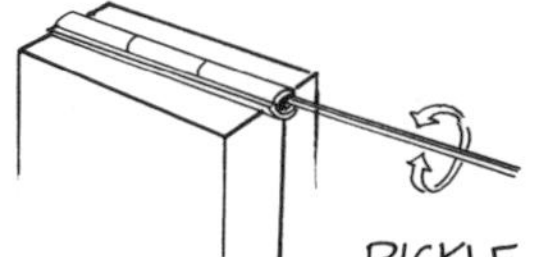

PICKLE, POLISH, WASH, AND SET HINGE BY FIRST REAMING WITH A BROACH. THIS IS NOT ESSENTIAL BUT MAKES A TIGHTER HINGE.

12

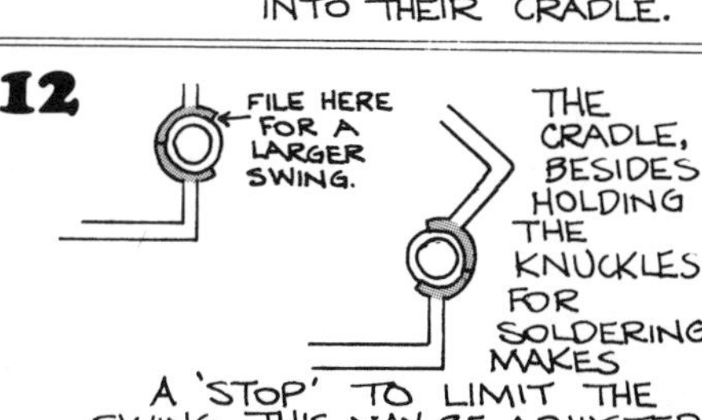

THE CRADLE, BESIDES HOLDING THE KNUCKLES FOR SOLDERING, MAKES A 'STOP' TO LIMIT THE SWING. THIS MAY BE ADJUSTED.

with tubing:

1. TABS ARE LAID OUT AND CUT. TO BE EFFECTIVE, THE FIT MUST BE EXACT.
2. WITH THE PIECES HELD TOGETHER A GROOVE IS SCRIBED AND FILED IN BOTH SECTIONS.
3. WITH THE TWO PIECES SEPARATED, A LENGTH OF TUBING IS SOLDERED INTO EACH GROOVE. AFTER SOLDERING, EXCESS TUBE IS CUT AWAY WITH A SAW.
4. TO PROVIDE A STOP TO KEEP THE LID FROM FALLING ALL THE WAY OPEN, SOLDER A LENGTH OF SQUARE WIRE ALONG ONE SIDE JUST IN FRONT OF THE TUBING: DON'T CUT ANY AWAY. THE HEIGHT OF THIS PIECE AND DISTANCE FROM THE TUBE WILL CONTROL THE SWING. HIGHER AND/OR CLOSER WILL LIMIT THE SWING.
5. FILE A BEVEL ON THE UNDERSIDE OF EACH COVE; BOTH PIECES. BE CAREFUL NOT TO FILE AWAY THE TOP EDGE, THUS RUINING THE ACCURACY OF #1.
6. ONE SECTION (WITH THE STOP) IS SOLDERED TO THE PIECE. A HOLE IS DRILLED IN THE SIDE OF THE OBJECT FOR THE PIN AND THE HINGE IS ASSEMBLED.

1.

2.

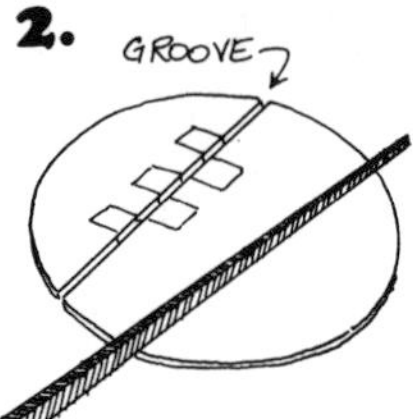

3.

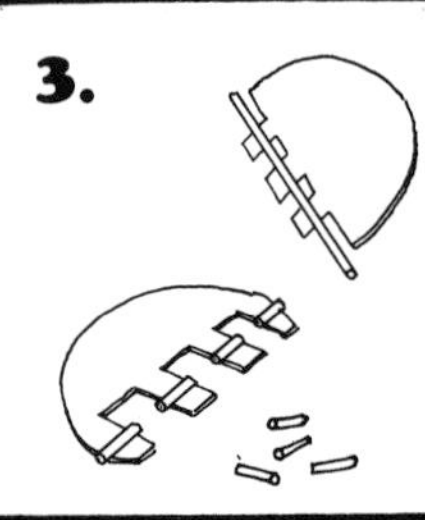

4.

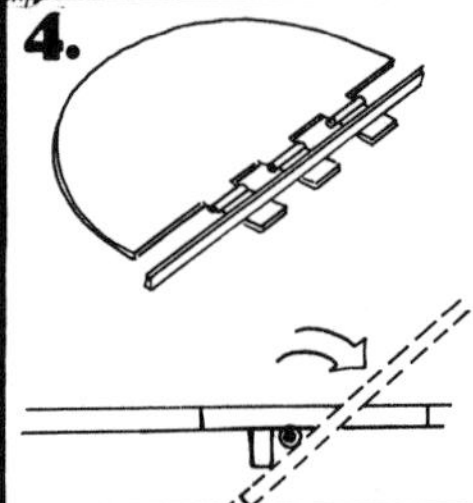

5.

6.

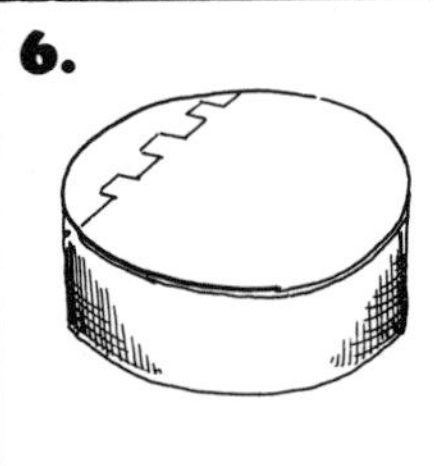

with sheet:

1. KNUCKLE UNITS ARE MADE OF HEAVY SHEET: 14-18 GAUGE. A STRIP IS CUT AND BENT AROUND ANOTHER PIECE OF THE SAME GAUGE. SQUEEZE WITH PARALLEL JAW PLIERS.

2. TO MAKE THE THIRD KNUCKLE OF THIS UNIT, A PIECE OF STRIP IS TEMPORARILY SOLDERED IN PLACE AND SQUEEZED IN THE SAME WAY.

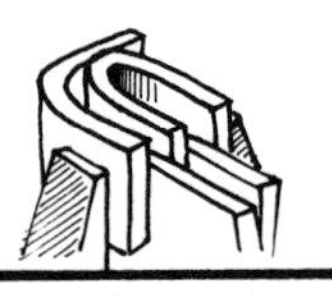

3. A SIMILAR DEVICE IS USED TO MAKE THE OTHER HALF OF THE HINGE, THIS ONE HAVING FOUR KNUCKLES. TWO UNITS LIKE #1 ARE MADE AND SOLDERED TEMPORARILY TO A BRACE TO HOLD THEM THE CORRECT DISTANCE APART (I.E. ONE THICKNESS AGAIN).

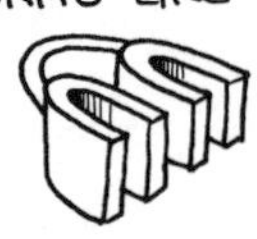

4. THIS UNIT IS THEN SOLDERED ONTO THE TOP OF THE CONTAINER AS SHOWN. WHEN SOLDERING IS COMPLETE, CUT AWAY EXCESS INCLUDING THE SHEET BETWEEN THE KNUCKLES.

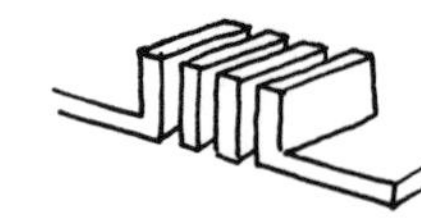

5. THE FIRST UNIT MUST OVERHANG ITS EDGE TO REACH INTO THE 4-KNUCKLE UNIT. TO KEEP THE TOP ON THE FINISHED HINGE FLUSH, FILE AS SHOWN. AFTER CHECKING THE SPACING BETWEEN THE KNUCKLES, SOLDER THIS UNIT TO THE LID. AFTER TRIMMING AWAY EXCESS, IT WILL LOOK LIKE THIS:

6. THE TWO UNITS ARE PUT TOGETHER AND A HOLE IS DRILLED THROUGH ALL. THE TOP IS THEN SOLDERED TO THE CONTAINER, THE PIN BEING SET BEFORE OR AFTER SOLDERING DEPENDING ON ACCESSIBILITY.

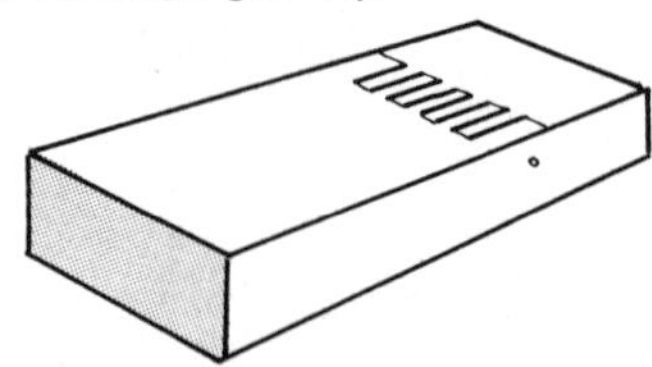

Coil Spring

THE SPRING IS PROVIDED BY A COIL OF HARD DRAWN WIRE. DEPENDING ON THE WEIGHT OF THE LID AND THE FINENESS OF THE PIECE THIS CAN BE GOLD, STERLING, BRASS, NICKEL SILVER, OR STEEL. THE STEEL MAY BE SALVAGED FROM A PEN SPRING.

THE HINGE IS MADE IN THE USUAL WAY EXCEPT THAT A SPACE IS LEFT TO BE OCCUPIED BY THE SPRING. THIS MAY BE MADE BY CUTTING AWAY ONE OF THE KNUCKLES BUT IT WILL BE NEATER IF YOU PLAN AHEAD.

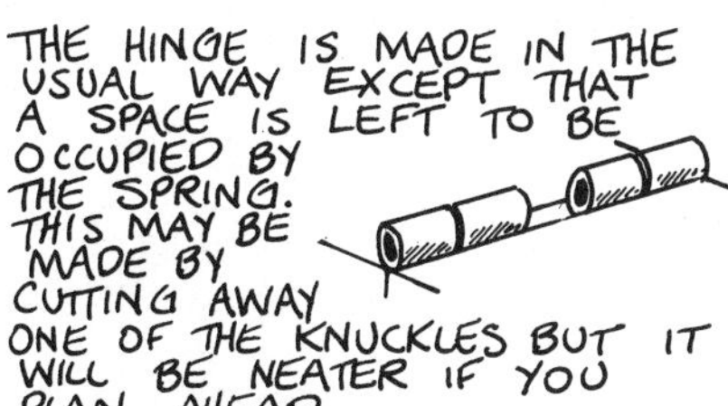

IN ASSEMBLING, THE SPRING IS COCKED INTO POSITION AND THE PIN IS RUN THROUGH IT. THIS CAN BE A TICKLISH OPERATION AND IS EASIER WITH TWO PEOPLE. THE TAILS OF THE COIL MUST PROTRUDE TO MAKE THIS SPRING WORK. DEPENDING ON WHERE YOU PUT THESE YOU CAN MAKE THE LID SPRING OPEN OR CLOSED.

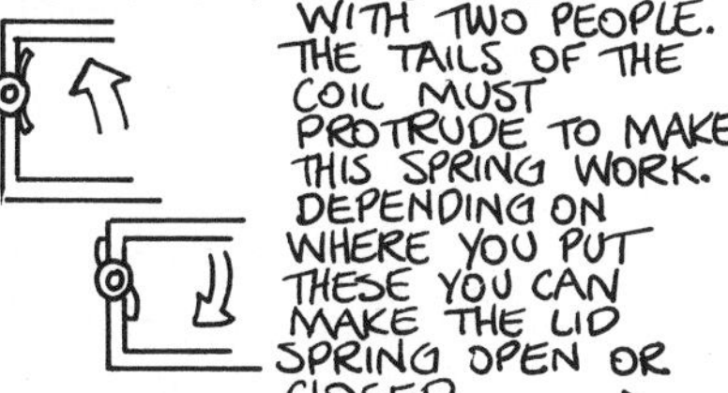

THE SPRING CAN BE MADE LESS OBVIOUS IF THE KNUCKLES ARE MADE OF COILED WIRE.

A Depression Spring

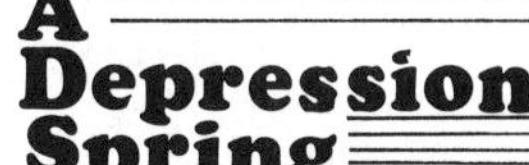

THIS IS BEST USED WHERE ONLY A SMALL PUSH IS NEEDED. IT IS COMMON ON THE COVERS OF POCKETWATCHES, FOR INSTANCE.

THIS SPRING, UNLIKE MOST, IS NOT IN THE HINGE AT ALL. SOMEWHERE NEAR THE HINGE IS A PIECE OF METAL THAT IS PUSHED DOWN WHEN THE LID IS CLOSED. WHEN THE CLASP IS RELEASED THE LITTLE TAB PUSHES UPWARD.

EACH APPLICATION WILL NEED ITS OWN DESIGN BUT HERE ARE A FEW EXAMPLES:

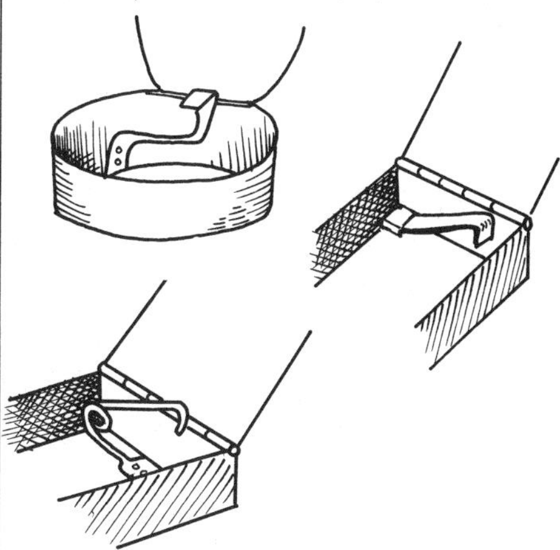

SPRING PIN

THIS IS THE MOST SUBTLE OF THE BUNCH BUT DOES NOT LEND ITSELF TO SHORT OR SLENDER HINGES.

1. MAKE A HINGE THAT IS CONVENTIONAL IN EVERY WAY EXCEPT THAT IT HAS AN EVEN NUMBER OF KNUCKLES. BOTH END KNUCKLES SHOULD NOT BE ATTACHED TO THE SAME UNIT.

2. AFTER POLISHING, SLIDE A TELESCOPING TUBE THROUGH THE HINGE, KEEPING IT A LITTLE SHORTER THAN THE HINGE PROPER. THIS WILL PROVIDE STRENGTH AND SMOOTH OPERATION.

3. YOU WILL NEED SEVERAL STRIPS OF FLAT SPRINGY METAL. WATCH MAINSPRINGS ARE BEST BUT IN A PINCH, HARD-DRAWN BRASS OR NICKEL SILVER CAN BE FLATTENED OUT AND USED. CUT 2 OR 3 PIECES ABOUT AN INCH LONGER THAN THE HINGE. SLIDE THESE THROUGH AND LOCK WITH A WEDGE, TAPPED INTO PLACE.

4. WITH PLIERS, GRIP THE EXTENDING SPRING PIECES AND GIVE THEM A TWIST OR TWO. LOCK IN PLACE WITH A SIMILAR WEDGE AND CHECK THE ACTION. IF SLACK, GIVE ANOTHER TWIST. DEPENDING ON THE DIRECTION OF TWIST THE SPRING WILL OPEN OR CLOSE. WHEN RIGHT, TAP WEDGE AND TRIM OFF EXCESS.

Friction

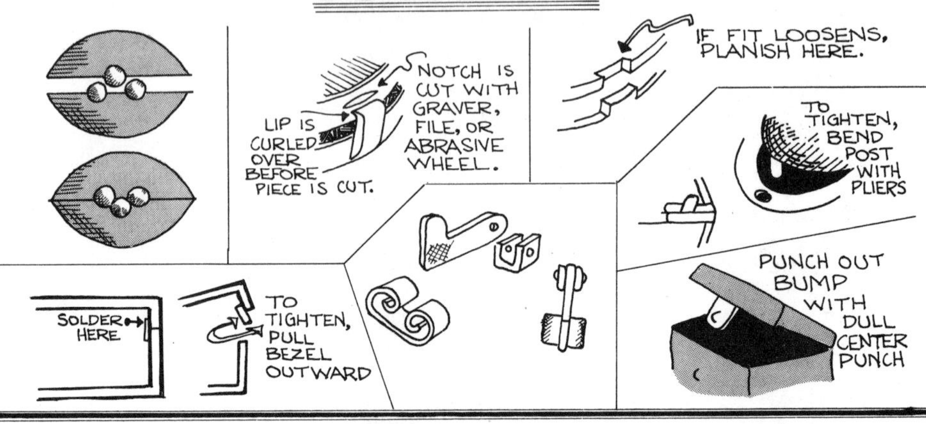

Hooks

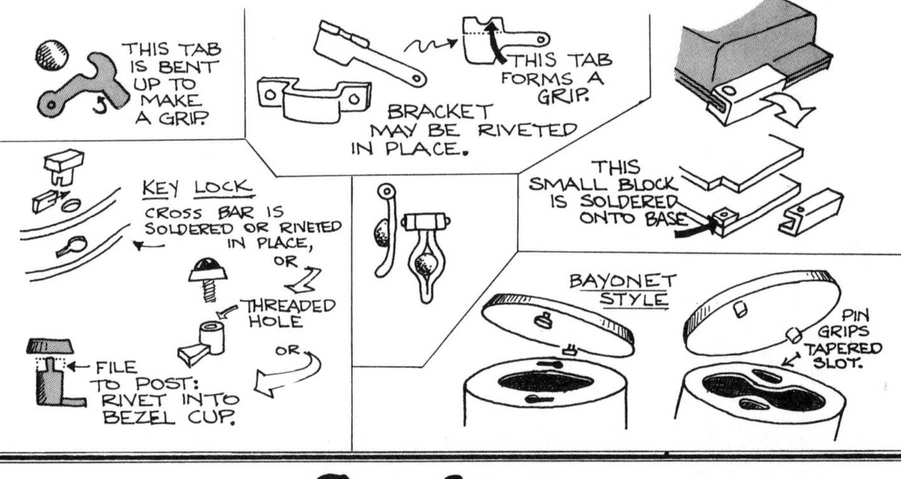

Spring

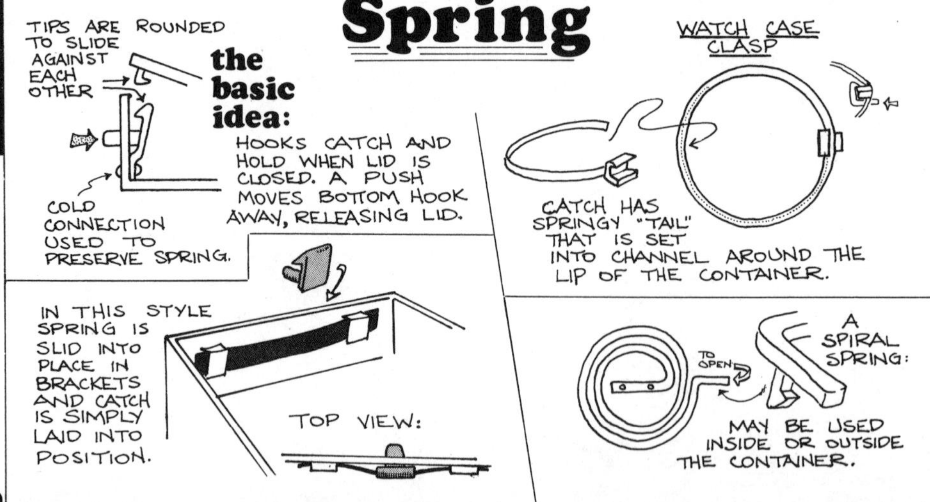

TOOLS

USES: SOME METALSMITHS, KNIFEMAKERS FOR INSTANCE, USE HARDENED STEEL IN THEIR PRODUCTS. ALL SMITHS SHOULD BE FAMILIAR WITH THESE TECHNIQUES, HOWEVER, SINCE THEY CAN BE USED TO MAKE TOOLS FOR THE SHOP. THESE WOULD INCLUDE PUNCHES, STAMPS, QUALITY STAMPS (E.G. "14K"), KNIVES, SCRAPERS, SPRINGS, AND DRILL BITS.

what happens

THE FIELD OF STEEL METALLURGY IS A LARGE AND COMPLEX TOPIC. FOR A CLOSER LOOK SEE PAGE 2. IN GENERAL TERMS, IRON THAT CONTAINS SMALL AMOUNTS OF CARBON HAS THE ABILITY TO TRANSFORM INTO SEVERAL STATES OR MOLECULAR CONFIGURATIONS. THESE EXHIBIT VARIOUS PROPERTIES LIKE HARDNESS, TOUGHNESS, AND SPRINGINESS. CONTROL OVER THE FORMATION OF THESE STATES DEPENDS ON THE ALLOY USED, TEMPERATURE AND DURATION OF HEAT USED, AND THE COOLING METHOD.

AS IT PERTAINS TO THE CRAFTSMAN THE PROCESS GOES LIKE THIS: TOOL STEEL IS GIVEN THE DESIRED SHAPE AND HEATED TO CHANGE ITS CRYSTALS TO HARD MARTENSITE. THIS IS GIVEN A SECOND HEAT TREATMENT TO RENDER IT AS TOUGH PARTICLES (CEMENTITE) BONDED IN A MATRIX OF RELATIVELY FLEXIBLE FERRITE.

CEMENTITE FERRITE

MARTENSITE (HARD, STRESSED) AFTER TEMPERING (TOUGH, RELAXED)

MILD STEEL (ALSO CALLED LOW CARBON STEEL) CONTAINS .15-.3 % CARBON. THIS AMOUNT IS INSUFFICIENT TO CAUSE HARDENING. TO TEST AN UNKNOWN PIECE OF MATERIAL HOLD AGAINST GRINDING WHEEL. TOOL STEEL THROWS WHITE STAR-LIKE SPARKS. MILD GIVES DULL, ROUND ORANGE SPARKS.

PROCESS

1. TOOL STEEL IS SOLD ANNEALED BUT IF YOU ARE REUSING A WORN TOOL THE FIRST STEP IS TO ANNEAL IT. HEAT TO BRIGHT RED AND COOL AS SLOWLY AS POSSIBLE. BURY STEEL IN SAND OR SET A FIREBRICK ON IT TO HOLD IN THE HEAT.
2. SHAPE TOOL BY FORGING, SAWING, GRINDING, AND FILING. FORGING MUST BE DONE WHILE THE STEEL IS RED HOT: DO NOT STRIKE AFTER THE COLOR HAS GONE OR THE STEEL MAY CRACK. WHEN MAKING A PATTERNED TOOL LIKE A STAMP, THE IMAGE IS CHECKED BY PRESSING INTO CLAY.
3. THE TOOL IS HARDENED BY HEATING TO BRIGHT, GLOWING RED AND QUENCHED IMMEDIATELY IN OIL OR BRINE. SMALL TOOLS MAY BE HELD IN THE TWEEZERS; LARGE PIECES ARE SET ON A BRICK OR MAY BE HEATED IN A FORGE OR FURNACE. PUNCHES ARE USUALLY HARDENED ONLY FOR THE INCH OR TWO UP FROM THE STAMPING END.
   - THE GOAL HERE IS TO CONVERT THE PEARLITE STAGE INTO MARTENSITE, WHICH IS NOT MAGNETIC. USE A MAGNET ON THE HOT STEEL TO BE SURE YOU GO TO A HIGH ENOUGH TEMPERATURE.
4. CHECK HARDNESS: A FILE STROKED ACROSS THE TOOL SHOULD NOT CUT IN AND SHOULD MAKE A GLASSY SOUND.
5. REMOVE GRAY OXIDE SCALE WITH FINE SANDPAPER, AT LEAST ON A SECTION OF THE HARDENED AREA. THIS WILL ALLOW BETTER PERCEPTION OF COLORS.
6. REDUCE BRITTLENESS BY HEATING IN A STEP CALLED DRAWING THE TEMPER (DRAWING; TEMPERING). THIS CAN BE DONE WITH A TORCH OR, FOR SMALL PIECES, ON A HOT PLATE. GO SLOWLY, LETTING HEAT TRAVEL FROM A THICK SECTION TO A THINNER ONE. THE HIGHER THE TEMPERATURE (LONGER HEAT) THE SOFTER, MORE FLEXIBLE THE STEEL WILL BECOME.

°C	°F	COLOR	PROPERTIES	USES
200-225	400-445	PALE YELLOW	HARD, LITTLE FLEXIBILITY	DRILL BITS
225-265	445-490	YELLOW	HARD, LESS BRITTLE	PUNCHES
265-300	490-535	GOLDEN YELLOW	HARD, INCREASED FLEX	CHISELS
300-325	535-580	BROWN-PURPLE	WILL HOLD GOOD EDGE BUT IS FLEXIBLE	THICK KNIFE BLADES
325-350	580-650	PURPLE	MEDIUM HARD, FLEXIBLE	THIN KNIFE BLADES
350-500	650-900	BLUE	NOT HARD	SPRINGS

THE JEWELER'S BENCH IS THE HUB OF THE SHOP. EVERY WELL-USED BENCH IS POPULATED WITH GADGETS THAT MAKE WORK THERE MORE EFFICIENT. HERE ARE SOME COMMON AIDS.

BENCH PIN

ANY HARDWOOD CAN BE USED. THESE SHAPES ARE COMMON STARTING PLACES: IN PRACTICE THE PIN IS FILED, DRILLED AND CARVED TO MEET SPECIFIC NEEDS.

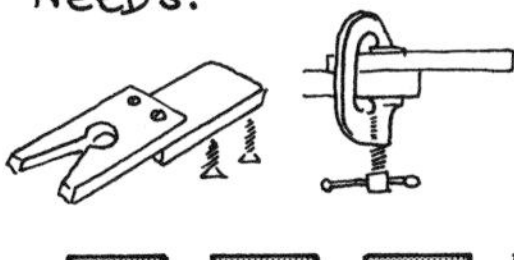

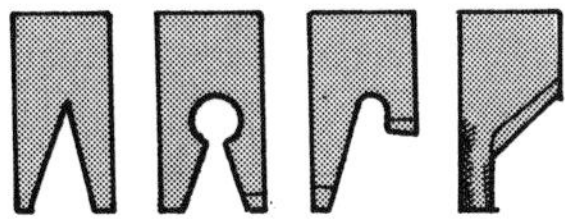

A SMALL SQUARE CAN BE MADE FROM STEEL OR BRASS BARS. ONE SIDE IS THICKER THAN THE OTHER TO ALLOW THE SQUARE TO REST AGAINST THE ITEM BEING MARKED. FILE AND SOLDER CAREFULLY, USING A STORE-BOUGHT SQUARE.

A PLIERS RACK CAN BE MADE FROM A PIECE OF COAT HANGER WIRE OR A ½" STRIP OF STEEL OR BRASS.

SAW BLADE HOLDER

bench knife

A KNIFE CAN BE IMPROVISED BY GRINDING AND RESHARPENING A USED KITCHEN PARING KNIFE. THESE CAN OFTEN BE BOUGHT AT A FLEA MARKET.

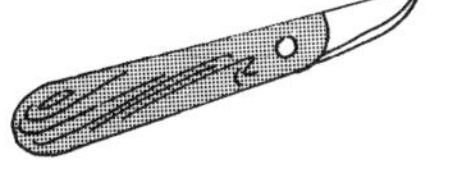

CHUCK KEYS

CAN BE MORE EASILY LOCATED IF THEY ARE FITTED INTO A FILE HANDLE OR SOLDERED ONTO A BROKEN SCREWDRIVER HANDLE.

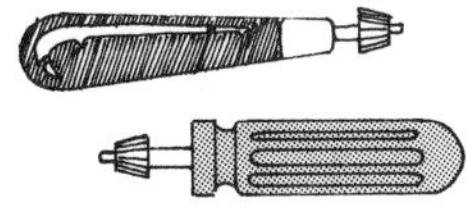

YOU WOULDN'T SAY AN AXE HANDLE HAS STYLE TO IT. IT HAS BEAUTY, AND AN APPROPRIATENESS OF FORM, AND A "THIS-IS-HOW-IT-SHOULD-BE-NESS." BUT IT HAS NO STYLE BECAUSE IT HAS NO MISTAKES. STYLE REFLECTS ONE'S IDIOSYNCRASIES.

CHARLES EAMES

Sanding Boards

MAKE UP THESE BOARDS BY GLUING PAPERS OF VARIOUS GRITS TO PANELS OF MASONITE. BOTH SIDES MAY BE USED SO THREE BOARDS WILL PROVIDE A THOROUGH RANGE OF GRITS. THESE ARE ESPECIALLY HANDY FOR TRUING UP FLAT AREAS.

SCRAPER

A SCRAPER CAN BE MADE BY BREAKING OFF AN OLD TRIANGULAR FILE AND GRINDING A POINT. FACES SHOULD BE GROUND SMOOTH AND POLISHED.

SWEEPS DRAWER

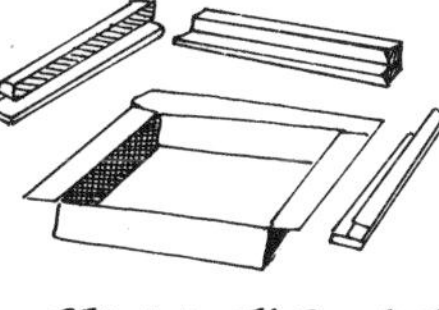

CUT A SMALL HOLE IN A BACK CORNER OF THE SWEEPS DRAWER AND COVER IT WITH A PIECE OF WINDOW SCREEN. BELOW THIS ATTACH A SHALLOW BOX THAT CAN BE REMOVED. THE BOX AND TRACK THAT HOLDS IT CAN BE MADE OF BRASS, TIN-PLATE, WOOD, OR PLASTIC.

BY SWEEPING SCRAPS OVER THIS THE LARGER PIECES CAN BE QUICKLY SORTED FROM FILINGS.

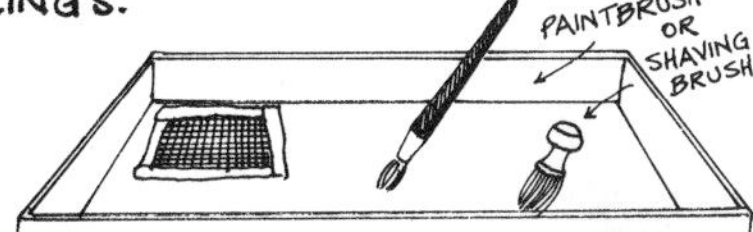

TOOLS

Hammers TOOLS

THE HEART OF THE METALSMITH'S SHOP IS IN THE HAMMERS. IN FACT THE WORD "SMITH" IS DERIVED FROM THE VERB "TO SMITE" OR TO HIT. MANY METALWORKING HAMMERS CAN BE BOUGHT NEW BUT ALONG THE WAY MOST SMITHS ALTER OLD HAMMER HEADS (AVAILABLE AT FLEA MARKETS, ETC.) TO SUIT THEIR NEEDS. SINCE METALWORKING HAMMERS ARE NOT USUALLY HARDENED THEY CAN BE MADE OF MILD STEEL.

Cast vs. Forged

THE ADVANTAGES OF CAST HAMMERHEADS ARE IN THEIR LOWER COST AND THEIR ABILITY TO BE RESHAPED EASILY BY GRINDING. FORGED HAMMERS ARE ABLE TO WITHSTAND HARDER USE BECAUSE THE CRYSTALS OF THE STEEL HAVE BEEN MORE DENSELY PACKED DURING MANUFACTURE. THESE ARE MADE IN A POWER HAMMER CALLED A DROP FORGE AND SO ARE CALLED "DROP FORGED HAMMERS."

care:

FACES THAT WILL CONTACT A WORKPIECE SHOULD BE FREE OF PITS AND SCALE. MANY SMITHS KEEP A PIECE OF CROCUS CLOTH AT HAND TO RUB THE FACE OF EACH HAMMER BEFORE USING IT. A MUSLIN BUFF (PREFERABLY STIFF) WITH A TOUGH ABRASIVE COMPOUND LIKE WHITE DIAMOND, SIMULCHROME, OR LEA COMPOUND WILL GIVE A GOOD POLISH.

FOR LONG TERM STORAGE THE FACE CAN BE PROTECTED WITH A LAYER OF VASELINE, WAX, OR OIL.

BY THE HAMMER AND HAND
ALL THE ARTS DO STAND.

handling

THE HANDLE IS FILED TO A TAPER THAT SNUGLY FITS THE EYE, WHICH HAS ITS LARGER OPENING UPWARD. MAKE A SAW CUT IN THE TOP OF THE HANDLE EQUAL TO THE LONG AXIS OF THE EYE. TAP HANDLE INTO PLACE AND CHECK ALIGNMENT. SLIDE A WOODEN WEDGE INTO PLACE, DAB SOME WHITE GLUE ON IT, AND POUND IT INTO POSITION. TRIM OFF EXCESS.

Mallets

TOOLS IN THIS FAMILY WILL BEND METAL WITHOUT STRETCHING OR MARRING IT. PROBABLY THE MOST POPULAR MATERIAL FOR MALLETS IS TREATED RAWHIDE. OTHER CHOICES INCLUDE WOOD, HORN, FIBER, PLASTIC, AND RUBBER.

THIS OLD ANVIL
LAUGHS AT MANY A
BROKEN HAMMER.
-CARL SANDBURG

styles:

PLANISHING

FORGING, CROSS PEEN

COLLET

EMBOSSING

RAISING

BALL PEEN

CHASING

HANDLES

THESE MUST BE STRONG WITHOUT BEING BULKY OR HEAVY. FIBROUS WOODS LIKE HICKORY AND ASH ARE COMMONLY USED. ANY DENSE WOOD CAN BE USED AS LONG AS THE GRAIN IS NOT CONVOLUTED.

A LONG HANDLED HAMMER DELIVERS MORE POWER BUT IS MORE DIFFICULT TO CONTROL THAN A SHORT-HANDLED ONE. A LENGTH THAT PROVIDES A COMFORTABLE MIX OF POWER AND CONTROL IS CORRECT. A ROUNDED END ON THE HANDLE WILL USUALLY BE MORE COMFORTABLE THAN A SQUARED OFF SHAPE.

NOT THIS

AN OVAL OR FACETED GRIP IS MORE EFFICIENT AND COMFORTABLE THAN A ROUND ONE. THOUGH RUBBERIZED COATINGS ARE AVAILABLE MOST PEOPLE PREFER THE FEEL AND GRIP OF SMOOTH UNTREATED WOOD. IT WILL ACQUIRE A HAND RUBBED OIL FINISH DURING USE.

THE EUROPEAN STYLE CHASING HAMMER IS MOST EFFECTIVE WITH A SPRINGY HANDLE. A FRUIT WOOD LIKE APPLE IS GOOD FOR THIS SINCE IT CAN BE MADE THIN ENOUGH FOR FLEXIBILITY WITHOUT IMPAIRING STRENGTH.

TOOLS

General:

FILES ARE GROUPED IN 4 CATEGORIES:
- HAND
- NEEDLE
- RIFFLERS
- RASPS

(HAND and NEEDLE:) THESE TWO ARE THE MOST OFTEN USED IN METALWORKING.

HAND FILES

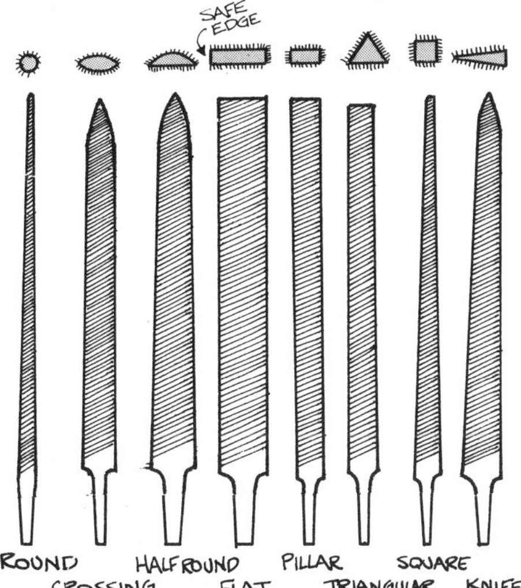

THIS IS THE LENGTH REFER TO WHEN BUYING THESE FILES, USUALLY 6, 8, OR 10 INCHES.

NEEDLE FILES ARE USUALLY SOLD BY TOTAL LENGTH.

FILES SHOULD BE EQUIPPED WITH HANDLES TO PROVIDE INCREASED LEVERAGE, WHICH MEANS FASTER CUTTING, AND TO PROTECT THE HAND FROM THE TANG.

HANDLES ARE USUALLY HELD ON BY FRICTION, THOUGH SOME HAVE A THREADING MECHANISM INSIDE FERRULE.

SINGLE CUT

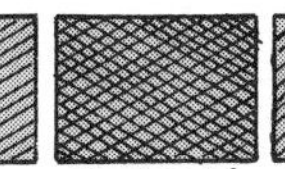

DOUBLE CUT

SOFT METAL (FLOATING)

A SINGLE CUT IS GENERALLY FINER AND WILL CUT SLOWER THAN A DOUBLE CUT, BUT THE COARSENESS OF THE FILE MUST BE CONSIDERED SINCE VARIOUS DEGREES OF COARSENESS ARE AVAILABLE IN EACH STYE.

FOREIGN MADE FILES ARE OFTEN INCORRECTLY CALLED "SWISS" FILES. THEY ARE USUALLY GRADED BY NUMBER FROM 00 (COARSEST) TO 8. AMERICAN MADE FILES USUALLY USE THE NAMES SHOWN →

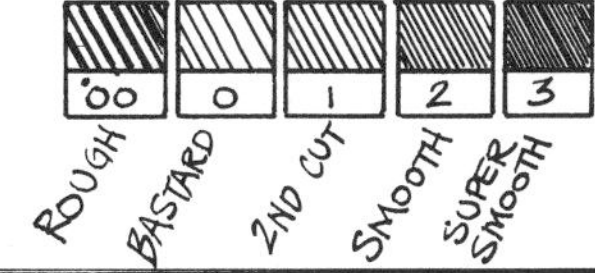

TIPS

THE TEETH ON ALL FILES POINT AWAY FROM THE HANDLE AND THEREFORE CUT ON THE PUSH STROKE. LIFT FILE OR EASE PRESSURE ON THE RETURN STROKE.

HOLD INDEX FINGER LIKE THIS TO PRESS DOWN WHILE FILING. WORK MUST BE STABILE: CUT NOTCHES IN THE BENCH PIN AS NEEDED. DON'T FILE WHILE WALKING AROUND.

DUST FILES WITH CHALK TO PREVENT CLOGGING.

KEEP FILES CLEAN WITH A FILE CARD (A WIRE BRUSH).

WITHOUT ANY DOUBT GOOD AND ACCURATE USE OF FILES COMES FROM PRACTICE AND MORE PRACTICE.
— CHARLES JARVIS

STORAGE

MOST DAMAGE IS DONE TO FILES WHEN THEY RUB ONE ANOTHER. ARRANGE SOME METHOD AT YOUR BENCH TO KEEP FILES APART. DRAWER DIVIDERS OR CARDBOARD TUBES CAN BE USED.

RIFFLERS, ALSO CALLED DIE SINKERS FILES, HAVE A SMALL CURVED TOOTH SECTION AT EACH END. EACH HAS A SPECIFIC AND LIMITED USE AND ARE NEEDED ONLY FOR EXACT WORK.

Measuring TOOLS

Rulers:

A RULER, LIKE ANY OTHER TOOL, REQUIRES SOME CARE IN USE TO DO ITS JOB WELL.

- WHEN MEASURING OR DRAWING A LINE, USE A SHARP PENCIL OR SCRIBE THAT CAN SLIDE EVENLY ALONG THE RULER'S EDGE. WORK IN LIGHTING THAT DOES NOT CAST SHADOWS.

- STEEL RULERS ARE BETTER THAN PLASTIC ONES.
- DO NOT TAKE MEASUREMENTS FROM THE END OF A RULER SINCE IT COULD BE WORN, AND THEREFORE INACCURATE.

WORN

THE SMALLEST DIVISION OF ANY RULER IS PRINTED NEAR ONE END. IN THIS CASE THE SMALLEST CALIBRATION IS TENTHS OF A MILLIMETER.

Degree Gauge:

ALSO CALLED A DOUZIEME GAUGE OR 72 GAUGE FROM THE FRENCH WATCHMAKERS MEASUREMENT CALLED A LIGNE.

12 DOUZIEMES = 1 LIGNE = .0888 INCH.

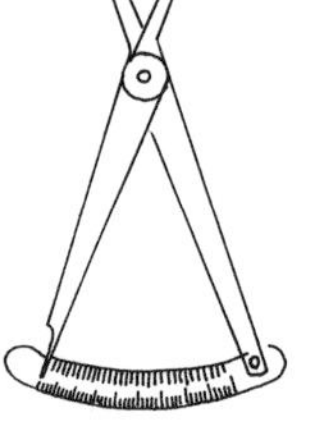

IN THIS SPRING-ACTIVATED TOOL THE SIZE OF THE OPENING AT THE TOP IS INDICATED BY THE SCALE AT THE BOTTOM.

Gauge Plate

THIS IS A THICK PIECE OF STEEL CUT WITH SLOTS OF A SPECIFIC SIZE. IT MEASURES BOTH SHEET AND WIRE IN THE BROWN AND SHARPE SYSTEM (ALSO CALLED AMERICAN STANDARD AND AMERICAN WIRE GAUGE, A.W.G.). THE OTHER SIDE SHOWS THOUSANDTHS OF AN INCH.

IN USING, FIND THE SLOT THAT MAKES A SNUG FIT BUT DON'T DISTORT THE METAL BY JAMMING IT IN. BE CAREFUL NOT TO MEASURE WHERE THE EDGE HAS BEEN THINNED BY PLANISHING, OR WHERE A BUR RAISED BY SNIPS WILL AFFECT ACCURACY.

ART IS NOT A THING; IT IS A WAY.

ELBERT HUBBARD

micrometer:

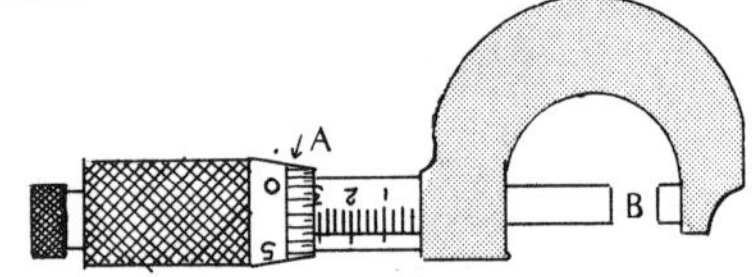

THIS IS A PRECISE AND ACCURATE TOOL USED FOR MEASURING THICKNESS, USUALLY IN THOUSANDTHS OF AN INCH. THE BARREL UNSCREWS ALONG THE SHANK, ROTATING THROUGH THE DIGITS 1-25 AT **A**. ON THE SHANK ARE MARKS INDICATING UNITS OF 25 THOUSANDTHS EACH. THE SMALL NUMBERS ON THE SHANK, THEN DEMARK HUNDREDTHS (I.E. FOUR UNITS OF 25).

AS SHOWN ABOVE, THE SPACE AT **B** IS .3035 INCHES.

DIVIDERS

IN ADDITION TO MAKING CIRCLES LIKE A COMPASS, THE DIVIDERS CAN BE USED TO "HOLD" A MEASUREMENT FOR QUICK REFERENCE.

ANOTHER USE IS TO LAY OUT PARALLEL LINES BY DRAGGING ONE LEG OF THE TOOL ALONG THE EDGE OF A PIECE OF METAL.

Vernier (VUR´NĒ·R)

THIS IS A SECONDARY GAUGE USED TO SUBDIVIDE THE SMALLEST UNITS ON A PRIMARY GAUGE. IT IS MOST OFTEN FOUND ON A SLIDING CALIPER.

TO USE IT →

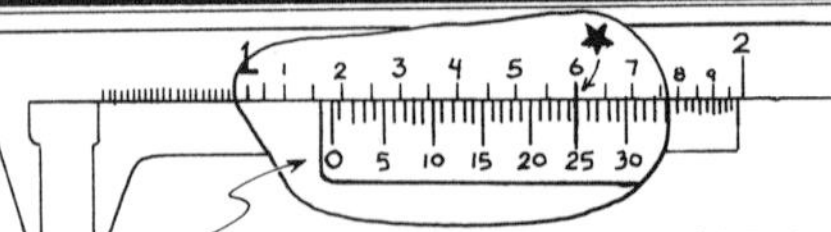

THE MEASUREMENT IS READ AT THE ZERO LINE. THE WHOLE INCHES ARE READ FIRST, THEN THE SMALL UNITS ON THE MAIN BAR WHICH HERE EQUAL .050" EACH. THE SECONDARY SCALE INDICATES UNITS OF .001" AND IS READ TO THE PLACE WHERE ONE OF ITS LINES COINCIDES WITH A LINE ON THE MAIN SCALE, AT ★. THIS SHOWS 1.175"

ANVILS

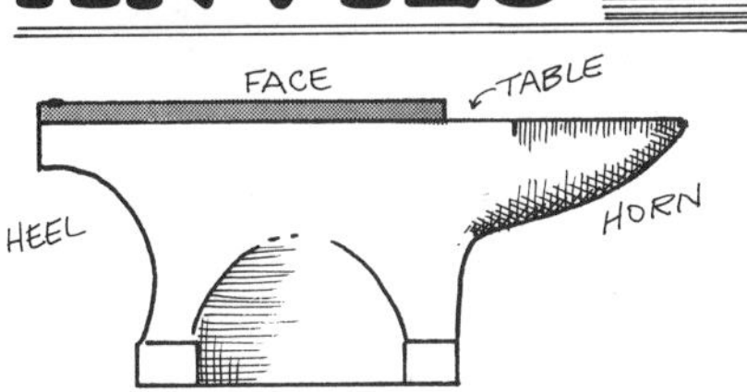

THERE WAS A TIME WHEN EVERY FARM HAD AN ANVIL, BUT NOWADAYS THEY ARE INCREASINGLY HARD TO FIND. USED ANVILS TURN UP AT AUCTIONS AND SCRAP DEALERS BUT EVEN THESE ARE RARE AND EXPENSIVE. FOR A ROUGHED UP ANVIL YOU CAN PAY ABOUT A DOLLAR A POUND.

THE FACE OF AN ANVIL CAN BE GROUND SMOOTH BY A MACHINE SHOP. ATTEMPTS TO WELD ON A NEW FACE OR FILL RECESSES WITH WELDING ROD ARE GENERALLY UNSUCCESSFUL. IN GRINDING THE FACE, TAKE CARE NOT TO CUT AWAY THE HARDENED STEEL PLATE THAT MAKES UP THE TOP ½" OF THE FACE, (SHADED ABOVE).

A GOOD ANVIL CAN BE MADE OF A PIECE OF RAILROAD TRACK. FIND THESE AT A JUNK YARD OR FOUNDRY. THE SURFACE SHOULD BE GROUND SMOOTH, EITHER BY A MACHINE SHOP OR WITH A BELT SANDER. A POINT CAN BE CUT WITH AN OXY-ACETYLENE TORCH BUT THAT IS NOT NECESSARY.

OTHER FLAT PIECES OF STEEL MAY BE USED AS ANVILS. THOUGH IT HELPS TO HAVE HARDENED STEEL IT IS NOT ESSENTIAL. KEEP IT HEAVY, SMOOTH AND WELL-ANCHORED, AND IT WILL WORK.

SOME ANVIL STANDS

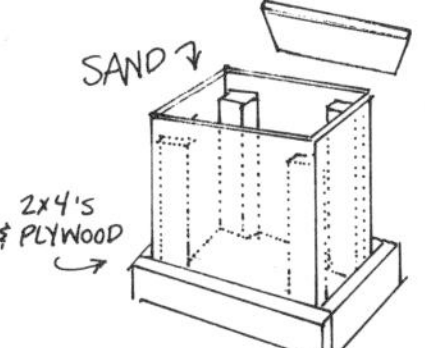

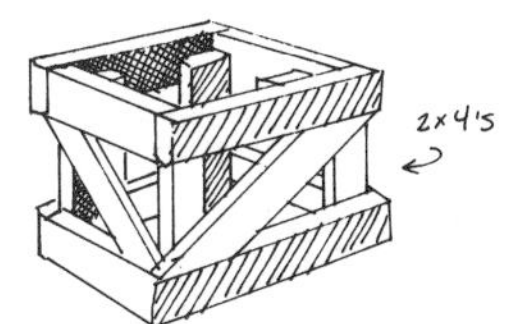

mandrels

IT WOULD SEEM IMPOSSIBLE TO HAVE TOO MANY MANDRELS. THE MOST COMMON VARIETIES ARE THE TAPERED MANDRELS: BEZEL, RING AND BRACELET. ANY HARD OBJECT THAT WILL LEND YOU ITS SHAPE WILL WORK — HERE ARE SOME ECONOMICAL SUBSTITUTES FOR CONVENTIONAL TOOLS:

ALSO, IF AN AREA VOCATIONAL OR TECHNICAL SCHOOL HAS A MACHINE SHOP YOU MIGHT FIND THAT THEY WILL CUSTOM MAKE MANDRELS TO YOUR SPECIFICATIONS AT A LOW COST.

stakes

SINCE MOST STAKES MUST SUIT A SPECIAL NEED IT IS LESS LIKELY THAT YOU WILL TURN UP A PIECE OF SCRAP STEEL WITH THE RIGHT SHAPE. SOME SMITHS FORGE THEIR OWN STAKES OR MAKE PATTERNS IN WOOD TO BE CAST BY A FOUNDRY. EITHER WAY IS TIME CONSUMING AND EXPENSIVE, BUT SOMETIMES NECESSARY.

HARD WOOD LIKE MAPLE CAN BE USED IN MANY CASES AND HAS THE ADVANTAGE OF NOT MARRING THE WORK PIECE; SEE PAGE 41 FOR A WOODEN SIDE STAKE FOR CRIMPING AND RAISING. SOME SMITHS HAVE BEEN EXPERIMENTING WITH STAKES MADE OF HARD PLASTIC LIKE DELRIN OR NYLON. THERMOPLASTICS THAT CAN BE SHAPED WHEN WARM OPEN MANY POSSIBILITIES.

molds

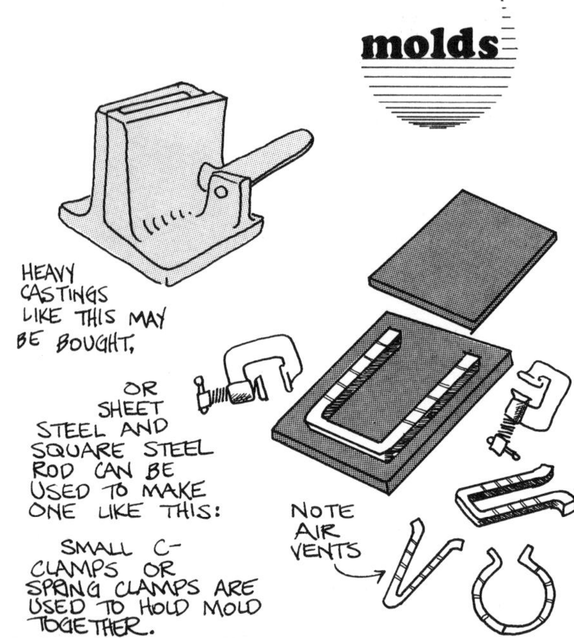

HEAVY CASTINGS LIKE THIS MAY BE BOUGHT,

OR SHEET STEEL AND SQUARE STEEL ROD CAN BE USED TO MAKE ONE LIKE THIS:

SMALL C-CLAMPS OR SPRING CLAMPS ARE USED TO HOLD MOLD TOGETHER.

investment

CAN BE USED TO MAKE A MOLD:

MIX UP SOME INVESTMENT, REMOVE BUBBLES, AND POUR INTO A MILK CARTON. WHEN THE BLOCK HAS HARDENED TEAR OFF THE CARTON AND CARVE THE DESIRED SHAPES.

POUR A SLAB OF INVESTMENT TO MAKE THE OTHER SIDE OF THE MOLD BY LAYING PIECES OF WOOD ON A SHEET OF GLASS.

BOTH MOLD PIECES SHOULD BY CURED BY HEATING THEM IN A KILN TO AROUND 1000°F.

TO USE, TIE THE TWO PIECES TOGETHER WITH BINDING WIRE, SET INTO A DISH OF SAND, AND PREHEAT FOR ABOUT HALF A MINUTE BEFORE POURING. THERE IS NO NEED TO LUBRICATE.

PROCESS:

① LUBRICATE THE MOLD WITH VASELENE OR ANIMAL FAT.

② HEAT MOLD UNTIL THE LUBRICANT STARTS TO SMOKE. MOLD MAY BE SET INTO A PAN OF SAND.

③ HEAT METAL IN A POURING CRUCIBLE, ADDING FLUX A COUPLE OF TIMES.

④ POUR METAL THROUGH THE FLAME IN A SINGLE EVEN FLOW. ALLOW THE RED COLOR TO FADE BEFORE REMOVING AND QUENCHING THE INGOT.

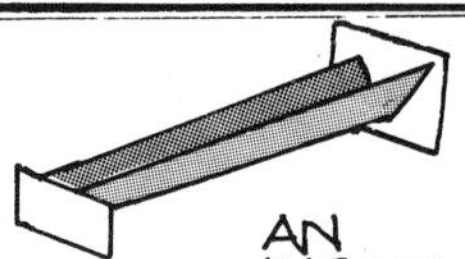

AN INGOT FOR POURING RODS CAN BE MADE BY WELDING CAPS ONTO BOTH ENDS OF A PIECE OF ANGLE IRON.

FOR LEAD, TIN, AND PEWTER THE END CAN BE CLOSED OFF WITH CLAY (NOT PLASTICENE).

WHEN YOU NEED A SHORT PIECE OF HEAVY WIRE AN INGOT MAY BE IMPROVISED BY DRILLING A HOLE OF THE DESIRED SIZE IN A CHARCOAL BLOCK. USE A POURING CRUCIBLE OR CARVE A MELTING RECESS IN THE BLOCK AND TILT TO POUR.

FIREBRICK MAY BE USED AS ABOVE FOR AN INGOT MOLD. THE DETAIL WILL NOT BE AS SMOOTH BUT THESE BUMPS WILL COME OUT IN SUBSEQUENT ROLLING, FORGING, OR DRAWING.

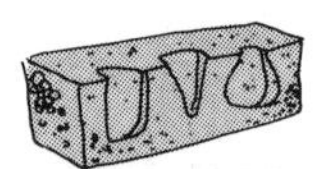

CERTAIN TRADITIONAL TOOLS OF THE MANUAL CRAFTS SCARCELY CHANGED IN FORM FOR TEN OR MORE CENTURIES BECAUSE THEY WERE PERFECTLY SUITED TO THE REQUIREMENTS OF THESE CRAFTS.

MAURICE DAUMAS

A HISTORY OF TECHNOLOGY & INVENTION

THIS BENCH AND THE ONE ON THE NEXT PAGE HAVE BEEN DESIGNED SO THEY CAN BE MADE FROM EASILY AVAILABLE MATERIALS WITHOUT SOPHISTICATED WOODWORKING EQUIPMENT. THOSE WITH SPECIAL NEEDS OR SKILLS MIGHT USE THESE IDEAS AS A POINT OF DEPARTURE.

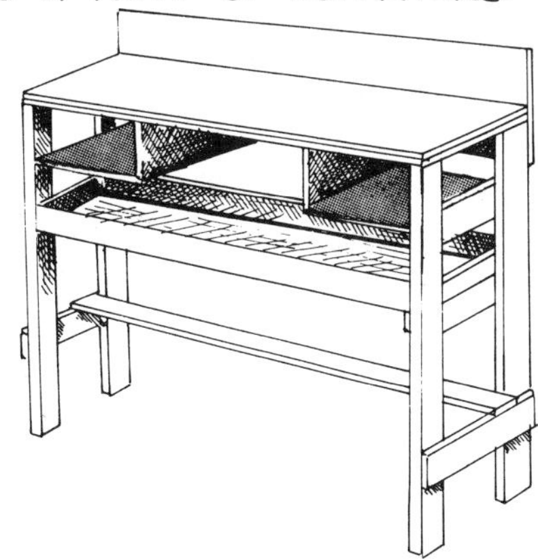

Cost: $20-25

Materials:

1	4x4'	SHEET OF ½" PLYWOOD, ONE SIDE SURFACED
2	8 FOOT	2x4's (STRAIGHT)
12'	1x3	
30	1½"	FLAT HEAD WOOD SCREWS
4	3"	" " "
1 PKG.	1"	FINISHING NAILS

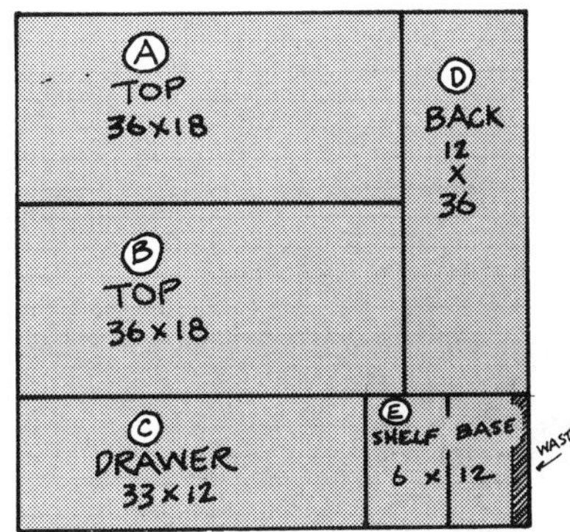

directions—

1. CUT PIECES. THE DIMENSIONS GIVEN DO NOT ALLOW FOR A SAW KERF. WHEN USING A HANDSAW THIS SPACE IS NOT IMPORTANT BUT THE ⅛" KERF LEFT BY A TABLE SAW MUST BE COMPENSATED FOR BY DECREASING THE GIVEN SIZES.
2. TO THICKEN THE TOP, TWO PIECES (A & B) ARE GLUED TOGETHER. USING A WHITE GLUE LIKE ELMER'S OR TITE-BOND, SET THE PIECES TOGETHER AND CLAMP OR WEIGHT THEM OVERNIGHT.
3. MAKE BOTH LEG ASSEMBLIES, USING A SQUARE TO BE SURE THAT ALL THE PIECES ARE PERPENDICULAR. GLUE AND SCREW EACH JOINT.
4. STAND UP THE LEGS WITH THE DOUBLE STRIP FACING INWARD. REST THE CROSS BRACE ON THE LOWER LEG BRACES AS SHOWN. GLUE AND SCREW. ATTACH THE BACK (D), ALLOWING HALF OF IT PROJECT ABOVE THE LEGS.

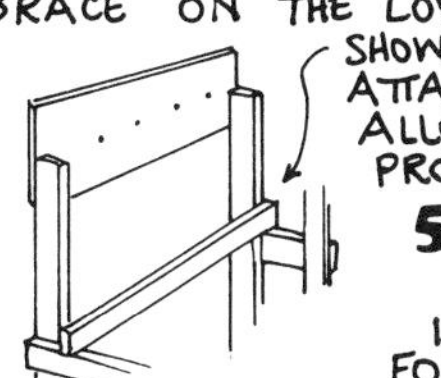

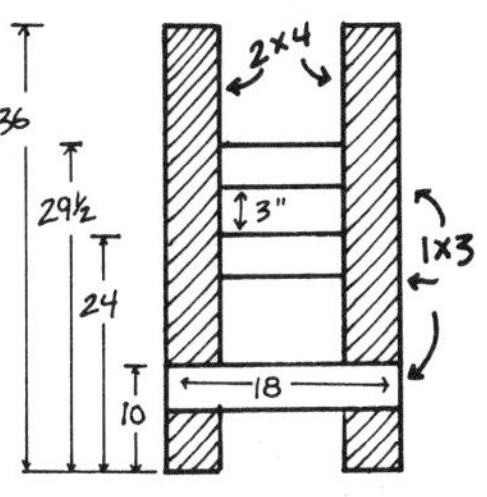

5. SET THE TOP IN PLACE AND USE THE 3" SCREWS TO JOIN IT TO THE LEGS. USE FOUR SCREWS ACROSS THE BACK TO ANCHOR IT TO THE TOP.
6. TO MAKE THE TOOL SHELVES, CUT A ONE FOOT STRIP OF 1x3 DOWN THE MIDDLE LENGTHWISE TO MAKE TWO CLEATS. TO EACH OF THESE IS SCREWED TWO 6½" PIECES OF 1x3 OR (NOT INCLUDED IN MATERIALS) A SINGLE PIECE OF WOOD, 6½ x 12." THE SHELF IS GLUED AND SCREWED ONTO THIS AND THE WHOLE ASSEMBLY IS SET INTO PLACE, RESTING ON THE LEG BRACE.

CLEAT

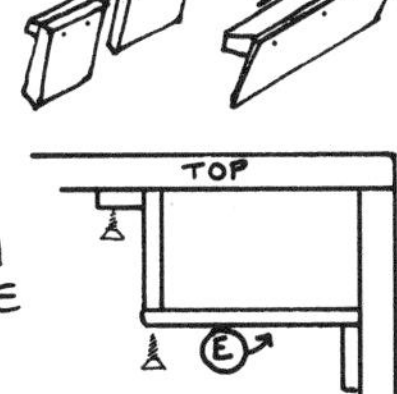

7. THE SWEEPS DRAWER FITS BETWEEN THE TWO UPPER LEG BRACES. IT IS MADE OF A 1x3 FRAME THAT IS SCREWED ONTO THE PLYWOOD PIECE, C. THE DRAWER AND BRACES CAN BE RUBBED WITH WAX OR SOAP TO MAKE THEM SLIDE EASIER.

SWEEPS DRAWER

THIS CAN BE CUT ON THE DOTTED LINE FOR EASIER ACCESS. DO NOT CUT AWAY MUCH OF THE SIDES.

TOOLS

THIS LARGER BENCH COSTS A LITTLE MORE AND IS A BIT MORE DIFFICULT TO MAKE THAN THE ONE ON THE PRECEDING PAGE.

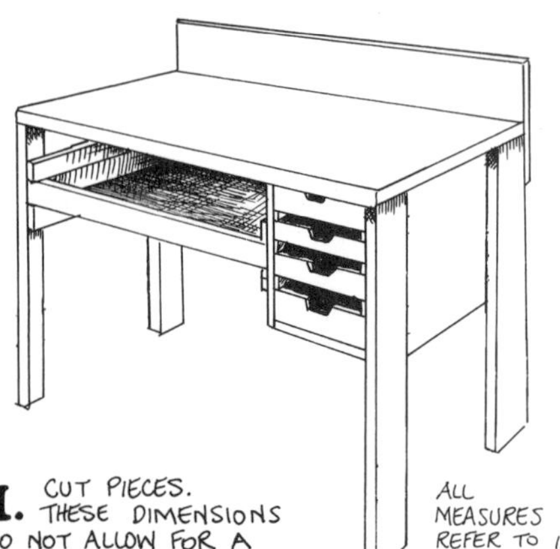

Cost: $45-50

Materials:

1	4×4×¾"	PLYWOOD CDX) (ONE SIDE SURFACED)
1	4×8×¼"	TEMPERED MASONITE
2	2×4×8'	(STRAIGHT ONES)
41'	1×3	RANDOM LENGTHS
18'	1×2	" "
6'	1×6	" "
4 LB.	1½"	PLASTER NAILS
4	3"	FLAT HEAD WOOD SCREWS

1. CUT PIECES. THESE DIMENSIONS DO NOT ALLOW FOR A SAW KERF. IF USING A TABLE SAW, PLAN AHEAD AND ALLOW FOR THIS.

ALL MEASURES REFER TO INCHES.

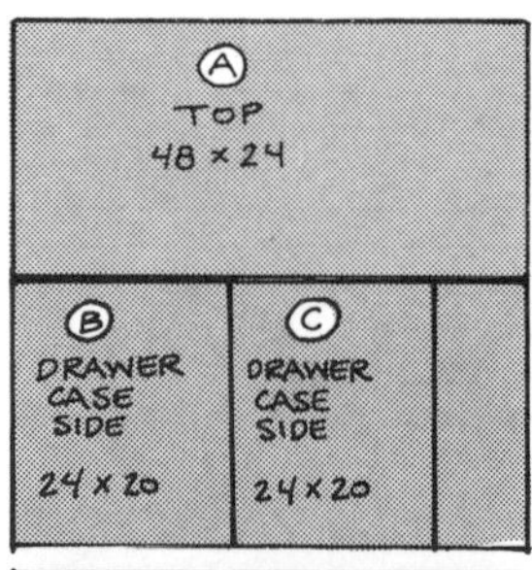

2. THE TOP IS MADE BY GLUING PIECES Ⓐ AND Ⓐ' TOGETHER. THEY SHOULD BE CLAMPED OR WEIGHTED WHILE THE GLUE DRIES. THE OPTIONAL CUTOUT BAY IS MADE WITH A BANDSAW OR SABRE SAW AFTER THE GLUE HAS DRIED.

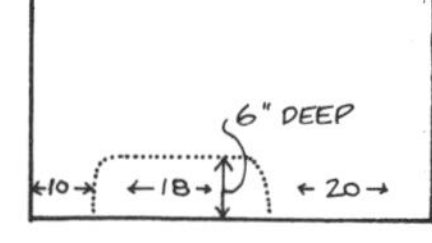

3. ASSEMBLE THE LEG UNITS. USE A SQUARE TO BE SURE ALL PIECES ARE PERPENDICULAR.

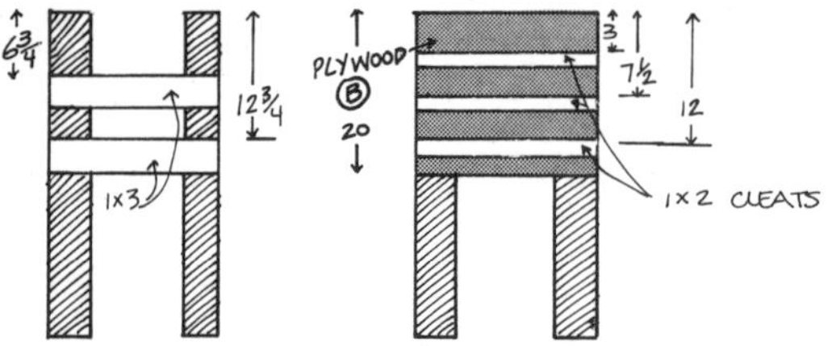

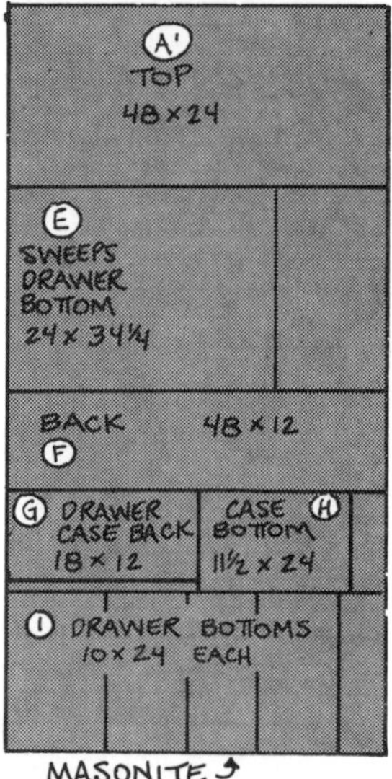

4. ASSEMBLE LEFT SIDE OF THE DRAWER CASE. THESE DIMENSIONS WILL DUPLICATE THOSE ON THE LEG UNITS.

THESE SHADED CLEATS ARE 1×3; THE OTHERS ON THIS PIECE ARE 1×2.

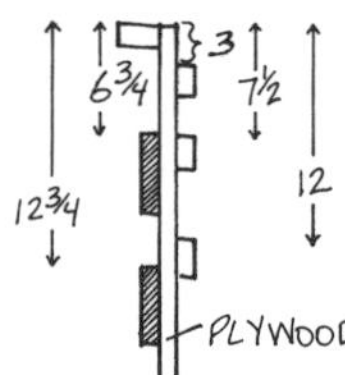

5. STAND UP LEG UNITS AND JOIN WITH A 48" PIECE OF 1×3, ABOUT 20" DOWN FROM THE TOP.

6. SET THE TOP INTO POSITION AND SECURE WITH THE 3" SCREWS. COUNTERSINK THESE SO THE TOP WILL BE FLUSH. SET THE MASONITE BACK INTO POSITION SO ABOUT HALF OF IT (6") PROJECTS ABOVE THE BENCH TOP.

7. THE LEFT SIDE OF THE DRAWER CASE IS FASTENED INTO POSITION BY THE CLEAT THAT JOINS THE UNDERSIDE OF THE TOP, AND BY THE MASONITE PANEL Ⓗ THAT SPANS THE BOTTOM EDGES OF THE PLYWOOD DRAWER CASE.

DRAWERS

BOTTOM DRAWER USES 1×6. 3 TOP DRAWERS USE 1×2. SWEEPS DRAWER USES 1×3 ON SIDES AND BACK, 1×2 ON FRONT.

APPENDIX

General Safety APPENDIX

WITH COMMON SENSE AND REASONABLE CAUTION METALSMITHING IS A SAFE AND REWARDING PURSUIT. IN ORDER TO CLEARLY MARK THOSE AREAS THAT DESERVE SPECIAL ATTENTION THIS BOOK USES A SYSTEM OF SAFETY ALERT LOGOS.

THEIR MEANINGS ARE SIMPLE. IT IS UP TO YOU TO HEED THE WARNINGS AND PROTECT YOURSELF.

TOXIC FUMES ARE BEING PRODUCED. ACTIVE VENTILATION IS NEEDED (SEE BELOW).

DANGEROUS MACHINERY. THIS INFORMATION IS MEANT TO SUPPLEMENT INSTRUCTION BY A QUALIFIED OPERATOR.

NOT A RECOMMENDED PROCEDURE. IT IS INCLUDED HERE PRIMARILY AS AN HISTORICAL REFERENCE.

Studio Dress

DO NOT WEAR LOOSE CLOTHING IN THE STUDIO. AVOID SYNTHETIC FABRICS BECAUSE OF FIRE HAZARD. WEAR GOGGLES AND HEAVY RUBBER GLOVES WHENEVER EVEN A SMALL CHANCE OF RISK IS PRESENT.

COMMON SENSE

THIS IS YOUR BEST PROTECTION. EVEN SAFE PROCEDURES CAN BE DANGEROUS IF ABUSED. REMEMBER THAT ACCIDENTS DON'T HAPPEN ONLY TO THE "OTHER GUY." IF YOU FEEL UNCERTAIN ABOUT A TOOL, GET HELP. IF YOU FEEL ILL OR DIZZY, STOP DOING WHAT YOU'RE DOING. IF ILLNESS PERSISTS CONTACT YOUR STATE HOSPITAL SYSTEM (DEPARTMENT OF OCCUPATIONAL SAFETY) OR

THE CENTER FOR OCCUPATIONAL HAZARDS, INC.
5 BEEKMAN STREET
NEW YORK, NY 10038
(212) 227-6220

Note:

WHILE MOST OF THE INFORMATION IN THIS BOOK CAN APPLY TO METALSMITHING ON ANY SCALE, KEEP IN MIND THAT IT IS WRITTEN PRIMARILY FOR WORK OF JEWELRY SIZE. IT IS NOT INTENDED AS A RESOURCE FOR LARGER STUDIOS OR INDUSTRY, WHERE OTHER SAFETY REQUIREMENTS MAY VERY WELL EXIST. FOR HELP IN THIS AREA CONTACT YOUR STATE OFFICE OF OCCUPATIONAL SAFETY AND HEALTH ADMINISTRATION (OSHA) OR THE INDUSTRIAL SAFETY DIVISION OF YOUR STATE LABOR DEPARTMENT.

*Ventilation

AS USED IN THIS BOOK, "VENTILATION" REFERS TO A POWERED MOVEMENT OF AIR. OPENING A WINDOW IS A PLEASANT THING TO DO ON A SUNNY DAY, BUT IT DOES NOT CONSTITUTE VENTILATION.

THE SIZE OF YOUR STUDIO AND THE TYPE AND VOLUME OF FUMES BEING PRODUCED WILL DETERMINE THE SCALE OF THE BLOWERS NEEDED. IN A SMALL SHOP A VENTED STOVE HOOD MAY BE ADEQUATE. THE CAN BE BOUGHT FROM LUMBER-YARDS OR KITCHEN REMODELING COMPANIES, WHO SOMETIMES HAVE USED OR SCRATCHED UNITS AT A REDUCED PRICE.

RESPIRATORS

RESPIRATORS FILTER AIR BEFORE IT ENTERS YOUR SYSTEM. THEY ARE GENERALLY CONSIDERED LESS EFFECTIVE THAN ACTIVE VENTILATION SINCE THEY CAN BE A LITTLE UNCOMFORTABLE.

A WORTHWHILE RESPIRATOR WILL HAVE A CANISTER OR CARTRIDGE FILTER TO CHEMICALLY REMOVE IMPURITIES. IT WILL COST $15-20.

1. LOOK FOR NIOSH SEAL OF APPROVAL.
2. CHOOSE A FILTER MADE FOR THE DANGER TO WHICH YOU ARE EXPOSED.
3. YOUR MASK MUST MAKE A TIGHT AND COMFORTABLE FIT.
4. CHANGE FILTERS AS NEEDED. YOU WILL NOTICE ODORS ENTERING MASK, OR WILL FEEL DIFFICULTY IN PULLING AIR IN.
5. IF YOU HAVE TROUBLE BREATHING OR A HISTORY OF RESPIRATORY ILLNESS, CONSULT YOUR DOCTOR.
6. BEARDS PROHIBIT AIRTIGHT CONTACT BETWEEN FACE AND MASK. HELP SEAL WITH VASELINE.

APPENDIX *Health & Safety*

Compound	Effects	Precaution
ACETONE	HEADACHE, DROWSINESS, SKIN IRRITATION. ONE OF THE LEAST TOXIC SOLVENTS.	ADEQUATE VENTILATION.
ACETYLENE	MILD NARCOTIC (INTOXICANT) IN SMALL DOSES. LARGE DOSES CAN CUT OFF OXYGEN.	USE CAUTION. CHECK EQUIPMENT REGULARLY FOR LEAKS. HAVE PROFESSIONALLY REPAIRED IF FOUND.
ACIDS · NITRIC · SULFURIC · HYDROCHLORIC · AQUA REGIA	THESE ARE STRONG OXIDIZERS – HIGHLY CORROSIVE IN ALL FORMS. LIQUIDS AND FUMES CAN CAUSE SERIOUS AND PAINFUL BURNS.	WEAR PROTECTIVE CLOTHING, GLOVES AND GOGGLES. WORK ONLY IN A WELL-VENTILATED AREA, NEAR A SOURCE OF RUNNING WATER. KEEP BAKING SODA HANDY TO NEUTRALIZE SPILLS.
ASBESTOS	MADE UP OF FIBERS THE BODY CANNOT DISSOLVE. A CARCINOGEN WHOSE EFFECTS TAKE 20-30 YEARS TO DEVELOP.	AVOID IT. AVOID IT. AVOID IT. AVOID IT. USE SUBSTITUTES.
BENZENE SOLVENT FOR PLASTICS	INTOXICATION, COMA, RESPIRATORY FAILURE.	USE ALTERNATIVE SOLVENT. AVOID IT!
CADMIUM SOLDER INGREDIENT	AFFECTS BRAIN, NERVOUS SYSTEM, LUNGS, KIDNEYS.	AVOID IF POSSIBLE; USE VERY GOOD VENTILATION.
CHLORINATED HYDROCARBONS EPOXY SOLVENT	DISSOLVES FATTY LAYER OF SKIN. CAUSES LIVER AND KIDNEY DAMAGE.	AVOID IF POSSIBLE; VENTILATE, WEAR NEOPRENE RUBBER GLOVES.
COPPER COMPOUNDS	OXIDES CAN IRRITATE LUNGS, INTESTINES, EYES, AND SKIN.	VENTILATE WHEN HEATING. WEAR GLOVES WHEN HANDLING A LOT, LIKE WHEN RAISING.
CYANIDES USED IN PLATING	MISTS INHALED OR FALLING ON SKIN ARE POISONOUS.	VENTILATE WELL, WEAR PROTECTIVE CLOTHING. NO NUDE PLATING.
FLUORIDES FLUX BASE	CAN FORM HYDROFLUORIC ACID IN THE LUNGS.	VENTILATE. AVOID BREATHING THE FUMES.
LEAD	DAMAGES BRAIN, CENTRAL NERVOUS SYSTEM, RED BLOOD CELLS, MARROW, LIVER, KIDNEYS. FUMES ARE ESPECIALLY DANGEROUS.	AVOID IF POSSIBLE. VENTILATE WELL. MINIMIZE HANDLING, WASH HANDS AFTER TOUCHING.
KETONES ACETONE, LACQUER THINNER, ETC.	SKIN, EYE, AND RESPIRATORY TRACT IRRITANTS. CAN CAUSE PERIPHERAL NERVE DAMAGE.	VENITILATE, WEAR APPROPRIATE RESPIRATOR. WEAR GLOVES.
LIVER OF SULPHUR (POTASSIUM SULPHIDE)	WHEN HEATED TO DECOMPOSITION IT RELEASES SULPHUR OXIDE FUMES THAT REACT WITH MOISTURE TO FORM HYDROGEN SULFIDE. IN HIGH CONCENTRATION THIS CAN CAUSE BRAIN DAMAGE AND SUFFOCATION.	DO NOT ALLOW MIXTURE TO COME TO A BOIL. ALL COLORING BENEFITS CAN BE OBTAINED FROM A WARM, NOT HOT, SOLUTION.
MERCURY	DAMAGES BRAIN, NERVOUS SYSTEM AND KIDNEYS.	AVOID FUMES AND SKIN CONTACT. VENTILATE AND WEAR GLOVES.

Compound	Effects	Precautions
PITCH	SKIN IRRITANT	WEAR GLOVES, AVOID HEATING TO A BOIL.
PLATINUM	METAL IS SAFE BUT FUMES WHEN MELTING CAN CAUSE LUNG AND SKIN IRRITATION.	VENTILATE.
POLYESTER RESINS	SKIN IRRITANTS. SOME RELEASE TOXIC FUMES ON MIXING WITH BINDERS. SOME ARE EXPLOSIVE.	WEAR GLOVES AND VENTILATE. STORE ACCORDING TO DIRECTIONS.
SILVER COMPOUNDS -SILVER CHLORIDE -SILVER NITRATE	ABSORBED INTO SKIN AS VAPOR OR DUST THESE CAN CAUSE A DISEASE CALLED ARGYRIA. SILVER DUST IN EYES CAN CAUSE NIGHT BLINDNESS.	WEAR GOGGLES, GLOVES AND A RESPIRATOR.
SULPHURIC ACID AND SPAREX (SODIUM BISULPHATE)	IRRITATES SKIN AND RESPIRATORY TRACT. DAMAGES CLOTHING.	VENTILATE. KEEP CONTAINER COVERED. DO NOT MIX STRONGER CONCENTRATION THAN NECESSARY. NEUTRALIZE WITH BAKING SODA AND WATER MIXTURE.
TELLURIUM	FUMES GENERATED IN REFINING GOLD, SILVER, COPPER, AND IN WELDING. IRRITATES SKIN AND GASTRO-INTESTINAL SYSTEM.	VENTILATE. EARLY SYMPTOM IS 'GARLIC BREATH' AND A METALLIC TASTE IN THE MOUTH.
TOLUENE TOLUOL SUBSTITUTE FOR BENZINE	CAUSES HALLUCINATION, INTOXICATION, LUNG, BRAIN, AND RED BLOOD CELL DAMAGE.	AVOID IF POSSIBLE. VENTILATE WELL.
TURPENTINE	SKIN IRRITANT. BRAIN AND LUNG DAMAGE POSSIBLE.	VENTILATE. WEAR GLOVES.
ZINC COMPOUNDS	DUST AND FUMES ATTACK THE CENTRAL NERVOUS SYSTEM, SKIN, AND LUNGS.	VENTILATE AND WEAR RESPIRATOR.

suppliers of safety equipment

AMERICAN OPTICAL CORP.
SAFETY PRODUCTS DIV.
100 CANAL STREET
PUTNAM, CT 06260

BINKS MFG. CO.
9201 W. BELMONT AVE.
FRANKLIN PARK, IL
60131

CESCO SAFETY PRODUCT
PO BOX 1237
KANSAS CITY, MO
64141

H.S. COVER CO.
107 E. ALEXANDER
BUCHANAN, MI
49107

DE VILBISS CO.
BOX 913
TOLEDO, OHIO
43692

EASTERN SAFETY
EQUIPMENT CO.
45-17 PEARSON
LONG ISLAND CITY,
N.Y. 11101

MINE SAFETY
APPLIANCES CO.
600 PENN CNTR BLVD
PITTSBURGH, PA
15235

SAFELINE PRODUCTS
BOX 550
PUTNAM, CT
06260

3M COMPANY
ST PAUL, MN
55101

SELLSTROM MFG. CO.
59 E. VAN BUREN
CHICAGO, IL 60605

U.S. SAFETY SERVICE
BOX 1237
KANSAS CITY, MO
64141

WELSH MFG. CO.
9 MAGNOLIA ST.
PROVIDENCE, RI
02909

WILLSON PRODUCTS
DIV. ESB, INC.
BOX 622
READING, PA 19603

CURTIN MATHESON
357 HAMBURG TPK
WAYNE, NJ
07470

FISHER SCIENTIFIC
711 FORBES AVE.
PITTSBURGH, PA
15219

GEN. SCIENTIFIC EQUIP.
LIMEKILN PIKE AND
WILLIAM AVE.
PHILADELPHIA, PA
19151

KERODEX PRODUCTS
AVERST LABS
685 THIRD AVE.
NEW YORK, NY
10017

MILBURN COMPANY
3246 E. WOODBRIDGE
DETROIT, MI
48207

To divide a circle into any number of equal parts

1. DRAW DIAMETER **AB**.
2. WITH **A** AS CENTER AND **AB** AS RADIUS DESCRIBE ARC. WITH **B** AS CENTER AND THE SAME RADIUS, DESCRIBE ANOTHER ARC CROSSING AT **C**.
3. WITH A RULER, DIVIDE **AB** INTO AS MANY PARTS AS YOU WISH TO MAKE; IN THIS CASE FIVE.
4. DRAW A LINE FROM **C** THROUGH THE <u>SECOND</u> DIVISION, REGARDLESS OF THE NUMBER OF PARTS BEING DIVIDED.
5. STEP THE DISTANCE **AD** AROUND THE CIRCLE WITH A COMPASS TO DETERMINE EQUIDISTANT POINTS.

Describing an Oval

1. DRAW MAJOR AND MINOR AXIS **ab** AND **cd** WITH **o** REPRESENTING THE CENTER OF EACH LINE.
2. WITH RADIUS **ao** AND CENTER **c** DESCRIBE AN ARC TO FIND **e** AND **f** (CALLED FOCI).
3. TAKE THREE PINS AND STICK THEM FIRMLY AT **c**, **e**, AND **f**. WRAP A STRING AROUND THESE PINS AND TIE IT TO FORM A LOOP.
4. REMOVE PIN AT **c** AND REPLACE IT WITH A SHARP PENCIL POINT. MOVE THE PENCIL AROUND, KEEPING THE STRING TIGHTLY STRETCHED. THE LINE FORMED WILL BE AN ELLIPSE.

making a cone pattern

① DRAW THE SIDE VIEW OF THE CONE EXACTLY AS YOU WANT IT.

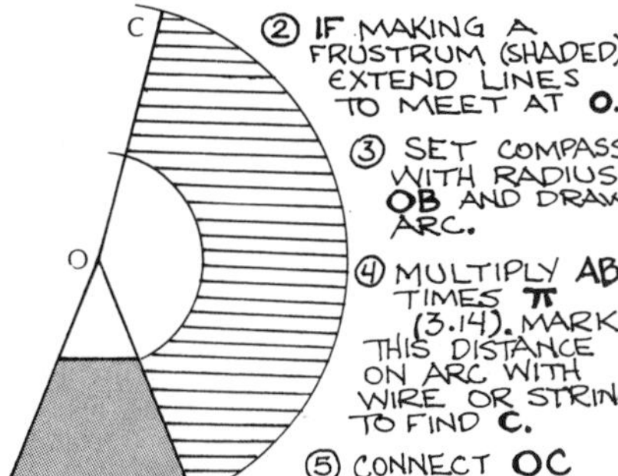

② IF MAKING A FRUSTRUM (SHADED) EXTEND LINES TO MEET AT **O**.

③ SET COMPASS WITH RADIUS **OB** AND DRAW ARC.

④ MULTIPLY **AB** TIMES **π** (3.14). MARK THIS DISTANCE ON ARC WITH WIRE OR STRING TO FIND **C**.

⑤ CONNECT **OC** STRIPED AREA IS PATTERN.

laying out a pyramid

① DRAW SIDE VIEW OF THE DESIRED FORM.

② TRACE PATTERN ON A SEPARATE SHEET OF PAPER AND CUT IT OUT.

③ LAY PATTERN WITH SIDE EDGE TOUCHING FIRST PYRAMID AS SHOWN.

④ DRAW AROUND PATTERN AND RELOCATE AS MANY TIMES AS NEEDED. SHADED PATTERN WOULD HAVE FOUR SIDES, ETC.

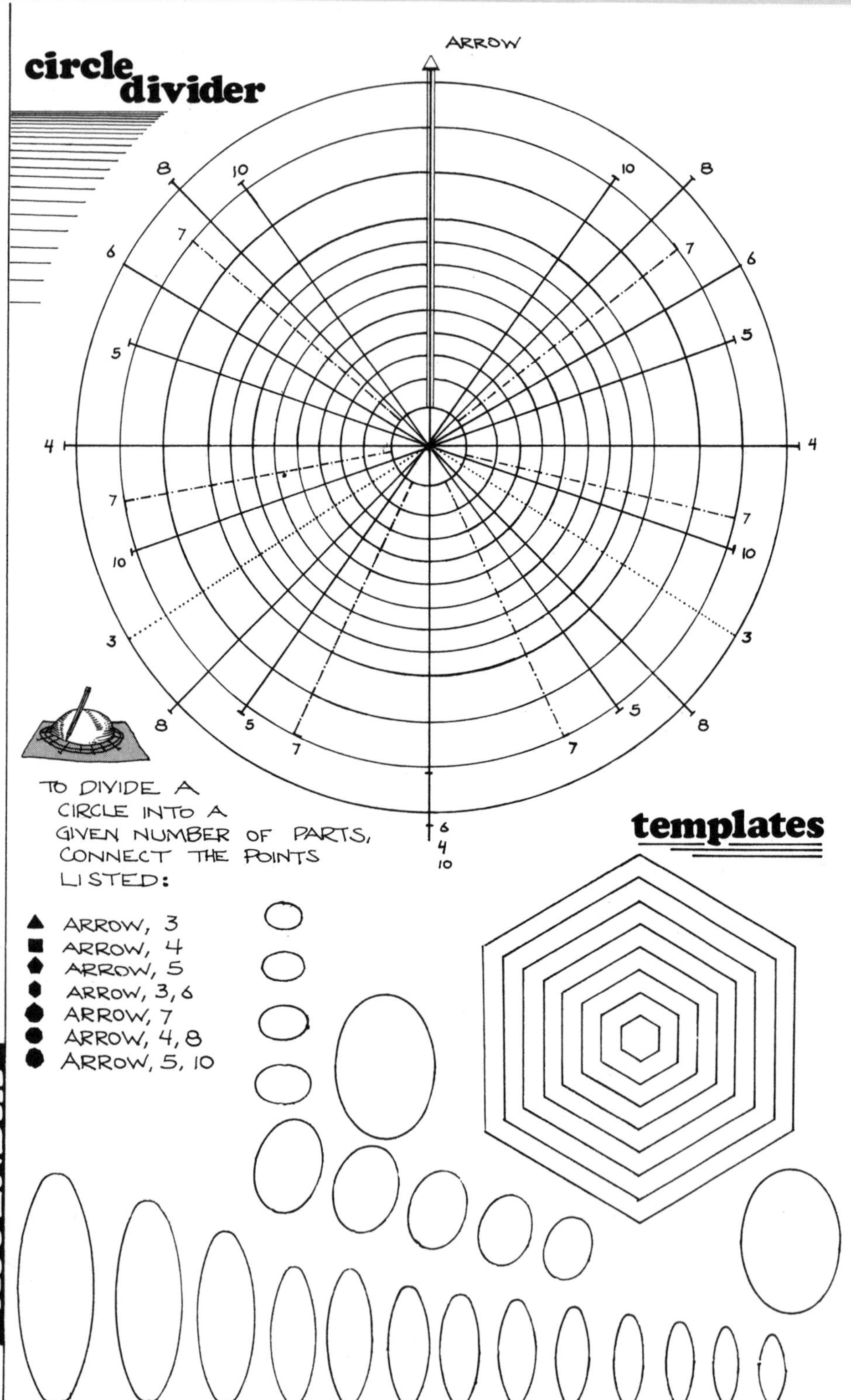
circle divider
ARROW
8
10
10
8
7
7
6
6
5
5
4
4
7
7
10
10
3
3
8
5
5
8
7
7
6
4
10
TO DIVIDE A CIRCLE INTO A GIVEN NUMBER OF PARTS, CONNECT THE POINTS LISTED:
▲ ARROW, 3
■ ARROW, 4
⬟ ARROW, 5
⬢ ARROW, 3, 6
ARROW, 7
ARROW, 4, 8
ARROW, 5, 10
templates

GENERAL

PHOTOGRAPHY IS COMPLEX, BOTH AS AN ART AND A SCIENCE. TO DEAL WITH IT COMPREHENSIVELY REQUIRES MUCH MORE SPACE THAN THIS. EXPERIMENTATION AND PROFESSIONAL HELP WILL COMPLEMENT THESE BASIC HINTS.

PHOTOGRAPHING SMALL REFLECTIVE OBJECTS PRESENTS TWO PROBLEMS. BECAUSE THE CAMERA IS CLOSE TO THE WORK, DEPTH OF FIELD BECOMES CRITICAL. TO KEEP BOTH THE FOREGROUND AND BACKGROUND IN FOCUS IT IS NECESSARY TO KEEP THE LENS OPENING (f-STOP) SMALL, LIKE 16. TO ALLOW ENOUGH LIGHT TO REACH THE FILM THROUGH SUCH A SMALL OPENING A LONG EXPOSURE IS NEEDED. SINCE IT IS IMPOSSIBLE TO KEEP THE CAMERA STEADY FOR THE LENGTH OF TIME REQUIRED A TRIPOD MUST BE USED. TO FURTHER REDUCE THE RISK OF JIGGLING THE CAMERA, A CABLE RELEASE IS RECOMMENDED.

WORK MAY BE SHOT IN DAYLIGHT (WITH THE APPROPRIATE FILM) BUT GREATER CONSISTENCY IS POSSIBLE WHEN USING LIGHTS. THESE MUST BE MATCHED TO THE FILM BEING USED AND ARE USUALLY MOUNTED IN ALUMINUM REFLECTORS. TO PROPERLY ILLUMINATE THE WORK AND AVOID SHADOWS LIGHT SHOULD COME FROM BOTH SIDES AND SOMETIMES FROM ABOVE AND BELOW. TO AVOID HOT SPOTS (BRIGHT REFLECTIONS) THE LIGHT SHOULD BE DIFFUSED AND/OR REFLECTED.

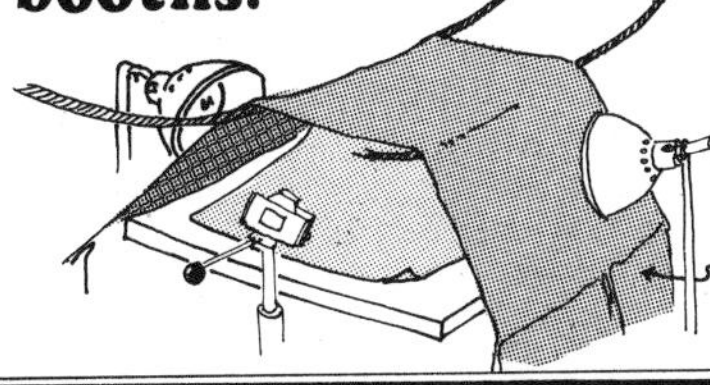

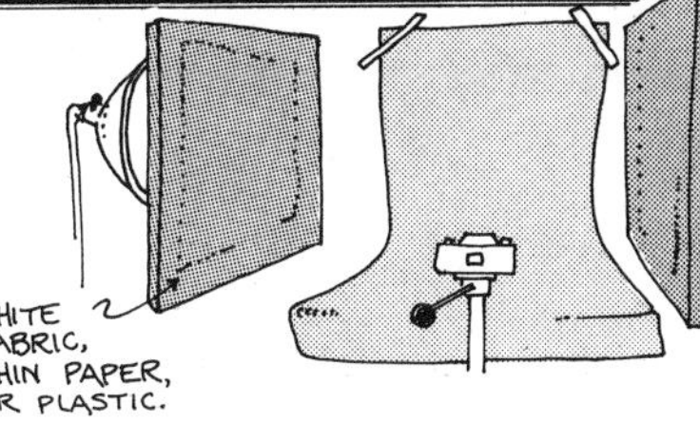

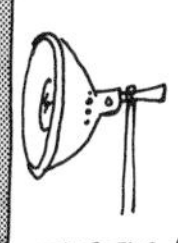

Background Material

IT'S EASY TO FORGET THAT A CLOSE UP PHOTO IS LIKE A MAGNIFYING GLASS. MATERIALS THAT LOOK GOOD TO THE NAKED EYE, LIKE BURLAP, VELVET, WOOD GRAIN, ETC. WHEN VIEWED CLOSE UP BECOME A JUNGLE OF LINT, LINES, AND FLAWS. MEDIUM VALUES OF COLORED PAPER MAKE GOOD BACKGROUND SURFACES. COLOR-AID PAPER (AVAILABLE AT AN ART SUPPLY STORE) IS ESPECIALLY RICH LOOKING.

SCALE

SCALE IS IMPORTANT IN TRANSLATING AN OBJECT INTO A PHOTOGRAPH. IN RARE CASES A COMMON OBJECT LIKE A COIN CAN BE SET BESIDE THE PIECE BUT THIS USUALLY CREATES A DISTRACTION. INSTEAD THE RELATIONSHIP BETWEEN THE OBJECT AND THE PICTURE AREA IS USED TO PROVIDE A SENSE OF SCALE.

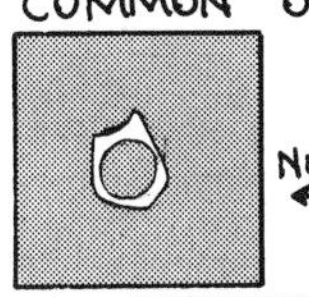

YES!

CHECKLIST

1. SET OBJECT IN PLACE AND ARRANGE FOR BEST ANGLE AND PROPER FRAMING.
2. ADD CLOSE UP RINGS OR LENS TO ALLOW FOCUSING.
3. TURN ON LIGHTS; CORRECT SHADOWS AND HOT SPOTS.
4. LOOK FOR REFLECTIONS. FIGURE OUT WHAT THEY ARE AND HOW TO GET RID OF THEM.
5. SET GRAY CARD (AVAILABLE FROM A PHOTO DEALER) INTO FIELD AND TAKE A READING. SET APERTURE AT HIGHEST NUMBER AND ADJUST SHUTTER SPEED AS REQUIRED.
6. DOUBLE CHECK THE IMAGE. IF IT'S GOOD, MAKE THE EXPOSURE.

VERY REFLECTIVE OBJECTS ARE DIFFICULT TO PHOTOGRAPH BECAUSE THEY MIRROR THE OBJECTS AROUND THEM. MAKE A WHITE ENCLOSURE (ABOVE) TO SOLVE THIS. TO AVOID THE REFLECTION OF THE CAMERA AND PHOTOGRAPHER, USE A PIECE OF WHITE CARDBOARD OR SHEET WITH A HOLE CUT OUT FOR THE LENS. A SMALL PIECE OF CARDBOARD CAN BE HAND-HELD TO "CATCH" REFLECTIONS.

*slightly condensed

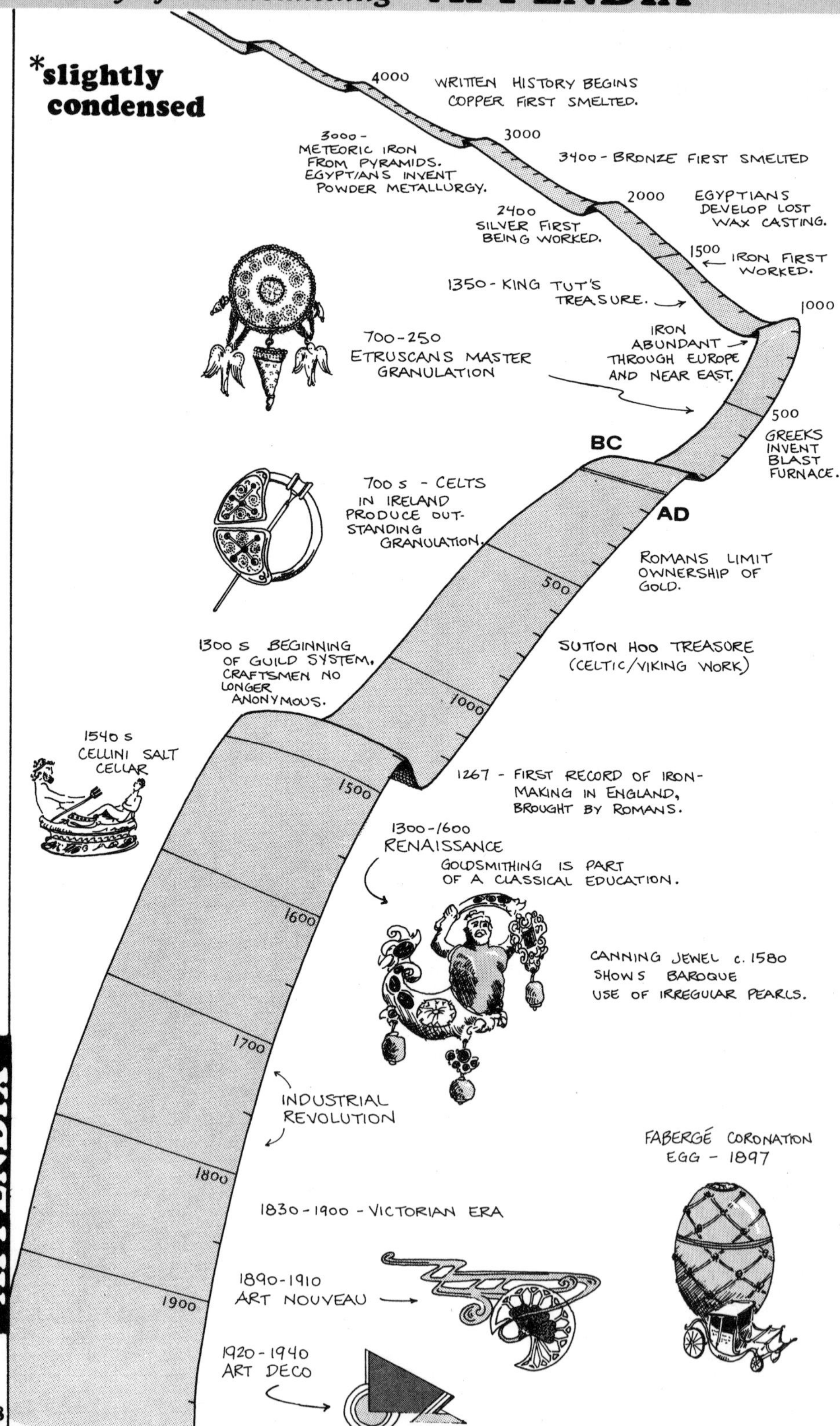

temperature conversion

centigrade to fahrenheit:

- MULTIPLY °C BY 9
- DIVIDE BY 5
- ADD 32

fahrenheit to centigrade:

- SUBTRACT 32 FROM °F
- MULTIPLY BY 5
- DIVIDE BY 9

°C	°F	°C	°F	°F	°C	°F	°C
0	32	650	1202	32	0	1300	704
50	122	675	1247	100	38	1350	732
75	167	700	1292	150	66	1400	760
100	212	725	1337	200	93	1450	788
125	257	750	1382	250	121	1500	816
150	302	775	1427	300	149	1550	843
175	347	800	1472	350	177	1600	871
200	392	825	1517	400	204	1650	899
225	437	850	1562	450	232	1700	927
250	482	875	1607	500	260	1750	954
275	527	900	1652	550	288	1800	982
300	572	925	1697	600	316	1850	1010
325	617	950	1742	650	343	1900	1038
350	662	975	1787	700	371	1950	1066
375	707	1000	1832	750	399	2000	1093
400	752	1025	1877	800	427	2050	1121
425	797	1050	1922	850	454	2100	1149
450	842	1075	1967	900	482	2150	1177
475	887	1100	2012	950	510	2200	1204
500	932	1125	2057	1000	538	2250	1232
525	977	1150	2102	1050	566	2300	1260
550	1022	1175	2147	1100	593	2350	1288
575	1067	1200	2192	1150	621	2400	1316
600	1112	1225	2237	1200	649	2450	1343
625	1157	1250	2282	1250	677	2500	1371

relative characteristics of common metals

heat conduction

1. SILVER
2. COPPER
3. GOLD
4. ALUMINUM
5. NICKEL
6. IRON PLATINUM
7. TIN
8. LEAD
9. ZINC
10. TITANIUM

malleability

THE ABILITY TO BE HAMMERED OR ROLLED OUT WITHOUT BREAKING.

1. GOLD
2. SILVER
3. ALUMINUM
4. COPPER
5. TIN
6. PLATINUM
7. LEAD
8. ZINC
9. IRON

electrical conduction

1. SILVER
2. COPPER
3. GOLD
4. ALUMINUM
5. ZINC
6. NICKEL
7. IRON
8. PLATINUM
9. TIN
10. LEAD

ductility

THE ABILITY TO BE DRAWN INTO WIRE.

1. GOLD
2. SILVER
3. PLATINUM
4. IRON
5. COPPER
6. ALUMINUM
7. NICKEL
8. ZINC
9. TIN
10. LEAD

tensile strength

RESISTENCE TO STRESS AND RUPTURE IN THE DIRECTION OF LENGTH.

1. IRON
2. COPPER
3. PLATINUM
4. SILVER
5. ZINC
6. GOLD
7. ALUMINUM
8. TIN
9. LEAD

Sizes & Weights **APPENDIX**

B&S	mm	thousandth of an inch	inch fraction	drill size
0	8.25	.325		
1	7.35	.289	9/32	
2	6.54	.258	1/4	
3	5.83	.229	7/32	1
4	5.19	.204	13/64	6
5	4.62	.182	3/16	15
6	4.11	.162	5/32	20
7	3.67	.144	9/64	27
8	3.26	.129	1/8	30
9	2.90	.114	7/64	33
10	2.59	.102		38
11	2.30	.090	3/32	43
12	2.05	.080	5/64	46
13	1.83	.072		50
14	1.63	.064	1/16	51
15	1.45	.057		52
16	1.29	.050		54
17	1.15	.045	3/64	55
18	1.02	.040		56
19	.912	.036		60
20	.813	.032	1/32	65
21	.724	.029		67
22	.643	.025		70
23	.574	.023		71
24	.511	.020		74
25	.455	.018		75
26	.404	.016	1/64	77
27	.361	.014		78
28	.330	.013		79
29	.279	.011		80
30	.254	.010		

STERLING STATISTICS

sheet			round			half-round			square		
B&S GA.	APPROX. SIZE PER OUNCE	SQUARE INCHES PER OUNCE	B&S GA.	LENGTH PER OUNCE	OUNCE PER FOOT	B&S GA.	LENGTH PER OUNCE	OUNCE PER FOOT	B&S GA.	LENGTH PER OUNCE	OUNCE PER FOOT
10	6"x¼"	1.78	4	5"	2.14	2	8"	1.65	4		
12	6"x⅜"	2.27	6	9"	1.35	4	13"	.94	6	7½"	1.75
14	6"x½"	2.86	8	15"	.85	6	1½'	.68	8	11"	1.09
16	6"x⅝"	3.61	10	2'	.54	8	2⅓'	.424	10	1½'	.682
18	6"x¾"	4.55	12	3'	.337	10	3½'	.28	12	2⅓'	.429
20	6"x1"	5.75	14	5'	.212	12	6'	.168	14	3¾'	.269
22	6"x1⅜"	7.25	16	7½'	.133	14	9½'	.106	16	6'	.169
24	6"x1½"	9.09	18	12'	.084	16	15'	.061	18	9⅓'	.107
26	6"x2"	11.5	20	19'	.053	18	24'	.042	20	15'	.067
28	6"x2½"	14.5	22	30'	.033	20	38'	.027	22	23.8	.042
30	6"x3¼"	18.2	24	48'	.021	22	60'	.017	24	37	.027

APPENDIX *Alloy Chart*

SYMBOL	NAME	GOLD	SILVER	COPPER	ZINC	OTHER	Melting Point		Specific Gravity
		Au	Ag	Cu	Zn		°C	°F	
Al	ALUMINUM					100 Al	660	1220	2.7
Sb	ANTIMONY					100 Sb	631	1168	6.6
Bi	BISMUTH					100 Bi	271	520	9.8
260	BRASS (CARTRIDGE)			70	30		954	1749	8.5
226	NU-GOLD (JEWELRY BRONZE)			88	12		1030	1886	8.7
220	RED BRASS			90	10		1044	1910	8.8
511	BRONZE			96		4 Sn	1060	1945	8.8
Cd	CADMIUM					100 Cd	321	610	8.7
Cr	CHROMIUM					100 Cr	1890	3434	6.9
Cu	COPPER			100			1083	1981	8.9
Au	GOLD (FINE)	100					1063	1945	19.3
920	22 KARAT YELLOW	92	4	4			977	1790	17.3
900	22 KARAT (COINAGE)	90	10				940	1724	17.2
750	18 KARAT YELLOW	75	15	10			882	1620	15.5
750	18 KARAT YELLOW	75	12.5	12.5			904	1660	15.5
750	18 KARAT GREEN	75	25				966	1770	15.6
750	18 KARAT ROSE	75	5	20			932	1710	15.5
750	18 KARAT WHITE	75				25 Pd	904	1660	15.7
580	14 KARAT YELLOW	58	25	17			802	1476	13.4
580	14 KARAT GREEN	58	35	7			835	1535	13.6
580	14 KARAT ROSE	58	10	32			827	1520	13.4
580	14 KARAT WHITE	58				42 Pd	927	1700	13.7
420	10 KARAT YELLOW	42	12	41	5		786	1447	11.6
420	10 KARAT YELLOW	42	7	48	3		876	1609	11.6
420	10 KARAT GREEN	42	58				804	1480	11.7
420	10 KARAT ROSE	42	10	48			810	1490	11.6
420	10 KARAT WHITE	42				58 Pd	927	1760	11.8
Fe	IRON					100 Fe	1535	2793	7.9
Pb	LEAD					100 Pb	327	621	11.3
Mg	MAGNESIUM					100 Mg	651	1204	1.7
	MONEL METAL			33		60 Ni 7 Fe	1360	2480	8.9
Ni	NICKEL					100 Ni	1455	2651	8.8
752	NICKEL (GERMAN) SILVER			65	17	18 Ni	1110	2030	8.8
Pd	PALLADIUM					100 Pd	1549	2820	12.2
	BRITTANIA			7		85 Sn 2 Sb 6 Bi	244	471	7.7
	OLD PEWTER					80 Pb 20 Sn	304	580	9.5
Pt	PLATINUM					100 Pt	1774	3225	21.4
Ag	SILVER (FINE)		100				961	1762	10.6
925	STERLING		92.5	7.5			920	1640	10.4
800-900	COIN SILVER		80-90	10-20			890	1634	10.3
	STEEL					99 Fe 1 C	1511	2750	7.9
	STAINLESS STEEL					91 Fe 9 Cr	1371	2500	7.8
Sn	TIN					100 Sn	232	450	7.3
Ti	TITANIUM					100 Ti	1800	3272	4.5
Zn	ZINC				100		419	786	7.1

mm	inches	B & S	weight per square inch						
			FINE SILVER	STERLING	FINE GOLD	10 K	14 K	18 K	PLATINUM
			OUNCES	OUNCES	DWTS.	DWTS.	DWTS.	DWTS.	OUNCES
6.54	.2576	2	1.42	1.41	52.5	31.4	35.5	42.3	2.91
5.19	.2043	4	1.12	1.12	41.6	24.9	28.1	33.6	2.31
4.11	.1620	6	.894	.884	33.0	19.8	22.3	26.6	1.83
3.26	.1285	8	.709	.701	26.2	15.7	17.7	21.1	1.45
2.59	.1019	10	.562	.556	20.8	12.4	14.0	16.7	1.15
2.05	.0808	12	.446	.441	16.5	9.85	11.1	13.3	.913
1.63	.0641	14	.354	.350	13.1	7.81	8.82	10.5	.724
1.29	.0508	16	.281	.277	10.4	6.21	7.00	8.35	.574
1.02	.0403	18	.223	.220	8.20	4.91	5.55	6.62	.455
.813	.0320	20	.176	.174	6.51	3.90	4.40	5.25	.361
.643	.0253	22	.140	.138	5.16	3.09	3.49	4.16	.286
.511	.0201	24	.111	.110	4.09	2.45	2.77	3.30	.227
.404	.0154	26	.088	.087	3.24	1.94	2.19	2.62	.180
.330	.0126	28	.070	.069	2.58	1.54	1.74	2.08	.143
.254	.0100	30	.055	.055	2.04	1.22	1.38	1.65	.113
.203	.0080	32	.044	.043	1.62	.969	1.09	1.31	.090
.157	.0063	34	.035	.034	1.29	.768	.868	1.03	.071
.125	.0050	36	.028	.027	1.02	.610	.689	.821	.057

mm	inches	B&S	weight per foot						
			FINE SILVER	STERLING	FINE GOLD	10 K	14 K	18 K	PLATINUM
			OUNCES	OUNCES	DWTS.	DWTS.	DWTS.	DWTS.	OUNCES
6.54	.2576	2	3.45	3.41	128	76.3	86.1	104	7.07
5.19	.2043	4	2.17	2.14	80.1	48	54.2	64.6	4.45
4.11	.1620	6	1.36	1.35	50.4	30.2	34.1	40.6	2.80
3.26	.1285	8	.856	.848	31.6	19	21.4	25.6	1.76
2.59	.1019	10	.541	.534	20	11.9	13.5	16.1	1.11
2.05	.0808	12	.339	.335	12.6	7.5	8.47	10.1	.695
1.63	.0641	14	.214	.211	7.87	4.72	5.33	6.36	.437
1.29	.0508	16	.135	.132	4.96	2.97	3.35	4.0	.275
1.02	.0403	18	.085	.084	3.11	1.87	2.11	2.51	.173
.813	.0320	20	.053	.053	1.96	1.17	1.33	1.58	.109
.643	.0253	22	.034	.033	1.23	.738	.833	.994	.068
.511	.0201	24	.021	.021	.775	.464	.524	.625	.043
.404	.0154	26	.013	.013	.488	.292	.330	.393	.027
.330	.0126	28	.008	.008	.306	.184	.207	.247	.017
.254	.0100	30	.005	.005	.193	.115	.130	.155	.010
.203	.0080	32	.003	.003	.122	.073	.082	.098	.007
.157	.0063	34	.002	.002	.076	.046	.051	.061	.004
.125	.0050	36	.001	.001	.048	.029	.032	.039	.003

FOR SQUARE WIRE MULTIPLY THESE FIGURES BY 1.2732

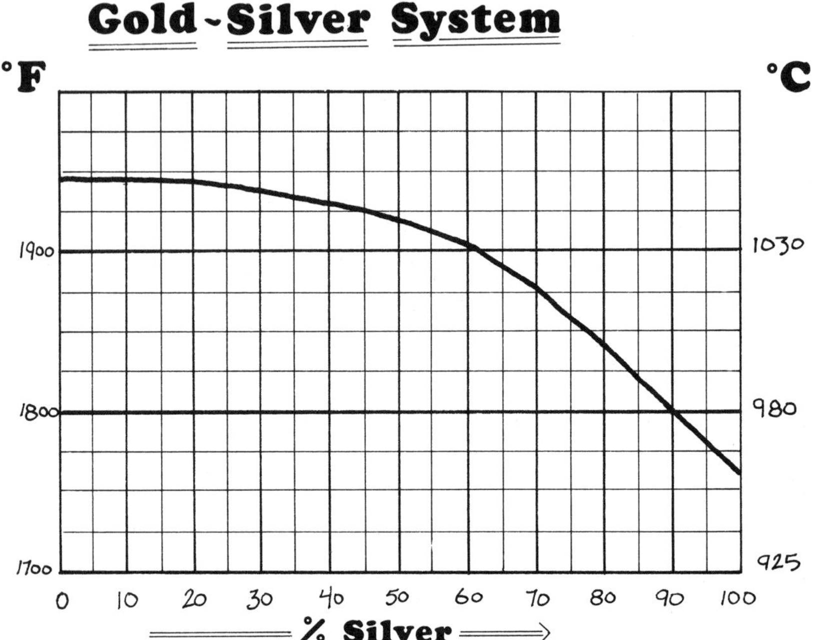
Gold~Silver System
°F
°C
1900
1800
1700
1030
980
925
0
10
20
30
40
50
60
70
80
90
100
% Silver

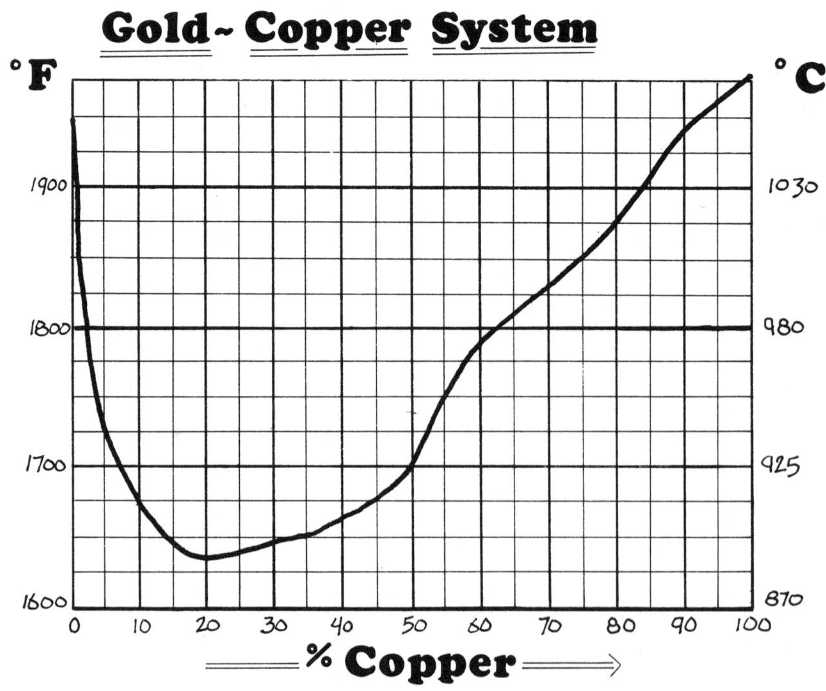
Gold~Copper System
°F
°C
1900
1800
1700
1600
1030
980
925
870
0
10
20
30
40
50
60
70
80
90
100
% Copper

Copper~Silver System

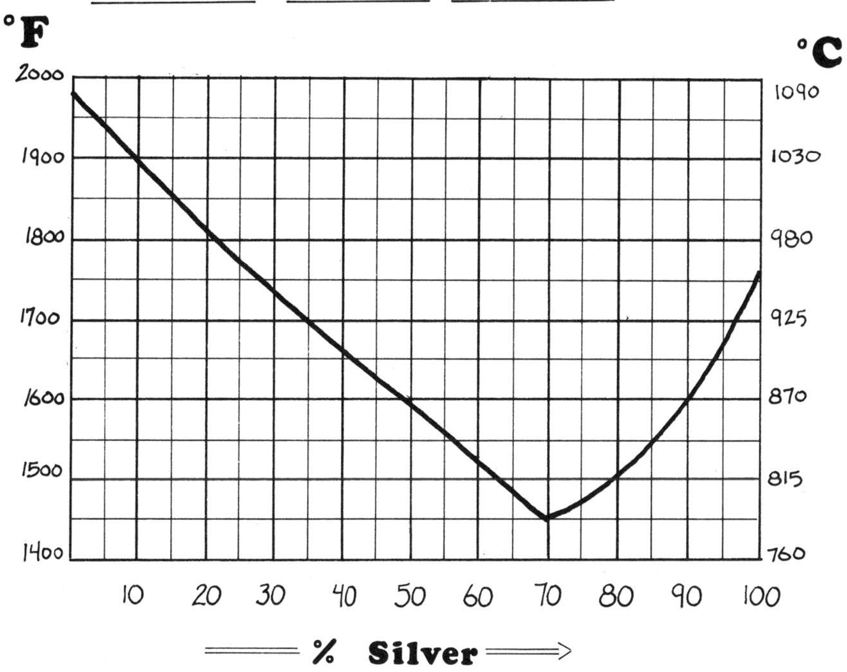

Copper~Zinc System

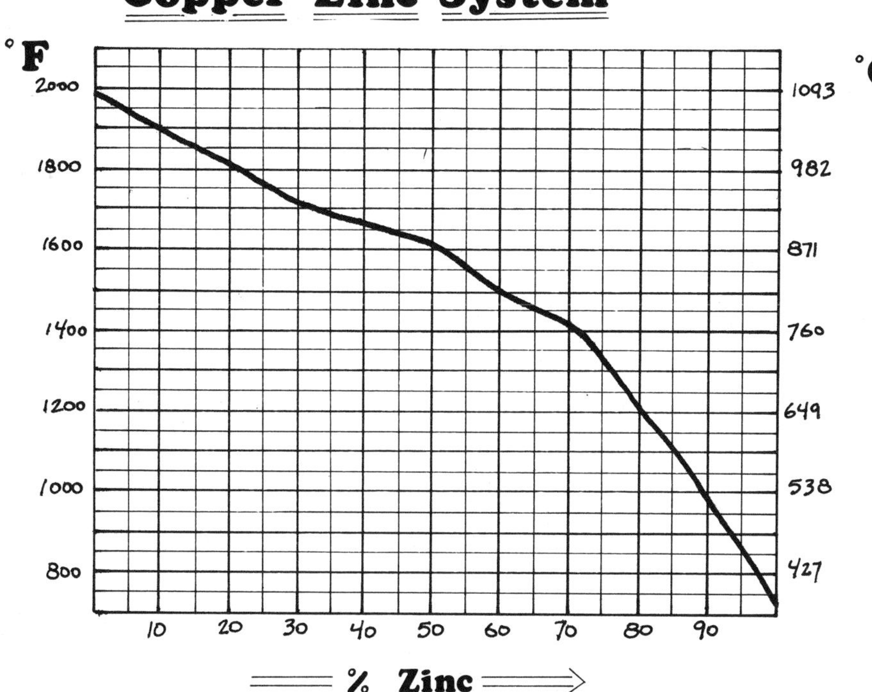

Conversion Formulas

MULTIPLY KNOWN UNIT BY FIGURE TO THE RIGHT OF THE MEASUREMENT YOU DESIRE.

CARATS	TO GRAINS	X 3.0865
	TO GRAMS	X .2
	TO MILLIGRAMS	X 200
GRAINS	TO CARATS	X .324
	TO GRAMS	X .0648
	TO MILLIGRAM	X 64.799
	TO OZ, AVOIR.	X .002286
	TO OZ, TROY	X .00208
	TO DWT	X .04167
GRAMS	TO CARAT	X 5
	TO GRAINS	X 15.4324
	TO OZ, AVOIR	X .03527
	TO OZ, TROY	X .03215
	TO DWT	X .64301
KILOGRAMS	TO OZ, A.	X 35.274
	TO OZ, TROY	X 32.1507
	TO DWT	X 643.015
	TO LB, AVOIR	X 2.20462
	TO LB, TROY	X 2.6792
MILLIGRAMS	TO GRAINS	X .015432
OZ, AVOIR	TO GRAINS	X 437.5
	TO GRAMS	X 28.3495
	TO OZ, TROY	X .91146
	TO DWT	X 18.2291
	TO LB, TROY	X .07595
OZ. TROY	TO GRAINS	X 480
	TO GRAMS	X 31.1035
	TO OZ, AVOIR	X 1.0971
	TO DWT	X 20
	TO LB, AVOIR	X .06857
OZ, U.S. FLUID	TO CU. CM.	X 29.5737
	TO CU. INCHES	X 1.80469
	TO LITERS	X .02957

PENNYWEIGHTS	TO GRAINS	X 24
	TO GRAMS	X 1.5551
	TO OZ. AVOIR	X .05486
POUNDS AVOIR.	TO GRAINS	X 7,000
	TO GRAMS	X 453.59
	TO KILOGRAM	X .4536
	TO OZ. TROY	X 14.5833
POUNDS TROY	TO GRAM	X 373.242
	TO KILOGRAM	X .3732
	TO OZ, AVOIR	X 13.165
	TO LB, AVOIR	X .82286

FEET	TO CENTIMETERS	X 30.48
	TO METERS	X .3048
METERS	TO FEET	X 3.2808
	TO INCHES	X 29.37
	TO YARDS	X 1.0936
MILLIMETERS	TO FEET	X .00328
	TO INCHES	X .03937
INCHES	TO CENTIMETER	X 2.54
	TO METERS	X .0254
	TO MILLIMETERS	X 25.4

CUBIC CM	TO CU. IN.	X .061
	TO U.S. FL. OZ.	X .0338
CUBIC IN.	TO CUBIC CM	X 16.387
	TO LITERS	X .01639
	TO U.S. FL. OZ.	X .554
GALLONS (U.S.)	TO LITER	X 3.785
	TO CUBIC IN.	X 231
	TO CUBIC FEET	X .1337
LITERS	TO GALLON US.	X .2642
	TO QUART US.	X 1.0567

Comparative Measures

1 LB TROY = 12 OZ TROY = 13.165 OZ AVOIR. = 5,760 GRAINS = 373.2 GRAMS

1 LB AVOIR. = 16 OZ AVOIR. = 14.58 OZ TROY = 7000 GRAINS = 453.6 GRAMS

1 KILOGRAM = 2.2 LB AVOIR. = 2.67 LB TROY = 35.274 OZ AVOIR. = 32.151 OZ. TROY

1 OZ AVOIR. = 18.229 DWT. = .9114 OZ. TROY = 28.35 GRAMS

1 OZ TROY = 20 DWT. = 1.097 OZ. AVOIR. = 31.103 GRAMS = 480 GRAINS

1 DWT. = 24 GRAINS = 1.555 GRAMS = .0548 OZ. AVOIR.

1 GRAM = .035 OZ. AVOIR. = .032 OZ. TROY = .643 DWT. = 5 CARATS

1 CARAT = .2 GRAMS = 200 MILLIGRAMS = 3.087 GRAINS

1 METER = 100 CENTIMETERS = 3.28 FEET = 29.37 INCHES

1 FOOT = 12 INCHES = 30.48 CENTIMETERS = .3048 METERS

1 INCH = 25.4 MILLIMETERS = 2.54 CENT. = .0277 YARDS = .083 FEET

1 GALLON = 4 QTS. = 8 PINTS = 128 FL. OZ. = 231 CU. IN = 3.785 LITER

1 US FL. OZ. = 29.57 CU. CM = 1.8 CU. IN. = .0295 LITER

APPENDIX *Suppliers*

NOTE: IN MOST CASES, CATALOGUE CHARGE IS REFUNDED ON THE FIRST ORDER. TOLL-FREE LINES ARE FOR ORDERS ONLY.

Supplier	GOLD	PLATINUM	SILVER & STERLING	COPPER, BRASS, NICKEL	PEWTER	HAND TOOLS	CASTING EQUIP.	CUT STONES	FINDINGS	BOOKS	CATALOGUE	MINIMUM ORDER	VISA • MASTER CARD	SCHOOL DISCOUNT
ALLCRAFT TOOL & SUPPLY 100 FRANK RD, HICKSVILLE, NY 11801 (800) 645-7124 (516) 433-1660			•	•		•	•	•	•	•	3	10	•	•
ANCHOR TOOL & SUPPLY CO. P.O. BOX 265, CHATHAM, NJ 07928-0265 (201) 635-2094			•	•	•	•	•	•	•	•	2		•	•
ARE, INC. BOX 8, GREENSBORO BEND, VT 05842 (800) 451-5086 (802) 533-7007	•		•	•	•	•	•	•	•	•			•	•
JULES BOREL & COMPANY 1110 GRAND AVE, KANSAS CITY, MO 64106 (816) 421-6110	•		•	•		•	•	•	•	•	2		•	
CALIFORNIA CRAFTS SUPPLY 1096 N. MAIN ST, ORANGE, CA 92667 (714) 633-8891	•		•	•		•	•	•	•	•	2	10	•	
CHASELLE, INC. 9645 GERWIG LANE, COLUMBIA, MD 21046 (301) 997-9611			•	•	•	•	•		•	•		25	•	
GRIEGER'S, INC. P.O. BIN 41, PASADENA, CA 91109 (800) 423-4181 (213) 795-9775	•		•			•	•	•	•	•		5	•	•
T.B. HAGSTOZ & SON 709 SANSOM ST, PHILADELPHIA, PA 19106 (215) 922-1627	•	•	•	•	•	•	•		•	•				•
HAUSER & MILLER CO. 10950 LIN-VALLE, ST. LOUIS, MO 63123 (314) 487-1311	•	•	•	•	•		•					5		
C.R. HILL COMPANY 2734 W. ELEVEN MILE RD, BERKLEY, MI 48072 (313) 543-1555	•		•	•	•	•	•	•	•	•	2	10		
THE MAISEL COMPANY 1500 LOMAS N.W. ALBUQUERQUE, NM 87104 (800) 545-9172 (505) 836-3713	•		•	•	•	•	•	•	•	•	3	25	•	•
MARSHALL-SWARTCHILD COMPANY 2040 N. MILWAUKEE AVE, CHICAGO, IL 60647 (800) 621-4767 (312) 278-2300						•	•	•	•	•	3	15		•
RIO GRANDE JEWELERS SUPPLY 6901 WASHINGTON N.E, ALBUQUERQUE, NM 87109 (800) 545-6566 (505) 345-8511	•		•	•	•	•	•	•	•	•			•	
C.W. SOMERS & COMPANY 387 WASHINGTON ST. BOSTON, MA 02108 (617) 426-6880	•		•			•	•	•	•	•	1	5		
SWEST, INC. 10803 COMPOSITE DR. DALLAS, TX 75220 (800) 527-5057 (214) 350-4011	•	•	•		•	•	•	•	•	•	5	25	•	•
MYRON TOBACK 23 W. 47TH ST. NEW YORK, NY 10036 (800) 223-7550 (212) 247-4750	•	•	•						•			10	•	
TSI, INC BOX 9266, 101 NICKERSON ST. SEATTLE, WA 98109 (800) 258-0823 (206) 282-3040	•		•	•	•	•	•	•	•	•	1	25	•	•

APPENDIX

Books APPENDIX

AMULETS AND SUPERSTITIONS	E.A. WALLIS BUDGE	DOVER NEW YORK 1978, ORIG. 1930
BEGINNING JEWELRY	ROGER ARMSTRONG	STAR PUBLISHING PALO ALTO, CA 1977
CENTRIFUGAL OR LOST WAX JEWELRY CASTING FOR SCHOOLS, TRADESMEN, AND CRAFTSMEN	MURRAY BOVIN	BOVIN FOREST HILLS, N.Y. 1971, REV. 1977
CONTEMPORARY JEWELRY	PHILIP MORTON	HOLT, RINEHART, AND WINSTON NEW YORK 1970, REV. 1976
CREATIVE CASTING	SHARR CHOATE	CROWN NEW YORK 1966
CREATIVE JEWELRY TECHNIQUES	HAROLD O'CONNOR	DUNCONOR LAKE CITY, CO. 1978
DESIGN & CREATION OF JEWELRY	ROBERT VON NEUMANN	CHILTON RADNOR, PA 1961, REV. 1972
RINGS FOR THE FINGER	GEORGE KUNZ	DOVER NEW YORK 1973, ORIG 1917
FORM EMPHASIS FOR METALSMITHS	HEIKKI SEPPÄ	KENT STATE UNIV. KENT, OHIO 1978
THE JEWELER'S BENCH REFERENCE	HAROLD O'CONNOR	DUNCONOR LAKE CITY, CO 1977
THE JEWELRY ENGRAVER'S MANUAL	HARDY & ALLEN	VAN NOSTRAND REINHOLD NEW YORK 1976, ORIG. 1954
JEWELRY MAKING FOR SCHOOLS, TRADESMEN & CRAFTSMEN	MURRAY BOVIN	BOVIN FOREST HILLS, N.Y. 1967, REV. 1971 & 1979
JEWERY MANUFACTURE AND REPAIR	CHARLES JARVIS	BONANZA NEW YORK 1978
THE CURIOUS LORE OF PRECIOUS STONES	GEORGE KUNZ	DOVER NEW YORK 1971, ORIG. 1913
THE MAKING OF TOOLS	ALEXANDER WYGERS	VAN NOSTRAND REINHOLD NEW YORK 1973
METALSMITH PAPERS	S.N.A.G. MEMBERS	SOCIETY OF NORTH-AMERICAN GOLDSMITHS CLINTON, OHIO 1981
METALSMITHING FOR THE ARTIST-CRAFTSMAN	RICHARD THOMAS	CHILTON RADNOR, PA 1960
METALWORK & ENAMELLING	HERBERT MARYON	DOVER NEW YORK 1971, ORIG. 1912
METALWORKING FOR JEWELRY	TIM MCCREIGHT	VAN NOSTRAND REINHOLD NEW YORK 1979
METAL TECHNIQUES FOR CRAFTSMEN	OPPI UNTRACHT	DOUBLEDAY GARDEN CITY, N.Y. 1968
STEP-BY-STEP KNIFEMAKING	DAVID BOYE	RODALE EMMAUS, PA 1977
TEXTILE TECHNIQUES IN METAL	ARLINE FISCH	VAN NOSTRAND REINHOLD NEW YORK 1975

INDEX

INDEX